HUMAN—COMPUTER INTERACTION AND OPERATORS' PERFORMANCE

OPTIMIZING WORK DESIGN WITH ACTIVITY THEORY

Ergonomics Design and Management: Theory and Applications

Series Editor

Waldemar Karwowski

Industrial Engineering and Management Systems
University of Central Florida (UCF) – Orlando, Florida

Published Titles

Ergonomics in Developing Regions: Needs and Applications
Patricia A. Scott

Ergonomics and Psychology: Developments in Theory and Practice
Olexiy Ya Chebykin, Gregory Z. Bedny, and Waldemar Karwowski

Human–Computer Interaction and Operators' Performance: Optimizing Work
Design with Activity Theory
Gregory Z. Bedny and Waldemar Karwowski

Trust Management in Virtual Organizations: A Human Factors Perspective
Wiesław M. Grudzewski, Irena K. Hejduk, Anna Sankowska, and Monika Wańtuchowicz

Forthcoming Titles

Ergonomics: Foundational Principles, Applications and Technologies
P. McCauley-Bush

Knowledge Service Engineering Handbook
Jussi Kantola and Waldemar Karwowski

Handbook of Human Factors in Consumer Product Design, 2 vol. set
Waldemar Karwowski, M. M. Soares, and Neville A. Stanton

> Human Factors Interaction: Theories in Consumer Product Design
> Human Factors Design: Case Studies in Consumer Product Design

Organizational Resource Management: Theories, Methodologies, and Applications
Jussi Kantola

HUMAN—COMPUTER INTERACTION AND OPERATORS' PERFORMANCE

OPTIMIZING WORK DESIGN WITH ACTIVITY THEORY

EDITED BY
GREGORY Z. BEDNY
WALDEMAR KARWOWSKI

CRC Press
Taylor & Francis Group
Boca Raton London New York

CRC Press is an imprint of the
Taylor & Francis Group, an **informa** business

CRC Press
Taylor & Francis Group
6000 Broken Sound Parkway NW, Suite 300
Boca Raton, FL 33487-2742

Library of Congress Cataloging-in-Publication Data

Human-computer interaction and operator's performance : optimizing work design with
    activity theory / editors, Gregory Z. Bedny and Waldemar Karwowski.
        p. cm. -- (Ergonomics design and management. Theory and applications)
    Includes bibliographical references and index.
    ISBN 978-1-4398-3626-2 (alk. paper)
        1. Human-computer interaction. 2. Human-computer interaction--Psychological
aspects. 3. Human engineering--Psychological aspects. I. Bednyi, G. Z. (Grigorii
Zakharovich) II. Karwowski, Waldemar, 1953- III. Title. IV. Series.

    QA76.9.H85H8562 2010
    004.01'9--dc22   2010005496

Visit the Taylor & Francis Web site at
http://www.taylorandfrancis.com

and the CRC Press Web site at
http://www.crcpress.com

Contents

Section I   Activity Theory in Studying Performance

Section II   Human–Computer Interaction

Section III   Evaluation of Computer Users' Psychophysiological Functional State

Preface

In recent times, activity theory (AT) has increasingly attracted the attention of scientists and practitioners in various countries. Numerous articles and books in the field in English, published in the West including the United States, are based primarily on the works of the Russian scholars Vygotsky and Leont'ev. However, these publications do not reflect any important data from the applied activity theory (AAT). Furthermore, a majority of the scientists from the former Soviet Union who have contributed to this field are unknown in the West.

There is a fundamental difficulty in the translation of texts containing the data obtained in the field of AT from Russian into English. Thus, inadequate translations have led to incorrect descriptions and interpretations of the original meaning of the basic terms and concepts from the perspective of general AT. In addition, the terminologies and basic principles of AAT are practically unknown in the West. Contradictions between general AT and AAT, in addition to the existence of different schools of psychology, create additional difficulty in the interpretation of the terminologies and principles of both general AT and AAT. The words and technical terms used in AAT have different meanings and interpretations in Western psychology. In this context, this book provides readers with new and important theoretical and applied concepts in the study of human work from the perspective of AT. The conceptual apparatus of AAT and its relationship with systemic-structural activity theory (SSAT) are discussed in this book.

According to AT, human mental development is the result of the acquisition of sociohistorical experience. Human beings are born into a social world where they interact with other humans and artifacts. In AT, artifacts are physical tools, used as both means of work and sign systems. Together, artifacts and culture have shaped mental development and the nature of human thought. Work is a critical form of human activity, whereas technology (means, tools of work, and sign systems) is an important class of artifacts that influences the structure of human thought. Technological progress drives human history and determines its progress. The nature of technology influences the types of activity that humans engage in. For example, an average American currently spends hours every day using the Internet, which only became popular in the 1990s. Technology thus influences human activity, and conversely, human activity influences technology. This interdependence of technology and human work activity was the departure point for the development of general and applied AT in Russian psychology.

Vygotsky (1978), the founder of the cultural–historical theory of human mental development, was the first to introduce the concept of tools and signs in psychology as an important source of mental development. He used these notions to explain the origins of consciousness and cognition in general.

Vygotsky strove to understand human behavior and consciousness through a historical analysis of human labor. Technology and human labor are considered the basis for the development of AT and its applications. The relationship between technology and human activity is a major object of study in ergonomics and work psychology.

Because AT is specifically suited to the understanding of the interaction of human activity and technology, there is an ever-increasing interest in the application of AT to the study of human work (Engestrom 2000; Kaptelinin and Nardi 2006; and others). For the English-speaking world, access to important works in AAT has been relatively fragmental until now. The Russian scholars Rubinshtein (1957, 1959) and Leont'ev (1978) developed general AT. The parts of the work of Leont'ev are relatively well known through the collection of works edited by Wertsch in 1981 and the later translation of his work. However, an important branch of general AT founded by Rubinshtein is practically unknown in the West. Although Leont'ev's and Rubinshtein's schools share some similarities, they differ significantly. Leont'ev emphasized the importance of the internalization process in mental development, whereas Rubinshtein argued that a person does not simply internalize ready-made standards. According to Rubinshtein, a person's external world acts on the mind through the mind's own internal conditions. "External influences on mental development always act through internal conditions" (Rubinshtein 1959). Some of the data from Rubinshtein's works and a comparison of his work with those of Vygotsky (1978) and Leont'ev (1978) can be found in the works of Bedny and Karwowski (2007) and Chebykin, Bedny, and Karwowski (2008).

General AT, as developed by Leont'ev and Rubenshtein, is not easily applicable to the study of human work. This fact has motivated the development of AAT or operationalized AT (Zarakovsky 2004). Unlike general AT, AAT has various methods for describing the structure of work activity and utilizes not only qualitative, but also quantitative methods of analysis. (For overviews of the latter in English, refer to Lomov 1969; Landa 1976; Bedny and Meister 1997; Ponomarenko 2004; and Zarakovsy 2004.)

During the last two decades, a new approach within AT has been developed (Bedny and Karwowski 2007). This approach is called "systemic-structural activity theory" and is derived from general AT and AAT (Bedny 1987). SSAT views activity as a goal-directed system, where cognition, behavior, and motivation are integrated and organized by the mechanisms of self-regulation toward achieving a conscious goal. Cognitive and behavioral actions, operations, function blocks, and members of human algorithms are considered the basic units of activity analysis. SSAT classifies, describes, and extracts the units of activity analysis from the processes of activity. It also proposes stages and levels of work-activity analysis, in addition to qualitative and quantitative methods of study. SSAT studies work activity at various levels of detail, depending on the purpose of study. It proposes simplified methods of analysis when they are appropriate to the purpose of the analyses. SSAT is a general psychological theory or framework that is theoretically articulated and has been successfully applied in a wide range of situations, in addition to examining human work at a very detailed level.

AAT represents itself in a number of relatively independent theoretical concepts. In some cases, the terminology and the theoretical principles are not sufficiently coordinated and are not in agreement with each other. All these factors create considerable difficulties in describing the original meaning of data obtained in AAT.

In this context, it should be noted that the data in some articles have been obtained from the AAT field and interpreted from an SSAT perspective. The latter (SSAT) allows the presentation of a more accurate description of experimental data. Another important factor for using SSAT is that English-speaking readers can more easily overcome terminology barriers and contradictions that exist between the different schools of psychology of AT used in the former Soviet Union. The term "systemic-structural analysis" often utilizes both general AT and AAT. Nevertheless, there is a considerable divergence between the general philosophical discourse on the systemic-structural analysis of activity and its implementation.

At present, with just a few notable exceptions (Galaktionov 1978; Konopkin 1980; Kotik 1978; Ponomarenko 2004, 2006; Zarakovsky and Pavlov 1978, 2004), we can discern only general philosophical discussions about the systemic-structural analysis of activity, rather than its real development in the study of work activity.

Moreover, the authors mentioned above (as exceptions) have described only some aspects of systemic-structural analysis of activity. Under the systemic-structural analysis of activity, we understand the principles and methods that allow us (1) to create a standardized language for describing the structure and measurements of activity and the units of analysis and (2) to develop various methods of creating mutually interdependent models of activity. Such an approach should provide clearly described stages and levels of analysis of activity, methods of qualitative and quantitative analysis, and so on. It seems to us that at present, such an approach has been developed only in SSAT. Specifically, the SSAT approach allows the unified translation and interpretation of data in the field of AAT. SSAT must hence be considered a new theoretical and practical direction or framework, which is closely linked to general AT and AAT. The basic ideas in SSAT should not be confused with the system concept in ergonomics, which studies man–machine systems. Here, we discuss activity as a structurally organized system that interacts with a machine system. There are probabilistic interrelationships between the structure of the technical components and the structure of activity.

In this volume, we present the works of different authors describing new data obtained in relation to general AT, AAT, and SSAT. However, the terminology used in this volume is unified and standardized from the SSAT perspective (Bedny and Karwowski 2007).

This book contains five sections. Section I, "Activity Theory in Studying Performance," includes two chapters. Chapter 1, "An Introduction to Applied and Systemic-Structural Activity Theory," by W. Karwowski and G. Bedny, introduces readers to AT. There is some information in it about general AT, AAT, and SSAT; it also discusses the relationships between these theories, which could help readers to better understand the material presented later in the book. Chapter 2, "The Relationship between External and Internal Aspects of Activity Theory and Its Importance in the Study of Human Work," by G. Bedny, W. Karwowski, and F. Voskoboynikov, describes the relationship between external (behavioral) and internal (mental) components of activity. The importance of studying the interdependence of external and internal activity in the analysis and design of human work is examined in this chapter. A comparative analysis of the works of Vygotsky, Rubinshtein, and Leont'ev in this area of psychology is presented. The issue of external and internal components in SSAT is considered.

Section II, "Human–Computer Interaction," includes four chapters. It begins with Chapter 3, by W. Karwowski and G. Bedny, entitled "Task Concept and Its Major Attributes in Ergonomics and Psychology." Task analysis is at the core of studies in human–computer interaction (HCI). Hence, the concept of task has a fundamental significance in HCI studies. Currently, however, specialists in the field of HCI do not have a clear understanding of the concept of task. The authors of this chapter review the basic concept of task and analyze it in conjunction with other basic concepts of AT. In particular, the fundamental importance of concepts such as goals in the analysis and description of a task is shown. The difference in understanding the concept of goals in terms of cognitive psychology and AT is demonstrated. The authors of this chapter analyze the relationship between motivation and goals. They also show the importance of concepts such as actions, operations, strategy, and so on in the description and analysis of the task. The presented data are essential to understanding the concept of task. Task analysis not only involves the field of HCI but the general study of human work.

In Chapter 4, "Task Concept in Production and Nonproduction Environments," by G. Bedny and W. Karwowski, the authors analyze HCI and demonstrate the importance of the task concept in studies of HCI in a nonproduction (entertainment) field, where emotional motivational aspects of task analysis become important. The motivations behind the risk-taking attitude of people addicted to gambling are described. Computer-based tasks are problem-solving tasks; therefore, studying the mechanisms of the thinking process is critical to understand the nature of computer-based tasks and strategies of performance. The thinking process during the performance of computer-based tasks is considered, and a model of operative thinking is suggested in this chapter. The described concepts include gnostic dynamics, verbalized and nonverbalized meanings, and conscious and unconscious strategies and their mutual transformations. The relationship between stages of motivation and operative thinking is outlined. Analysis of task performance in the nonproduction environment and classification of tasks are discussed.

Chapter 5, "Microgenetic Principles in the Study of Computer-Based Tasks," by T. Sengupta and I. Bedny, presents a new method of usability evaluation of computer-based tasks. The method is derived from the genetic principles of activity study. The performance of various tasks is studied during activity formation and development. A method for activity analysis during acquisition of computer-based tasks is suggested. Complexity of tasks and their ability to be learned are evaluated during analysis of skill-acquisition processes. New measures of usability of computer-based tasks are suggested.

Chapter 6, "Abandoned Actions Reveal Design Flaws: An Illustration by a Web-Survey Task," which concludes Section II, has been prepared by I. Bedny, W. Karwowski, and G. Bedny. In many situations, even experienced users have to discover some new details of task performance; to some extent, they have to explore how to perform a new task. Users who perform tasks gradually learn the required task sequence through exploratory activity. The explorative stage of task performance aims to examine the existing situation and the consequences of one's own actions. For studying the explorative aspects of task performance, the authors have introduced the concept of "abandoned actions." Cognitive and motor explorative actions that give undesirable results are called abandoned actions. In this chapter, the authors describe formalized and quantitative methods of analysis of explorative activity during the performance of computer-based tasks. A Web-survey task is selected as an illustration in this study. This applied study proves AT that links motivation, cognition, and behavior into a holistic self-regulated system has more advantages compared to the traditional cognitive analysis of computer-based tasks. Evaluation of the effectiveness of the Web-survey task is organized into three basic stages of analysis: qualitative, algorithmic, and quantitative stages. The last stage includes a new method for the quantitative analysis of a user's explorative activity.

Section III, "Evaluation of Computer Users' Psychophysiological Functional State," contains two chapters. This section presents data about users' psychophysiological states during the performance of computer-based tasks. The functional state describes the characteristics of those psychological and physiological functions that may change during a work shift or a particular period. Chapter 7, "Optimization of Human–Computer Interaction by Adjusting the Psychophysiological State of the Operator," is by A. Karpoukhina. She introduces readers to the little-known field of regulation of users' psychophysiological functional state during their interaction with computers. The objective of this chapter is to demonstrate the possibility of increasing the efficiency and the potency of the interface between humans and computers during a work shift. For this purpose, she utilizes laser-based acupuncture to improve operators' psychophysiological state. The advantage of this method is that the influence of laser-based acupuncture is below the sensitive threshold of a user. She shows that this method does not constitute a diversion; that is, it does not interfere with operators' activities and, at the same time, meets health and safety requirements.

Chapter 8, "Psychophysiological Analysis of Students' Functional States during Computer Training," is by D. Yakovetc and I. Bedny. The authors describe the basic types of functional states considered in AT. Psychic tension is the leading functional state that accompanies any goal-directed activity. The authors demonstrate that functional state can be utilized as a criterion for optimization of a student's interaction process with computers. Different instrumental and subjective methods for the analysis of students' functional states during interaction with computers are presented in this work.

Three chapters are included in Section IV, "Work Activity in Aviation." It introduces readers to studies involving the work activities of pilots and air-traffic controllers. The study of pilots' performances in these chapters was conducted based on the functional analysis of activity. Functional analysis of activity should be distinguished from study of the functional state of a subject during work. Functional analysis considers activity as a goal-directed self-regulated system. The major units of analysis in such a study are function blocks. Analysis of activity is performed based on models of self-regulation of activity developed in SSAT. Self-regulative models include the following types: goal-formation process, self-regulation of orienting activity, and general self-regulation of activity. Orienting activity is involved in comprehension of a situation. It describes activity that precedes execution.

AT has accumulated a wealth of information in the field of self-regulation. The studies have originated from the works of two outstanding physiologists, Anokhin and Bernshtein, and have been the theoretical basis for the development of various concepts of self-regulation in AT. In this book, we utilize the theory of self-regulation developed in SSAT, which is very different from the concepts of self-regulation derived from mechanistic models based on homeostatic principles. The proposed theory has clear practical applications. This book shows how this theory is applied in aviation, the study of attention, and so on.

The first chapter of this section, Chapter 9, "Characteristics of Pilots' Activities in Emergency Situations Resulting from Technical Failure," has been prepared by V. Ponomarenko and G. Bedny. The purpose of this chapter is to study pilots' activities in emergency situations resulting from equipment failure. The authors describe the model of self-regulation of orienting activity, and the application of this model for the study of pilots' activities in emergency situations is considered. The relationship between verbally logical and imaginative components in pilots' activities is analyzed. The role of (1) instrumental and noninstrumental signals and (2) relevant and irrelevant information in pilots' activities in emergency situations is discussed.

Chapter 10, "Functional Analysis of Pilot Activity: A Method of Investigation of Flight Safety," is presented by V. Ponomarenko. The author considers some new aspects of the safety of flights based on a functional analysis of pilots' activities. The relationship between functional and cognitive analyses of pilots' activities is discussed, along with examples of the functional analysis of pilots' activities, with particular attention to mechanisms such as goals, stable and dynamic models of flights, the relationship between complexity and difficulty, and the significance of tasks. The data obtained can be used to develop training methods, to design pilots' activity and informational systems, and to create standard alarm systems. This work demonstrates how functional analysis can be utilized in practice.

Chapter 11, "The Methodology of Teaching Flight-Specific English to Non-Native English–Speaking Air-Traffic Controllers," has been prepared by R. Makarov and F. Voskoboynikov. The authors address a very important issue: development of the communication skills of nonnative English–speaking air-traffic controllers. The English language is used all over the world as a means of communication between pilots of international flights and air-traffic controllers. Hence, the ability of pilots to use flight-specific English in the process of flight monitoring is of great importance to ensure the safety of international flights. Air-traffic controllers have to use foreign languages under stressful conditions. They must translate the messages received from their screens, the pilots, and other air-traffic controllers into actions immediately. Based on functional analysis of the controllers' activities and an analysis of their operative thinking, the authors have developed a method of training that significantly improves the controllers' ability to use the English language in emergency situations.

Section V, "Special Topics in the Study of Human Work from the Activity Theory Perspective," contains five chapters and begins with Chapter 12, "Functional Analysis of Attention," by G. Bedny and W. Karwowski. The focus of this chapter is on the study of mechanisms of attention. In the study, the authors utilize both cognitive and functional analyses of activity, which are considered interdependent stages of analysis in SSAT. In this experimental study, subjects receive various tasks in sequence; the second task immediately follows the first task. The complexity of each task is changed from one series of experiments to another. Based on the original method of study developed by the authors, new models of attention are suggested. According to the obtained data, attention is considered a goal-directed self-regulated system that contains a number of interdependent functional mechanisms. Time-sharing strategies and allocation of attention resources are discussed. The ability to adapt and tune various features of attention to a specific task's requirements is considered in detail.

Chapter 13, "Real and Potential Structures of Activity and Its Interrelationship with Features of Personality," is authored by G. Zarakovsky and W. Karwowski. The authors consider the relationship between real and potential structures of activity and an activity's interrelationship with features of personality. Particular attention is paid to the relationship among activity, personality, and professions. General psychophysiological schema of human activity and human potentials and their relationships to personality subsystems are analyzed. A new principle of occupational classification is suggested based on the obtained data.

In Chapter 14, "Application of Laser-Based Acupuncture to Improve Operators' Psychophysiological States," authored by A. Karpoukhina and O. Kokun, the psychophysiological method of activity analysis is demonstrated. The authors describe a possibility of utilizing laser-based acupuncture to improve operators' psychophysiological states. For this study, they utilized the following methods: laser-based acupuncture, measurement of simple sensory–motor reactions, mobility and flexibility of nerve processes, indexes of the cardiovascular system, self-rating, encephalography and so on. A positive effect of laser-based acupuncture on the operators' performance during training processes was detected. In particular, the authors discovered a positive effect on the operators' cerebral blood flow, an improvement in their static muscular endurance, and an increase in their muscle strength and hand-movement coordination. This study demonstrates the possibility of utilizing biologically active skin points not only for diagnoses, but also for regulation of the operators' functional states.

In Chapter 15, "Information Processing and Holistic Learning and Training in an Organization: A Systemic-Structural Activity Theoretical Approach," authors K. Synytsya and H. von Breven consider learning and training processes from the SSAT perspective. A case study of the work of technical support agents is described. It shows a lack of any close connection between learning and training on one side and professional information processing on the other, which not only leaves a serious gap in the agents' skills, but also limits their work efficiency. The need for a holistic view regarding human–computer systems is emphasized in this chapter and is supported by a context schema of agent's activity. An example of such a model that integrates triadic schema, instructional events identified by instructional theories, and some of the agents' tasks is presented. For this purpose, SSAT is suggested because it serves as an effective and all-embracing methodological base for the research and implementation of holistic learning and training systems.

Chapter 16, "Effort, Fatigue, Sleepiness, and Attention Networks Activity: A Functional Magnetic Resonance Imaging Study," by T. Marek and his colleagues, analyzes the relationship between fatigue and sleepiness as a potential source of drivers' errors. The neural basis of the human attention system and its stability over time have become a matter of major concern. The research focuses on the relationship between attention, workload, effort, fatigue, and performance and the opportunities to preview critical events of driving. An operator's functional state, that is, the capacity for sustaining effective task performance under constraints imposed by environmental factors, fluctuates with the time of day the tasks are performed. The major method of study includes MR registration of brain activity in simulated conditions between performances of driving tasks. The role of the brain's attention network in driving is described.

The purpose of this volume is to offer a balanced picture of theoretical and applied issues in the study of human work from the perspectives of general AT, AAT, and SSAT. This book provides readers with state-of-the-art information in AT by emphasizing its application to the study of human work while humans interact with advanced technology.

Gregory Bedny
Waldemar Karwowski
Editors

Contributors

G. Bedny
Institute for Advanced Systems
  Engineering
University of Central Florida
Orlando, Florida

I. Bedny
Institute for Advanced Systems
  Engineering
University of Central Florida
Orlando, Florida

E. Beldzik
Department of Neuroergonomics
  Institute of Applied Psychology
Jagiellonian University
Krakow, Poland

A. Domagalik
Department of Neuroergonomics
  Institute of Applied Psychology
Jagiellonian University
Krakow, Poland

M. Fafrowicz
Department of Neuroergonomics
  Institute of Applied Psychology
Jagiellonian University
Krakow, Poland

K. Golonka
Department of Neuroergonomics
  Institute of Applied Psychology
Jagiellonian University
Krakow, Poland

A. M. Karpoukhina
Academy of Pedagogical Sciences
G. S. Kostuk Institute of Psychology
Kiev, Ukraine

W. Karwowski
Department of Industrial
  Engineering and Management
  Systems
University of Central Florida
Orlando, Florida

O. Kokun
Academy of Pedagogical Sciences
G. S. Kostuk Institute of Psychology
Kiev, Ukraine

R. Makarov
Kirovograd Aviation University
Kirovogradska, Ukraine

T. Marek
Department of Neuroergonomics
  Institute of Applied Psychology
Jagiellonian University
Krakow, Poland

J. Mojsa-Kaja
Department of Neuroergonomics
  Institute of Applied Psychology
Jagiellonian University
Krakow, Poland

H. Oginska
Department of Ergonomics and
  Exercise Physiology
Collegium Medicum
Jagiellonian University
Krakow, Poland

V. Ponomarenko
All-Russian Scientific Research
  Institute of Aviation and Space
  Medicine
Moscow, Russia

T. Sengupta
Microsoft Corporation
Redmond, Washington

K. Synytsya
Kiev State University
Kiev, Ukraine

K. Tucholska
Department of Neuroergonomics
    Institute of Applied Psychology
Jagiellonian University
Krakow, Poland

A. Urbanik
Collegium Medicum
Jagiellonian University
Krakow, Poland

H. von Brevern
School of Informatics
City University of London
London, United Kingdom

F. Voskoboynikov
Baltic Academy of Education
St. Petersburg, Russia

D. A. Yakovetc
Astrakhan Engineering University
Astrakhan, Russia

G. Zarakovsky
All-Russian Institute of Ergodesign
Moscow, Russia

Section I

Activity Theory in Studying Performance

1

Introduction to Applied and Systemic-Structural Activity Theory

G. Bedny and W. Karwowski

CONTENTS

1.1 Introduction

Cognitive psychology studies various cognitive processes. Cognitive processes do not exist independently; they influence each other and are included into unitary human activity (*deyatel'nost'* in Russian). Activity is a socially historical phenomenon that has evolved during the historical development of society. Human labor plays an important role in the formation of activity. Through human labor and its historical evolution, the human mind and consciousness are being developed. Consciousness and human practice are closely interconnected. Human work drives history and determines our social world and values. According to activity theory, you are what you do. Human activity exists in various artifacts, which can be in the form of technology and sign systems developed in a society. Artifacts are created by humans to achieve specific goals of activity. Hence, artificial tools produced by humans have specific purpose and value in activity. It is important to know the kind of goals can be achieved with this technology, the cost of producing particular technology, the efficiency of the technology, and so on. An activity-based approach to the study of human work and technology is a major purpose of applied and systemic-structural activity theory (SSAT). In activity theory, technology is a tool and means of work.

Activity theory studies cognition from different perspectives. It emphasizes the role of activity in the development of cognition and considers the influence of cognition on human work. "Activity" can be defined as conscious, intentional, goal-oriented, and socially formed behavior of human beings and is specific to humans, in contrast to the term "behavior," which is widely used in psychology and covers such phenomena as acts, responses, reactions, movements, and other activities of both animals and humans. Activity theory emphasizes that there is a great difference between human and nonhuman psychic processes. Human psychic processes are unique with respect to their social aspects. The psychic processes of animals are developed according to the laws of biological evolution, whereas the psychic processes of humans are influenced by the laws of social–historical evolution. Furthermore, activity cannot be reduced to reactive behavior. It is an intentional goal-oriented system that includes cognitive and behavioral components. General activity theory is a fundamental theoretical approach from which applied theory and SSAT are derived.

Engestrom (1993), for whom Leont'ev's version of activity theory was a major point of departure, wrote that activity theory does not offer "ready-made techniques and procedures" for the study of human work. According to Engestrom, the conceptual tools of activity theory must be adapted to specific purpose of study. Leading specialists in the former Soviet Union who studied work psychology clearly defined general activity theory as simply a philosophical framework that cannot be directly applied to the study of human work. In the 1970s, they began developing applied activity theory, which can be used in study of human work. Among those were Bedny (1987), Gordeeva and Zinchenko (1982), Galaktionov (1978), Kotik (1974), Konopkin (1980), Landa (1976), Zarakovsky and Pavlov (1987), Zarakovsky (2004), Zinchenko et al. (1974), Zavalova et al. (1986), Platonov (1970), Pushkin (1978), and Lomov (1966). Zarakovsky (2004) called applied activity theory the operationally psychological concept of activity theory. In this chapter, we will consider in brief the general, applied, and SSAT theories.

1.2 General Activity Theory Concepts

In his textbook on general psychology, Petrovsky (1986) defined activity as internal (cognitive) and external (behavioral) processes regulated by a conscious goal and developed under conditions of social cooperation. General activity theory creates a theoretical framework for studying different forms of human praxis. This can be explained by the fact that work activity plays a fundamental role in the development of human psychic processes.

The origin of the term *deyatel'nost'* (translated in English as "activity") in the Russian language is not known. However, in everyday language, it is close to the words for the work, labor, and behavior. The scientific importance of this concept was formed in philosophy, physiology, sociology, and psychology. In Russian physiology, the word "activity" is used to describe the work of the nervous system. The term activity was introduced in Russian psychology by Grot (1879). Activity has been used as the basic concept of psychology since the Russian Revolution in 1917, when labor and work became crucial to the Soviet psychology. Thanks to the work of Rubinshtein (1922, 1986), a new scientific direction for Soviet psychology, which became known as activity theory, was developed.

According to Rubinshtein (1957, 1958, 1959), human consciousness has been formed through the process of work activity. Tools made by people determine the actions performed by people during their work activity. One generation is able to transfer its experiences to another through actions or operations, which are to be carried out by means of specific tools. This suggests that human psychic processes are formed by means of work activity.

Activity theory has its roots in the Marxian philosophy. Marx (1969) stated that historical changes in society and material life trigger changes in human consciousness and behavior. Labor not only modifies nature but also modifies people, who are the agents of activity. Engels (1959) elaborates this view by arguing that human history is the history of the development of human tools, which changes human work activity and human consciousness. Therefore, work is the basis for the development of human mental processes. Since activity theory in the Soviet Union was used to explain the emergence and functioning of consciousness in terms based on Marxist philosophy, it played a fundamental role in the Soviet psychological sciences. That kind of ideology led to the practical conclusion that a socialist society would be able to create a new generation of people, different from the people of a capitalist society.

Psychologists in the Soviet Union have had to prove that they are in line with Marxist philosophy. However, this does not diminish the fact that work plays a huge role in the historical development of humans. Here, we are faced with the usual situation where science and ideology can be used for very different purposes. Despite the certain negative influences of this philosophical basis on the development of science in general, it still had some positive effect on the development of activity theory. Activity theory emphasizes the importance of the study of human labor from psychological perspectives more than other branches of psychology. This theory was developed in the Soviet Union at approximately the same time as that when the behavioral approach was introduced in the United States. Activity theory is particularly useful for work psychology and ergonomics because of its connection with human work. Vygotsky (1971, 1978), Rubinshtein (1957, 1959), and Leont'ev (1978) played a leading role in the development of activity theory.

The cultural–historical theory of development of higher mental functions, developed by Vygotsky, had a significant impact on the development of activity theory. His major idea was that signs as mental tools are a major factor in human mental development. Leont'ev (1978), in contrast to Vygotsky, emphasized the importance of material activity and its interaction with material objects in mental development rather than social interaction. The process of internalization also plays an important role in mental development. Rubinshtein rejected the concept of internalization and proposed the principle of unity of consciousness and behavior (cognition and behavior). This idea was particularly important at a time when psychology was considered either a science of the mind or a science of behavior. The subject–object relationship is critical in Rubinshtein's subject-oriented activity approach. Objects can exist not only as a physical phenomenon but also as a sign system. Nature becomes objects only during interaction with subjects. An object, which can be mental and physical, is transformed according to the goal of the activity. Subjects always operate objects rather than symbols, and they can operate signs and symbols only when they designate some objects or phenomena (Brushlinsky 1979). Rubinshtein together with Leont'ev developed major units of activity analysis to which cognitive and behavioral actions and operations belong. Rubinshtein refined the concepts of motive, goal, and task in activity theory. The feeling of need becomes connected to the goals of activity and thus becomes motive. Rubinshtein did not accept the concept of internalization in Vygotsky's or Leont'ev's interpretation. At the same time, Rubinshtein and Leont'ev did not sufficiently consider the semiotic aspects of human activity. All of this will be considered in more detail in Chapter 2.

The relationship between needs, motives, and goals is an important aspect of activity. Needs may be transformed into motives only in those cases in which they acquire the capacity to induce a person to achieve a particular goal. One goal may cause different motives. Motivation is considered to be a hierarchically organized system of motives. The goal of activity is a conscious image or logical representation of the desired future result of the activity. Hence, needs and motives must be differentiated from goals. A goal is a conscious desired result of activity. In contrast, motives are not always understood. Conscious or unconscious motives appear only in the form of our experienced desires, wishes, intentions, or our striving toward a goal. Activity consists of actions that could be cognitive and behavioral. Actions are directed toward the achievement of conscious goals. Each action is composed of operations. However, general activity theory has not proposed a clear system of classification of motor and mental actions and a method of their extraction in the flow of activity.

The goals of activity have a hierarchical and logical organization. In a hierarchical organization of goals, it is assumed that achieving a high-order goal requires the achievement of subgoals. The logical organization of goals implies that after achievement of one or several interdependent goals, attention shifts to the achievement of other goals defined by the logical structure of activity. In this case, the logic rules for the selection of goals are important. The process of selecting goals and planning their achievement can occur both consciously and unconsciously. Goals of task or subgoals of tasks are higher-order hierarchical goals in comparison to goals of individual actions.

They are developed consciously more often. The goals of individual actions are often formed unconsciously or automatically. Such goals could be quickly forgotten after their achievements. If a person has more than one actual goal, his or her achievement is based on a process of switching from one goal to another. Such goals usually can be kept in the working memory, but the number of such goals is limited by the capacity of working memory. However, if additional goals are unconscious, they are transformed into an unconscious set (Bedny and Karwowski 2007).

Another important concept of activity is mediation. For example, subject and object do not act on each other directly. Tools, which could be external or material, also internal or mental, mediate the relationship between the subject and the object. We will not go further in our discussion of general activity theory here, because it is not the purpose of this book; some additional data in relation to this topic can be found in the works by Bedny and Meister (1997), Bedny, Karwowski, and Bedny (2001), Bedny and Karwowski (2007), Bedny, Karwowski, and Sengupta (2008), and Chebykin, Bedny, and Karwowski (2008). Here, we will outline the following major aspects of general activity theory, which are critical in the study of human work:

- Human activity cannot be reduced to animals' behavior; it is specifically a human kind of activity, which is social in nature.
- Cognition, external behavior, and motivation should be considered components of a unitary system of activity.
- One important principle of unity of cognition and external behavior is that consciousness and practice are interdependent, and work activity therefore emphasizes the role of naturalistic study and the role of culture and history in mental development and human work.
- Activity is not a reactive process. Rather, it is a goal-directed, voluntary regulated process. People can articulate their intention and consciously develop their plans to achieve the goals of activity.
- Activity can be considered a process of transformation of an object (material or mental) according to the goal of activity. There is a regulatory feedback, an intermittent and final result of activity, and a goal of activity. Hence, activity can be considered a self-regulative system.
- Human goals are the consciously desired result of activity in the future. Hence, goal and motives should be considered different components of activity.
- Activity is a historically developed phenomenon that evolves over time within a culture. Hence, activity has socially historical features and therefore should be described only in the context of the community in which it is developed.
- Activity has a complicated structure and consists of cognitive and behavioral actions and operations.

At the same time, all these principles are not sufficiently adapted for the study of human work. For example, the relationship between goal and motive was not precisely described in Leont'ev's concept of activity. There is no developed system of classification or method of extraction of behavioral and cognitive actions in the flow of activity. General activity theory does not develop systemic models of self-regulation of activity. Hence, general activity theory does not suggest principles of description of activity as a structure or as a system, allowing us to conclude that systemic principles of activity study are not developed in general activity theory. There are no precisely described units of analysis of work activity or practical methods or procedures of work activity study. Here we will outline some basic deficiencies of general activity theory for the study of human work. These aspects of activity theory were later elaborated upon in applied theory and SSAT. Thus, general activity theory should be considered only a general philosophical and theoretical background in the study of human work.

General activity theory has become popular in the West, and it has a substantial impact on a wide range of research in psychology. However, the ideas of the theory are not always used properly. We will demonstrate this in some examples. At present, the idea of the theory of activity is used in the study of mental development. One such theory that uses the idea of activity theory is called "embodied cognition." Cognition is embodied when it arises from bodily interactions with the world (Thelen 1995). Here Thelen replaces activity theory terminology. Instead of the term "subject $\rightarrow$ object" he uses the term "body $\rightarrow$ environment." From the activity perspective, the subject interacts with the outside world but not his or her own body. Moreover, the subject does not simply interact with the outside world or environment; he or she interacts with objects that exist in the outside world (Rubinshtein 1959; Leont'ev 1978). The subject has consciousness and language; he or she can develop conscious goals, acquire a certain culture, and so on. An object is something that is transformed in accordance with the goal of activity. The object may be material and ideal. Each of these terms is of fundamental importance in the theory of activity. Thelen uses the theory of activity in the study of mental development in a hidden manner by distorting the theory of activity terminology.

Suchman (1987) developed the *situated action concept*. In this concept, the term "action" is similar to the concept of activity to some degree. The study focused on situated aspects of activity. Situated aspects of cognition were also considered in a similar way by Lave (1988). Both of them emphasized the improvisatory nature of human activity. The work of these authors was correct when they criticized traditional cognitive science approach. However, according to functional analysis, activity and actions are constructed or adapted to a situation in accordance with the mechanisms of self-regulation. Description of activity as a self-regulative system is the purpose of a functional analysis of activity (Bedny and Karwowski 2007). Activity is a combination of components planned in advance and adapted to a situation.

Without understanding of the mechanisms of self-regulation, we cannot correctly describe situated aspects of cognition, actions, or activity.

1.3 Applied Activity Theory

Significant theoretical data was obtained from applied studies of activity theory in the former Soviet Union between 1970 and 1990. Zarakovsky defined general activity theory as a conceptually verbalized theory (Zarakovsky and Pavlov 1987; Zarakovsky 2004). He called the activity theory that was elaborated from the general theory of activity *operationalized* or *operational-psychological activity theory*. In further discussion, we will use the term "applied activity theory." In contrast to general activity theory, applied activity theory offers operationally defined units of analysis not only for externally observable behavior but also for internal, cognitive processes. Substantial research has been done in this area by Soviet scientists Anokhin (1962, 1969) and Bernshtein (1966). They introduced and began to develop the concept of self-regulation in activity theory in 1935. The concept of self-regulation allows scientists to present activity as a self-regulative system.

Applied and general activity theories are interconnected and influence each other. These connections of applied branches of activity theory with general activity theory are not accidental. In general activity theory, the concept of human labor is also critical.

One specific characteristic of activity theory is that not only does general activity theory influence the development of applied activity theory, but applied activity theory also influences general activity theory. One important requirement for psychological studies in the former Soviet Union was a possibility to utilize psychology in practical applications and particularly in the study of human work and learning. The goal concept becomes critical in applied activity theory. "Goal" includes not only verbally logical but also imaginative components. According to the latter, Lomov (1977) introduced the concept "image-goal" in applied activity theory. Goals as requirements of task accepted by a subject can be formulated not only verbally, but also in an imaginative form. Hence, the goal is broader than its verbal equivalent. Another important concept introduced in the applied field was a "motives → goal" vector. In SSAT, motive or motivation is energetic and goal is a cognitive component of activity. This issue will be considered further in Chapters 3 and 4.

Another important aspect within the frame of applied activity theory was an attempt to study cognitive processes from systemic analysis perspectives. For example, even the simplest psychic process, such as sensation, is considered an element of a complex activity system that includes goal, motivation, and strategies. Sensation is considered a process that interacts with other psychic processes. Psychophysical tasks, like any other tasks, require problem solving. This means that even a simple psychophysic task can be interpreted by subjects in different ways. Based on this concept, Zabrodin (1985) introduced the notions of "sensory space" and "space of decisions." Sensory space includes possible alternative images of the situation. The space of decisions includes alternative possible responses that may accomplish the goal of psychophysical task. Thus, in the simplest situation tied to the extraction of the weak signals from a noise background, task is considered from the perspectives of problem solution and strategies of performance.

Studies in the field of applied activity theory demonstrate that the efficiency of cognitive processes functioning depends on goal, motivation, and strategies of self-regulation. For example, the organization of a material may assist in its utilization and enhanced memorization. However, how subjects use this material is critical. Zinchenko (1961) suggested that memorization was dependent on the goal of activity. Zinchenko et al. (1980) demonstrated that short-term memory cannot be presented as a rigid, totally involuntary programmed processes. Goal, consciousness, motivation, and so on can influence short-term memory. Features of short-term memory depend not only on the features of our brain but also on voluntary regulated human activity. Particular attention was paid in applied activity theory to studies of practical thinking (Pushkin 1978; Tichomirov 1987). One important type of practical thinking is working or operative thinking. The role of emotional components in this type of thinking was shown.

Scientists in applied activity theory were able to obtain some interesting theoretical and applied data in studies of human work, particularly in the studies of pilots' work activity. Some general principles of such studies are presented in the works of Dobrolensky et al. (1975), Ponomarenko (2006), and Zarakovsky and Pavlov (1987).

In this book we pay particular attention to systemic-structural activity approach. The topic of systemic analysis of work activity has been very popular in Soviet psychology. Systemic analysis as an interdisciplinary field should not negate the need for the development of proper systemic psychological methods of activity study. Implementing such an approach becomes possible when activity can be described as a complex structure evolving over time. Therefore, we will discuss the systemic-structural analysis of work activity with higher precision. Creation of such an approach is possible only when we can develop methods of analysis to describe the activity as a systemic-structural entity. In this case, a description of activity as a system, which consists of the elements that are in specific relations and interaction with other elements of activity, is at the forefront of the research.

Transition from a general philosophical discussion about systemic analysis to its practical application is not so easy. Existing methods of systemic analysis of activity are important and useful from the theoretical and practical points of view. However, they are fragmented and cannot substitute for a unified and, to some extent, standardized approach to systemic analysis of work activity.

From Schedrovitsky (1995), we believe that the nature of systemic-structural method is determined by what methods and tools we use to describe an object as a system. This means that the same object can be presented as a system in different ways, depending on the methods used. Hence, we need to develop standardized methods, research procedures, and operations to provide the possibility of developing various types of models of the same object in the form of a system and structure.

We will now concentrate our efforts primarily on the application of the systemic-structural approach to an area of study that is related to the task analysis in ergonomics, work psychology, and labor economics in which the application of systemic-structural analysis of activity is the most important. Systemic-structural analysis raises a number of issues that need to be addressed. First of all, there are needs for development of standardized units of analysis of activity and language of its description, the selection of stages and levels of analysis, development of methods for constructing models of activity, analysis of their relations, identifying the relations between qualitative, formalized, and quantitative research methods, and so on. In this regard, we believe that the systemic-structural theory of activity is the most promising (Bedny and Karwowski 2007). Introduction to this theory will be discussed in the next section.

1.4 Systemic-Structural Activity Theory

Applied activity theory is not a unitary theory. The terminology used by different authors is not the same, and therefore, it does not unify the theory. There are no standardizing units of analysis and standardized procedures of work activity analysis. We have made an attempt to present these data in a more unified way (Bedny and Meister 1997) by developing an SSAT, described thoroughly in the work of Bedny and Karwowski (2007). An SSAT is a unique and independent approach in procedures of work activity analysis, and is distinct from those that preceded it. This approach is derived from general and applied activity theory. Let us briefly consider some aspects of this theory.

The SSAT views activity as a structurally organized self-regulated system rather than as an aggregation of responses to multiple stimuli, or linear sequence of information stages as described in behavioral or cognitive psychology. Furthermore, it views activity as a goal-directed rather than as a homeostatic self-regulative system. A system is considered goal-directed and self-regulated if it continues to pursue the same goal under changed environmental conditions. The system can reformulate the goal while functioning. Activity as a self-regulated system integrates cognitive, behavioral, and motivational components. A simplified schema of activity is presented

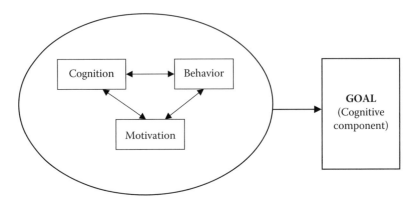

FIGURE 1.1
Simplified schema of activity as a system.

in Figure 1.1. The figure demonstrates that cognition, behavior, and motivation influence each other. Figure 1.1 shows that it is not only cognition that regulates behavior, but that behavior through a feedback also regulates mental processes and their development. We will consider this issue in a more detailed manner in Chapter 9.

The systemic approach is utilized in work psychology for describing a man–machine system. Here, we talk about activity as a system. Activity as a system is a set of independent elements (units of activity) that are organized and mobilized around a specific goal. Hence, here we utilize a systemic-structural analysis of activity. This analysis not only differentiates elements of activity in terms of their functionality but also describes their systemic interrelations and organization.

A systemic-structural approach depends not so much on the qualities and specificity of the object under consideration but rather on the perspectives and methods of analysis. It follows that considering the object as a system to a significant degree depends on the abilities of scientists to describe activity as a system. The object was not considered a system until the requisite tools and methods of study were developed. In the absence of the appropriate units of analysis the whole systemic-structural analysis of activity was impossible. In SSAT, which has standardized units of analysis of activity and their precise description, systemic-structural analysis is possible. For example, in SSAT, cognitive and behavior actions have precise descriptions. Method of their extraction from holistic activity is described in SSAT. There are also other units of analysis, such as cognitive and mental operations, functional macroblocks and microblocks, and members of human algorithm, which consists of one or several interdependent actions integrated by high-order goal. Members of algorithm have a logical organization and describe the logic of activity performance. In contrast, in general activity theory and even in applied activity theory, these units of analysis are not precisely described and unified. There are no principles for their organization into a holistic system and no standardized stages and levels of analysis. Therefore, from our point of view, general activity theory cannot be adopted for practical purposes, however some Western scientists have attempted to utilize activity theory for practical purposes. At the same time, applied activity theory has a number of powerful methods in a study of human work.

The cognitive approach, which is dominating currently, treats cognition and behavior as a process and makes it difficult to study activity and behavior from systemic-structural perspectives. The notion of the process does not allow us to describe the activity as a structure. Introducing standardized and unified units of analysis in SSAT helps us to describe activity as a structure that unfolds over time. SSAT does not reject the cognitive approach, but uses this approach as a stage of activity analysis.

SSAT invites and empowers the study of the same activity from a different point of view and from distinct aspects, thereby legitimizing the use of multiple approaches for the description of a single object. This implies that in applied research, an adequate description of the same object of study requires multiple interrelated and supplemented models and languages of description as presented in Figure 1.2.

In Figure 1.2, we see X as an object of study, but A, B, and C are different and interdependent supplementary presentation of the same object. Therefore, the activity system cannot be described by one best method as is done in cognitive psychology. The activity system calls for different stages and levels for description of the same activity. These stages are qualitative analysis, analysis of logical organization of activity (algorithmic description of activity), analysis of time structure of activity, and quantitative description of activity (evaluation

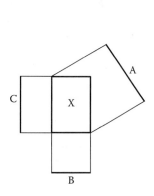

FIGURE 1.2
Systemic representation of an object.

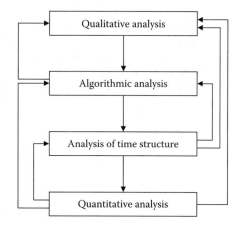

FIGURE 1.3
Four stages of systemic-structural design.

of task complexity or reliability of task performance; see Figure 1.3). These stages have a loop structure organization, that is, sequential stages of analysis can require reconsideration of the preliminary stages. Each stage of analysis can be performed with different levels of decomposition. Macrostructural and microstructural analyses determine the levels of analysis.

At the microstructural stage of analysis, cognitive and behavioral actions are described as a system of operations or functional microblocks. For example, motor actions can be described as a system of motions. Decomposition of activity into logically organized system of actions is related to morphological analysis of activity. Similarly, cognitive actions can be subdivided into mental operations. The starting point of an action is the initiation of conscious goal of action (goal acceptance or goal formation), and the action is completed when the actual result of the action is evaluated (Figure 1.4). A more detailed analysis of the method of extraction as well as description of action can be found in the work of Bedny and Karwowski (2006, 2007).

Through morphological analysis, cognition is considered not only a process but also a logical organization of cognitive actions and operations. Identification of motor actions and cognitive actions can be complex. For example, during analysis of computer-based tasks, it may be necessary to use eye movement registration to extract and classify cognitive actions (Bedny, Karwowski, and Sengupta 2006). Depending on the strategies of activity performance, the same task can contain different actions and their logical organization. The concept of tool and object depends on a goal of actions and a goal of task. According to the goal of the task, during transformation of the initial screen situation into the final screen situation, the initial situation can be considered an object of activity. However, during the analysis of separate actions, different objects can be extracted. For example, when a subject performs the action "reach and grasp the mouse with right hand," he or she performs a simple motor action with an approximately 30-centimeter distance. This action includes two motions (reach "R25A" and grasp "G1A"). Separate motions descriptions can be found in the MTM-1 system. However, their combination into motor actions is described in SSAT (Bedny and Karwowski 2007). The goal of the described action is the mental image of the result (grasp the mouse). In this action, the object of the action is the mouse. When the subject performs the second action "move the cursor to the required position and depress the left mouse button with the index finger for activation

FIGURE 1.4
Simplified model of action as a loop-structured system.

of particular icon" (move object to exact location "M7C" and apply pressure "AP1"), the mouse is not an object of action but is a tool of action. The pointer is an element associated with the mouse and also the tool of the action. At the same time, the icon on the screen is an object of the action.

There is another important aspect of action classification in SSAT. It is important to differentiate between activity elements (psychological units) and task elements (technological units) of analysis. Psychological units of analysis represent standardized elements of activity. Anyone who is familiar with these units of analysis can clearly understand what a subject does when he or she performs these actions. For example, in the MTM-1 system, R25A means reach an object in a well-known location when the distance is 25 centimeters. Note that the MTM-1 system does not distinguish psychological and technological units. This classification was introduced in SSAT and similar classification is applied for cognitive actions. When we attempt to design the time structure of activity and evaluate task complexity, we need to convert technological units of analysis into psychological ones. In the process of task analysis at the first two stages, a specialist usually utilizes technological units and then transfers them into psychological units.

Activity is a combination of normative or standardized and variable components. Therefore, it can be described with some approximation. Normative and variable components depend on the rules and prescribed procedures of task performance. They depend on existing technology at the time, safety requirements, and prescribed methods of task performance. Periodically, these normative requirements can be changed. However, task performance practices cannot be changed voluntarily according to the individual desire of a worker; written standardized practice must be maintained and preserved. New methods of task performance can be introduced only after careful analysis of existing methods and conclusion by professionals, who are responsible for the analysis and design of the job process. At the same time, normative methods have some flexibility, that is, a worker can utilize different individual strategies for a task performance. However, possible strategies of task performance are restricted by existing technological constraints. Individual strategies can be considered acceptable if they do not violate existing technological constraints. Such strategies in activity theory are known as individual styles of activity performance (Bedny and Karwowski 2007; Bedny and Seglin 1999). Any activity includes variable components that cannot be totally eliminated. Variability is a feature of not only human activity but also of many complex systems. For example, in mass production and manufacturing, each manufactured part is unique in its size and shape. However, if the variation of sizes and shapes is within an established range of tolerance, all parts are considered to have the "same" size and shape parameters. Similarly, if the same strategy of activity performance falls within an established range of tolerance, the strategy is considered the "same." Strategies, constraints, individual styles of performance, variability of activity, tolerance, and so on help us to understand relations between normative (standardized) and variable components of activity in the process of activity design. That is why we do not agree with the constraint-based approach of design suggested by Rasmussen and Pejtersen (1995) and Vicente (1999). Their approach ignores the relationship between normative (prescribed) and variable components of activity. The understanding of relationships between normative or standardized and variable elements of activity eliminates contradictions between ideal task models and those that represent real task instances, which are often discussed in the human–computer interaction field. If variations in task instances do not exceed the existing range of tolerance, then this strategy of task performance can be considered the expected standardized method of task performance.

1.5 Activity as a Self-Regulated System

In SSAT, functional analysis of activity is the systemic qualitative analysis of activity during task performance. In the process of functional analysis, activity is considered a self-regulative system. The purpose of this analysis is to describe possible strategies of activity performance, while the major units of analysis are functional mechanisms or function blocks. Functional blocks represent a coordinated system of cognitive processes that have a specific purpose in the regulation of activity. For example, functional blocks can be responsible for the interpretation and acceptance of a goal, formation dynamic model of situation, evaluation of difficulty and significance of situation, development program of performance, subjective criteria of successful result, and so on. A general model of self-regulation of activity includes 20 functional blocks. Any functional block that is introduced in a model of self-regulation, is the product of experimental and theoretical studies. During the process of functional analysis of activity, we attempt to describe the activity not in terms of cognitive processes or cognitive and behavioral actions but more in terms of complicated integrative functional mechanisms. The length of this chapter does not allow us discuss this model in detail. However, here we will briefly discuss the model of a goal-formation process (Figure 1.5). It is the simplest self-regulative model in the functional analysis of activity.

The content of each functional block can change depending on the specificity of task performance; however, the purpose of each functional block in the self-regulation of activity will remain the same. We can examine each functional block and their interrelationship during a task performance. A model of self-regulation of activity can be interpreted as an interdependent system of windows (functional blocks) from which we can observe human activity during the task performance. For example, a researcher can open a window called "Goal" and at this stage of analysis can direct his or her attention to aspects of activity such as goal interpretation, goal reformulation

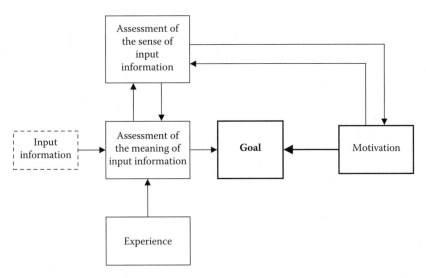

FIGURE 1.5
Self-regulation model of the goal formation process.

or formation, goal acceptance, the relationship between verbally logical and imaginative aspects of goal, possibility to understand the goal at different stages of task performance in more detail, influence of goal on interpretation of input information, and so on. In the next step, she or he can open another box called "Assessment of the meaning of input information." At this stage, it is important to find out how objectively presented information about task are being transformed into subjective situational meaning associated with goal.

In contrast, when we open the block "Assessment of the sense of input information," we concentrate our attention on aspects such as emotionally evaluative aspects of situation related to the goal. Therefore, the concept of "significance" becomes the major focus of analysis. We should note here that objectively important information is not the same as subjective evaluation of this information. The factor of significance influences cognitive interpretation of the situation. This block influences motivation. The more significant information is to the subject, the more the subject is motivated to reach the goal. Similarly, other windows or function blocks can be opened selectively, depending on the activity peculiarities. Depending on the relative importance of other windows or function blocks, a specialist can selectively open another block. Some functional blocks might be skipped altogether during functional analysis. Through the feed-forward and feedback between functional blocks, we can find the interdependence of the obtained data. The existence of blocks such as "Assessment of the sense of input information" and "Motivation" demonstrates that the specificity of the processing of information depends on emotionally motivational aspects of activity. A subject can interpret the information in a totally different way depending on the factor of significance. Cognitive psychology ignores this factor. Functional analysis of activity helps us to introduce emotionally motivational factors in cognitive psychology. This model demonstrates that vector "motives → goal" is a result of self-regulative process, and SSAT can explain this phenomenon in a much more detailed manner than general activity theory. Thus, each functional block determines different aspects of the study of the same task. These aspects also depend on the specificity of the task. For example, in one situation it is very important to pay attention how subjectively accepted goal deviates from objectively formulated goal and how this influences the interpretation of information. In another situation, it is important to find out the relationship between verbally logical and imaginative components of goal. These questions become common in the study of pilot's tasks.

A more detailed description of functional analysis is presented in the work of Bedny and Karwowski (2007) and in Chapter 9 of this book. The concept of a functional block with some approximation can be compared with the concept of a module in cognitive psychology. This concept is discussed by Sternberg (2008) in more detail. According to Sternberg, the complex human information-processing system should be divided into modules or parts that are independent in some manner and have different functions. In cognitive psychology and SSAT, methods for decomposition of mental processes are different. However, the theoretical bases of these approaches have some similarities. Any complex information-processing system can be split into small subsystems. This is a very important principle of systemic analysis of any complex system. From a functional analysis perspective, it follows that objective presentation of task to the subject and subjective representation of the task are not the same. This is important from a practical and theoretical point of view. For example, even the simplest sensory task, "detection of signal," can be perceived subjectively in different ways, and even the most rigorous instructions cannot prevent subjective interpretation of the task. During the design of an experiment, psychologists should consider this factor. The task includes objective requirements that can be transferred into subjectively accepted goals in different ways. Task conditions can also be interpreted in a subjective way. Based on this, multiple strategies of task performance can be developed.

Analysis shows that models of self-regulation of activity consist of functional blocks. Each block is associated with a stage of activity analysis, and each stage includes the system of theoretical concepts, a set of procedures and techniques that are used in the theory of activity. All of the methods and procedures of study are divided into stages, which are associated with particular blocks. Consequently, each functional block requires a specific set of research methods. Connections between different functional blocks in the model of self-regulation of activity demonstrate connections between the stages of analysis, and reflect the ways of transition from one stage of activity analysis to another. Hence, suggested models of self-regulation determine the systemic analysis of a performer's activity. In SSAT, there are also models of self-regulation of orienting activity and general models of self-regulation.

We will consider models of self-regulation of orienting activity when we discuss pilot performance in Chapters 9 and 10.

There have recently been attempts to present the concept of self-regulation ignoring the achievements of activity theory in this area. The work of those psychologists who work in the area of industrial/organizational (I/O) psychology is interesting in this regard. Let us consider briefly some ideas of self-regulation proposed by Vancouver (2005). In his diagram of self-regulation, there are three hypothetical stages or functions: input function, comparator, and output functions. Input function is a perceptual process responsible for the perception of incoming information. The following stage is comparator, which compares input information with the goal or standard. The comparator stage gives information about the error. This information is taken into account by the third stage, called output function, which is simply a behavior. All three functions are connected in sequence by lines. These lines represent where signals can pass. The result of the behavior affects the variable. Input function perceives the state of the variable. This single closed-loop process is considered self-regulation.

These three stages of self-regulation do not have sufficient scientific justification. For example, the input function is reduced to the perceptual process. However, this is only possible in very simple cases. For example, when a well-perceivable and familiar signal is presented to a worker, he or she must respond to this signal. In reality, the situation may be quite complex, and the worker should identify the relevant and irrelevant information, structure information, combined obtained data with information stored in memory, and so on. Hence, receiving information includes the function of memory and thinking. Then, Vancouver argues that the input function specialized only for the perception of the state of the variable, which depends on the output function and disturbances. However, the worker does not only perceive changes in a situation as a result of his or her behavior. He or she actively selects information from the situation, compares what has changed as a result of his or her own actions with elements of situation that was unchanged, and combines the selected data with the data extracted from memory. Based on this, the worker creates a dynamic mental model of the situation. The worker does so in accordance with the goal of his or her activity, which he or she formed for himself or herself and the specifics of the situation in general. The worker does not interact with the variable in isolation from the general situation.

The term "variable" is suitable for experimental procedures but not for the process of receiving complex information about the situation. The complex information about the state of the system and environment cannot be considered to be variable.

The goal in Vancouver's diagram is depicted as an externally given standard that influences the comparator. However, the goal cannot be presented as an externally given standard in ready form for the subject (Bedny and Karwowski 2007). Externally given requirements of task must be subjectively accepted or formed as a goal by the subject. The goal and subjective standard of successful result are not always the same. We want to emphasize that Vancouver's diagram is homeostatic. Self-regulation in humans is goal-directed. There are a number of proposed models of self-regulation outside of activity theory; these are much more detailed than Vancouver's diagram (Norman 1986; Frese and Zapf 1994). For example, Norman introduced seven stages of self-regulation in his model. Even regular textbooks propose a more advanced model of self-regulation than Vancouver's diagram (see, for example, Hilgard, R. Atkinson, and R. C. Atkinson 1979, p. 284). These models, as well as the Vancouver's diagram, have only one self-regulation loop. However, the self-regulation of activity in general cannot be represented as a single loop of self-regulation. Self-regulation consists of multiple subloops. In activity theory, we can find different models of self-regulation (Bedny and Meister 1997; Bedny and Karwowski 2007). In general, Vancouver's self-regulation diagram is a clear deterioration in addressing this problem and shows a complete ignorance of the problem. We cannot analyze in detail Vancouver's diagram of self-regulation here. We consider the problem of self-regulation in more detail from the standpoint of activity theory in Chapters 9 through 12.

1.6 Goal Concept: Comparative Analysis

The material presented in Sections 1.2 through 1.5 demonstrates that to understand the activity as a goal-directed system, we should understand the goal of activity. In cognitive psychology, social and personal psychology, and in the field of motivation, a goal is considered a combination of cognitive and motivational components. There is also a general consensus among professionals in these areas of psychology that a goal can be both conscious and unconscious (see Austin and Vancouver 1996; Pervin 1989; Locke and Latham 1990). In contrast to an approach where goal has both cognitive and motivational features, in activity theory, goal is a cognitive mechanism connected to motives. Data analysis demonstrates that in activity theory, the concept of goal has significantly different characteristics and meaning. The main differences are that the concept of goal is related only to human behavior or activity, the goal always includes conscious components, the goal is cognitive representation of a future result of subject's own activity, and goal is connected with motive and creates a vector "motives → goal."

In contrast, representatives of I/O psychology such as Austin and Vancouver (1996) gave the following definition of goal: "Goal is an internal representation of the desired state of the system." In activity theory, the term "goal" is used only for the analysis of human activity. The desired future state can be applied not just to a human goal; it may be a result of events not directly related to human activity. The desired future result becomes a goal only if it directs human activity and the goal can be achieved as a consequence of such activity. Human goal and system goal are not the same. Vancouver operates with the most important categories of psychology in a very confident and at the same time irresponsible way. He insists that there are two types of goals (Vancouver 2005, p. 329): (1) a goal for the perceptual unit of behavior and (2) a goal for the internal unit of behavior. For the internal unit, goal is simply the desired level of errors. In this case, the author's arguments are contrary to all existing data in psychology. Goal cannot be considered a desired level of errors. There are no external perceptual and internal goals. Perceptual activity cannot be strictly isolated from other kinds of mental activity and emotionally motivational processes. Moreover, external behavior is closely linked with mental and emotional–motivational processes.

With great self-confidence, Vancouver has introduced other types of goals. Trying to link the goal with self-regulation, he introduced attainment goal and maintenance goal. The notion of "maintenance goal" from his point of view helps to connect the concept of goal with the concept of self-regulation. The author misinterprets the concept of self-regulation of human activity by reducing the process of self-regulation to elimination of deviations from a so-called maintenance goal. Understanding self-regulation as a process of elimination of deviation from the maintenance goal is a homeostatic principle of self-regulation. However, human activity is not limited to the elimination of errors that deviate from the standard. Such simple tasks are usually accomplished by technical systems. Typically, if a situation deviates from acceptable limits, the performer formulates a new goal and therefore a new task that helps to eliminate the deviation. At the next stage, the performer performs a logical system of actions aimed to achieve this new goal. Thus, the introduction of maintenance goal is totally unfounded. Moreover, self-regulation cannot be reduced to elimination of errors. According to Vancouver (2005), if a subject makes an error, this gives him or her an opportunity to eliminate it. Then, another error is made and corrected. Human behavior cannot be reduced to a process of error elimination or moving from one error to another. In SSAT and ergonomics, there are notions such as error, mistake, failure, range of tolerance, acceptable level of deviation, and so on. Vancouver demonstrates a complete misunderstanding of the concept of error in psychology. There is a lot of information in psychological literature about errors (see, for example, Norman 1981; Reason 1990; Senders and Moray 1991; Kirwan 1994; Bedny and Meister 1997).

To substantiate his arguments and theory, Vancouver utilized the following bizarre hypothetical example (Vancouver 2005, p. 305). In a maintenance context, a widget-maker performs the following task: he monitors the state of the shelves in his store. His goal is to keep the shelf full. When customers purchase widgets, the widget-maker must make more, but only enough to fill the shelf. The customers are considered to be a source of disturbance of the variable (the state of the shelf). A customer disturbs the variable and produces an error. In the attainment context, the widget-maker has a goal to produce the required number of widgets to fill the shelf and to keep it full.

Each workday begins anew with zero widgets made and ends when the goal is reached. Further, the author writes "as customers purchase widgets, an error is created between the goal and the widget maker's perception of the state of the shelf" (Vancouver 2005, 315).

The above example demonstrates that the author does not understand the meaning of production process, work process, task, errors, and so on. First of all, we need to understand that a widget maker performs incompatible functions. When a worker produces a widget, she or he can perform a number of production tasks. Transportation also can include a number of tasks and takes time. A customer cannot wait for the completion of the production process. The same person cannot be responsible for production, storing, and selling. Customers do not disturb businesses. The main goal of the work process is to sell a product, not to keep the shelves full. In this example, the widget-maker has multiple goals that are not compatible over time. Moreover, it is not reasonable to fill the shelves after the sale of each widget. Only when the number of widgets approaches the minimum required quantity does a worker formulate a task (i.e., goal of task) to go to the production room and bring the widgets to fill the shelf. When the number of widgets becomes lower than the minimum required and results in an inability to serve the customer in a timely manner, such events can be considered errors. In other situations, a lower number of widgets can be considered a permissible level of deviation. Bringing the widgets to shelves is a particular stage of the work process that might include a number of tasks. If there are a number of tasks, there are a number of goals for the tasks. Each task includes a number of cognitive and behavioral actions, which also have their own goals. Hence, self-regulation cannot be considered the elimination of so-called disturbance and errors. The self-regulation process allows not only the correction of errors but also the prediction and prevention of them. Self-regulation happens even when there is no disturbance and error as it is considered in the above-described example. Vancouver reduces self-regulative process to elimination of errors that are the result of disturbances. Our activity is a self-regulative system. Self-regulation is a complex process that regulates the entire activity, so the term "maintenance goal" is not an accurate one. Disturbances include danger, unanticipated events, emergencies, and so on. Subjects have to improvise and adapt to the contingency of such disturbances. Because of the disturbances, the self-regulation process becomes more complex, and the strategies for task performance change. There are strategies that are utilized in normal work conditions, in dangerous situations, and for other disturbances, and transitory strategies are utilized when a subject transfers from an existing strategy to a new one. The foundation for all these strategies is the process of self-regulation that involves goal formation. We will consider some examples of self-regulation in a pilot's activity during the performance of a variety of tasks in emergency conditions in Chapters 9 and 10.

Analysis of Vancouver's publications demonstrates that there is currently no clear understanding of goal, task, self-regulation, and other important concepts that are necessary for task analysis. I/O psychologists who study human work cannot use such primitive examples even in their theoretical discussion. We would not find this level of examples in ergonomics, where specialists such as psychologists, engineers, and computer scientists work together.

Note that in activity theory, the goal of a future desired result of performance can be represented as an image or a verbal or symbolic description. The goal also has various levels of specificity or precision and can be clarified and become more precise in the course of activity. However, if the goal is completely changed, the achievement of this new goal will mean that the subject is involved in a new activity. The image or mental representation of the future result becomes the goal of the task only during its interaction with the motives of the activity, which determines the directness of the activity to achieve the goal. Depending on the motives with which goal is connected, it acquires a different personal sense. Goal acts as a cognitive component, and motivation acts as an energetic component of activity. In view of the fact that the activity is often polymotivated, it is possible that the subject relates to the same goal differently in various circumstances. For example, if there are conflicts between positive and negative motives in the process of approaching the goal, the acceptance or the formation of goals and hence the task performance in general is conducted in a conflicting motivational background. This conflict of motives can be changed in the process of goal attainment.

In activity theory, the concept of goal is closely linked to the concept of task. The goal of the task determines the integrity of the activity during the task performance. The relationship between the goals of the tasks determines their logical organization. The goals of individual actions during task performance are of particular importance in the analysis of individual actions and the formation of a task performance program.

In cognitive psychology, social psychology, and motivation theories, goal is not considered in the context of the tasks performance. It is considered and analyzed in relation to the behavior or activity in general. Moreover, terms such as "activity" and "actions" are often not distinguished. This makes the notion of goal very amorphous and undetermined. The goal emerges as a motive, a cognitive entity, and a characteristic of personality. This makes it virtually impossible to use this concept in task analysis. Awareness of the goal of task as one of its most important characteristics is ignored. All of these drawbacks are eliminated by goal interpretation in activity theory. The goal of the task and its relation with motivation are central to the task analysis in activity theory.

It is also important to consider the relationship between the goal and consciousness. In various areas of psychology, the goal is seen as a conscious or unconscious element of human behavior or activity. For example, Austin and Vancouver (1996) wrote that goals are not limited to a conscious level.

Indeed, if we are talking about low-level goals, which are specific to biological systems, they can be unconscious. However, in the theory of activity, human goals are always conscious to some degree. This is not to deny the existence of the low-level unconscious anticipatory mechanisms of behavior. However, in the theory of activity, there are other terms for low-level goals such as "purpose" (Tolman 1932), "anticipatory results" (Anokhin 1969), "required future" (Bernshtein 1945), and "neural model of a stimulus" (Sokolov 1969). These unconscious anticipatory mechanisms, as well as conscious predictive mechanisms (human goals), manifest themselves at different levels of activity performance. All of these predictive mechanisms make up direction to the human activity. However, only human beings reach the more complex conscious level of reflection of the future desired result during their own activity. This level of reflection is always associated with conscious goal. Such goals include some verbalized components. This means that the subject can describe his or her goal with some approximation.

In activity theory, it is possible to distinguish various types of goals that may have different organizations and relationships. For example, each task has a final goal, the achievement of which is the completion of the task. However, an individual needs to perform a logical sequence of actions that have his or her own goals in order to achieve the goal of task. Goals of actions are often formed involuntarily. They can be conscious within a short period of time and quickly forgotten. The goal of the task can be formulated more consciously and stored in memory for a longer time. There are proximate and distal goals. The proximate goals can be achieved in a relatively short period of time. The distal goals are remote in time. Progress toward a distal goal requires achievement of a number of intermediate goals. As shown in Section 1.5 and Chapter 9, goals may have various levels of difficulty and significance.

There are also "potential" and actual goals. A potential goal is not a real goal. It is something that in particular circumstances could be a goal. In analyzing the goals that are connected with past experiences, it is useful to distinguish potential and actual goals. Potential goals are kept in memory and associated with existing needs, and they are not conscious. An actual goal is actualized in memory due to a higher level of needs potentially associated with it. Needs might change over time due to the activation of associative memory connections (Zarakovsky and Pavlov 1987). External situations can trigger the activation of these associations. If potential needs exceed the certain intensity threshold, the potential goal is transformed into an actual goal. Such goal formation is unconscious, and after the formation of the goal, the performance program is triggered almost automatically. The formation of the required level of intensity of potential needs and the level of intensity of motivation associated with it also depends on the nonconscious feelings of importance of the situation, feelings of danger, subjective assessment of the difficulty of attainment possible in the future, and the not always clearly defined result (Bedny and Karwowski 2006). The more significant the feelings, in general, the higher is the intensity of the need-motivational components of activity.

The described method of goal formation should be distinguished from voluntary goal formation, which is often associated with a willingness involved in goal formation. An involuntary goal formation process is more typical for the formation of the goals of separate and especially habitual actions. If we are talking about a task's goal, such a goal is often formed voluntarily. Voluntary goal formation process is particularly important for the study of human work. Based on the analysis of the situation, existing requirements, guidelines, and past experience, a worker accepts or formulates a goal of task. When considering the process of goal acceptance or formation, it is essential to evaluate the goal significance (Bedny and Karwowski 2007; Bedny and Karwowski 2006). In extreme situations, the role of volitional processes is elevated. For example, in case of the presence of danger, the operator can utilize volitional efforts to suppress the motive to escape from the danger and to increase the importance of social motives aimed at suppressing fear and rescuing people. In all these examples, concepts such as significance, motivation, and difficulty are considered functional mechanisms of self-regulation.

Many scientists are ignoring the fact that the goals of the activity and behavior in general have not only hierarchical but also logical organization. Most goals have no subordinate relations with the other goals. Their consistent achievement may be defined arbitrarily by the subject. Austin and Vancouver (1996) reduce behavior and relationship between goals only to the hierarchical organization. Simon (1996) introduced the concept of "span of control." If the level of goals' subordination exceeds the span of control, such a hierarchical system has limitations in being efficiently regulated. Moreover, Austin and Vancouver use the term "low-level goals." However, in activity theory, these are not considered goals because they are not conscious.

Some authors use terms such as "antigoal" or "work avoidant goal" (Carver and Scheier 2005; Pintrich 2005). Such goals do not exist in activity theory. The subject may or may not accept the goal formulated by instruction. Moreover, in response to the presented goal, a subject can formulate her or his own goal, which contradicts with the goal objectively presented by instruction. In activity theory, goal is always associated with motives and creates the vector "motives → goal." This vector defines the direction of activity. This vector is directed at achieving the required goal. "Antigoal," "work avoidant goal," and so on are just new goals, and thus the new vector that determines directions of activity is formed and the subject formulates a new type of activity or task. A lack of clear understanding of the goal leads to a situation in which some authors propose to eliminate the concept of goal in task analysis entirely (see, for example, Diaper and Stanton 2004 and Karat et al. 2004). This is due to the fact that a concept such as goal has both cognitive and motivational features in cognitive and social psychology. In other words, the goal and motives in task performance are not distinguished. In most cases, human behavior is polymotivated. Hence, according to these scientists, one task has multiple goals of task, but this is impossible because one task can have only one general goal of task.

1.7 Conclusion

This chapter provided an introduction to some issues that have arisen during the study of work activity theory. Applied theory and SSAT describe activity as a goal-directed and self-regulated system. This approach considers cognition, behavior, and motivation in unity. Hence, activity theory integrates different domains of psychology and sheds new light on the study of human work. We showed that general activity theory cannot be sufficiently adapted for the study of human work. Applied theory and SSAT were developed for this purpose. We described some data that were obtained by us and other authors in their research in the field of applied theory and SSAT. From the examples that we presented, we can see that these data not only have an important application in practical use but also have significant theoretical meaning for activity theory and psychology in general. It is interesting that fundamental data in the studies of cognition, motivation, and behavior were obtained not only in general but also in the applied field of activity theory. The development of general activity theory strongly depends on applied research; however, it is not a unitary theory. The terminology used by different authors is not the same, and there are no standardized units of analysis or standardized procedures for work activity analysis. Hence, we suggest using SSAT terminology for the purpose of studying human work. This helps to integrate applied theory and SSAT. Cognitive analysis is not rejected by this approach, and is considered a stage of analysis. The SSAT approach considers work activity to be a multidimensional system that should be studied from the point of view of systemic principles. This ideology should be distinguished from the systemic principles of the study of man–machine systems. Here, we consider human activity to be a system. For this purpose, we need to have precise units of analysis that could give an opportunity to describe the holistic activity not only as a process but also as a structure. This gives us an opportunity to solve important ergonomic and work psychology problems in a different way. For example, in cognitive ergonomics, design is first of all an experimental process. In contrast, in SSAT, principles of design are derived from the fact that the physical configuration of equipment in a probabilistic manner can change the structure of activity. Therefore, through the analysis of the structure of activity, which can be described by means of analytical procedures, we can predict the more efficient configuration of equipment. An analytical comparison of the structure of activity and physical configurations of the equipment becomes the central component of ergonomic design. However, such methods cannot be utilized without precisely developed units of analysis, which can be found in SSAT. These units of analysis are functional block of self-regulation, members of human algorithms, human cognitive actions, and operations. These units of analysis help us to present activity as a structurally organized system. The concept of goal is particularly important for the extraction of such units of analysis. Note that terminology such as activity, action, goal, and self-regulation has different meanings in comparison with other approaches of psychology. We understand the concept of task in a significantly different way.

SSAT helps us consider the relationship between normative or standardized and variable components of activity in a new way.

Activity is a self-regulative system and therefore has situated features. At the same time, it includes normative features that are derived from technological prescriptions. Understanding the relationship between normative and standardized elements of activity eliminates contradictions between ideal task models and those that represent real task instances. Concepts and notions such as individual style of activity, the more efficient and representative strategies of activity, and constraints help us solve problems, such as the relationship between normative and variable aspects of activity, more efficiently. Finally, SSAT suggests standardized terminology, units of analysis and measurement, and procedures for the description and development of analytical models of activity. This approach does not eliminate methods of work performance analysis that is derived from cognitive psychology, where the major concept is a process.

The material presented in this chapter demonstrates that activity theory has important distinguishing features not only from the behavioral perspectives but also from cognitive approaches. These basic distinguishing features are as follows: (1) activity is a goal-directed, self-regulated system that cannot be studied as a reactive behavior or a computer-like information-processing system, and (2) activity should be analyzed as a system and therefore not only parametric but also systemic methods of study are required.

References

Anokhin, P. K. 1962. *The Theory of Functional Systems as a Prerequisite for the Construction of Physiological Cybernetics.* Moscow: Academy of Science of the USSR.

Anokhin, P. K. 1969. Cybernetic and the integrative activity of the brain. In *A Handbook of Contemporary Soviet Psychology*, ed. M. Cole and I. Maltzman, 830–57. New York: Basic Books.

Austin, J. T., and J. B. Vancouver. 1996. Goal construct in psychology: Structure, process, and content. *Psychol Bull* 120(3):338–75.

Bedny, G. Z. 1987. *The Psychological Foundations of Analyzing and Designing Work Processes.* Kiev, Ukraine: Higher Education Publishers.

Bedny, G. Z., and W. Karwowski. 2006. The self-regulation concept of motivation at work. *Theor Issues Ergon Sci* 7(4):413–36.

Bedny, G. Z., and W. Karwowski. 2007. *A Systemic-Structural Theory of Activity. Applications to Human Performance and Work Design.* Boca Raton, FL: Taylor & Francis.

Bedny, G. Z., W. Karwowski, and T. Sengupta. 2006. Application of systemic structural activity theory to work design of human-computer interaction tasks. In *International Encyclopedia of Ergonomics and Human Factors*, 2nd ed., ed. W. Karwowski, 1:1272–86. Boca Raton, FL: CRC Press.

Bedny, G. Z., W. Karwowski, and T. Sengupta. 2008. Application of systemic-structural theory of activity in the development of predicted models of user performance. *Int J Hum Comput Interact* 24(3):239–74.

Bedny, G., W. Karwowski, and M. Bedny. 2001. The principle of unity of cognition and behavior: Implications of activity theory for the study of human work. *Int J Cogn Ergon* 5(4):401–20.

Bedny, G., and D. Meister. 1997. *The Russian Theory of Activity: Current Applications to Design and Learning*. Mahwah, NJ: Lawrence Erlbaum Associates.

Bedny, G., and M. Seglin. 1999. Individual style of activity and adaptation to standard performance requirement. *Hum Perform* 12(1):59–78.

Bernshtein, N. A. 1945. *About Nature and Dynamic of Coordinative Functions*. Issues 90. Moscow, Russia: Scientific Publications, Lomonosov University of Moscow.

Bernshtein, N. A. 1966. *The Physiology of Movement and Activity*. Moscow: Medical Publishers.

Brushlinsky, A. V. 1979. *Thinking and Forecasting*. Moscow: Thinking Press.

Carver, C. S., and M. F. Scheier. 2005. On the structure of behavioral self-regulation. In *Handbook of Self-Regulation*, ed. M. Boekaerts, P. R. Pintrich, and M. Zeidner, 42–85. San Diego, CA: Academic Press.

Chebykin, O. Y, G. Z. Bedny, and W. Karwowski, eds. 2008. *Ergonomics and Psychology: Development in Theory and Practice*. London: Taylor & Francis.

Diaper, D., and N. A. Stanton. 2004. Wishing on a sTAr: The future of task analysis. In *The Handbook of Task Analysis for Human-Computer Interaction*, ed. D. Diaper and N. Stanton, 603–20. Mahwah, NJ: Lawrence Erlbaum Associates.

Dobrolensky, U. P., N. D. Zavalova, V. A. Ponomarenko, and V. A. Tuvaev. 1975. *Methods of Engineering Psychological Study in Aviation*. Moscow: Manufacturing Publishers.

Engels, F. 1959. *The Role of Human Labor in Process of Transformation of Monkey into Human*. Collection of scientific papers of K. Marx and F. Engels, version 20. Moscow: Political Publishers.

Engestrom, Y. 1993. Developmental studies of work as a testbench of activity theory. In *Understanding Practice: Perspectives on Activity and Context*, ed. S. Chaklin and J. Lave, 64–103. Cambridge: Cambridge University Press.

Frese, M. and D. Zapf. 1994. Action as a core of work psychology: A German approach. In *Handbook of Industrial and Organizational Psychology*, ed. H. C. Triadis, M. D. Dunnette and L. M. Hough, 271–340. Palo Alto, CA: Consulting Psychologists Press.

Galaktionov, A. I. 1978. *The Foundations of Engineering: Psychological Design of Automotive Technological Systems*. Moscow, Russia: Energy Publishers.

Gordeeva, N. D., and V. P. Zinchenko. 1982. *Functional Structure of Action*. Moscow: Moscow University Publishers.

Grot, N. Ya. 1879–1880. *Psychology of Feelings*. Petersburg, Russia: Sankt-Petersburg University.

Hilgard, R., R. L. Atkinson, and R. C. Atkinson. 1979. *Introduction to Psychology*. New York: Harcourt.

Karat, J., C.-M. Karat, and J. Vergo. 2004. Experiences people value: The new frontier for task analysis. In *The Handbook of Task Analysis for Human-Computer Interaction*, ed. D. Diaper and N. Stanton, 585–603. Mahwah, NJ: Lawrence Erlbaum Associates.

Kirwan, B. 1994. *A Guide to Practical Human Reliability Assessment*. London: Taylor & Francis.

Konopkin, O. A. 1980. *Psychological Mechanisms of Regulation of Activity*. Moscow: Science Publishers.

Kotik, M. A. 1974. *Self-Regulation and Reliability of Operator*. Tallinn, Estonia: Valgus.

Landa, L. M. 1976. *Instructional Regulation and Control: Cybernetics, Algorithmization and Heuristic in Education*. Englewood Cliffs, NJ: Educational Technology Publication (English translation).

Lave, J. 1988. *Cognition in Practice*. Cambridge, UK: Cambridge University Press.

Leont'ev, A. N. 1978. *Activity, Consciousness and Personality*. Englewood Cliffs, NJ: Prentice Hall.

Locke, E. A., and G. P. Latham. 1990. Work motivation: The high performance cycle. In *Work Motivation*, ed. U. Kleinbeck, H.-H. Quast, H. Thierry, and H. Hacker, 3–26. Mahwah, NJ: Lawrence Erlbaum Associates.

Lomov, B. F. 1966. *Man and Machine*. Moscow: Soviet Radio.

Lomov, B. F. 1977. On way of construction theory of engineering psychology based on systemic approach. In *Engineering Psychology*, ed. B. F. Lomov, V. F. Rubakhin, and V. F. Venda, 31–55. Moscow: Science Publishers.

Marx, K. 1969. *Kapital*, 1:358. Moscow: Political Publisher.

Norman, D. A. 1981. Categorization of action slips. *Psychological Review* 88:1–15.

Norman, D. A. 1986. Cognitive engineering. In *User-Centered System Design*, ed. D. Norman and S. Draper, 31–61. Hillsdale, NJ: Erlbaum.

Pervin, L. A. 1989. The goal concept: Definition and essential features. In *Goal Concept in Personality and Social Psychology*, ed. A. Pervin, 473–80. Mahwah, NJ: Lawrence Erlbaum Associates.

Petrovsky, A. V., ed. 1986. *General Psychology*. Moscow: Education Publishers.

Pintrich, P. R. 2005. The role of goal orientation in self-regulated learning. In *Handbook of Self-Regulation*, ed. M. Boekaerts, P. R. Pintrich, and M. Zeidner, 452–531. San Diego, CA: Academic Press.

Platonov, K. K. 1970. *Problem of Work Psychology*. Moscow: Medical Publisher.

Ponomarenko, V. A. 2006. *Psychology of Human Factor in Dangerous Professions*. Krasnoyarsk, Russia: International Academy of Human Problems in Aviation and Astronautics.

Pushkin, V. N. 1965. *Operative Thinking in Large Systems*. Moscow: Science Publishers.

Pushkin, V. V. 1978. Construction of situational concepts in activity structure. In *Problem of General and Educational Psychology*, ed. A. A. Smirnov, 106–20. Moscow: Pedagogy.

Rasmussen, J., and A. Pejtersen. 1995. Virtual ecology of work. In *Global Perspectives on the Ecology of Human-Machine Systems*, ed. J. Flach, P. Hancock, J. Caird, and K. J. Vicente, 121–56. Hillsdale, NJ: Lawrence Erlbaum Associates.

Reason, J. 1990. *Human Error*. Cambridge, UK: Cambridge University Press.

Rubinshtein, S. L. 1922/1986. The principle of creative activity. *Quest Psychol* 4:101–7.

Rubinshtein, S. L. 1957. *Existence and Consciousness*. Moscow: Academy of Science.

Rubinshtein, S. L. 1958. *About Thinking and Methods of Its Development*. Moscow: Academic Science.

Rubinshtein, S. L. 1959. *Principles and Directions of Developing Psychology*. Moscow: Academic Science.

Schedrovitsky, G. P. 1995. *Selective Work*. Moscow: Culture and Politics Publisher.

Senders, J. W., and N. P. Moray. 1991. *Human Error: Cause, Prediction, and Reduction*. Mahwah, NJ: Lawrence Erlbaum Associates.

Simon, H. A. 1966. *The Science of Artificial*. Cambridge, MA: MIT Press.

Sokolov, E. N. 1969. The modeling properties of the nervous system. In *Handbook of Contemporary Psychology*, ed. M. Cole and I. Maltzman, 671–704. New York: Basic Books, Inc.

Sternberg, S. 2008. Modular processes in mind and brain. In *Ergonomics and Psychology*, ed. O. Y. Chebykin, G. Bedny, and W. Karwowski, 111–34. London: Taylor & Francis.

Suchman, L. 1987. *Plans and Situated Actions*. Cambridge, UK: Cambridge Univ. Press.

Thelen, E. 1995. Time-scale dynamics in the development of an embodied cognition. In *Mind in Motion*, ed. R. Port and T. van Gelder. Cambridge, MA: MIT Press.

Tolman, E. C. 1932. *Purposive Behavior in Animals and Men*. New York: Century.

Vancouver, J. B. 2005. Self-regulation in organizational settings: A tale of two paradigms. In *Hand-Book of Self-Regulation*, ed. M. Boekaerts, P. R. Pintrich, and M. Zeidner. San Diego, CA: Academic Press.

Vicente, K. J. 1999. *Cognitive Work Analysis: Toward Safe, Productive, and Healthy Computer-Based Work*. Mahwah, NJ: Lawrence Erlbaum Associates.

Vygotsky, L. S. 1971. *The Psychology of Arts*. Cambridge, MA: MIT Press.

Vygotsky, L. S. 1978. *Mind in Society. The Development of Higher Psychological Processes*. Cambridge, MA: Harvard University Press.

Zabrodin, Y. M. 1985. Methodological and theoretical problems of psychophysics. In *Psychophysics of Discrete and Continual Tasks*, ed. B. F. Lomov and Y. M. Zabrodin, 3–26. Moscow: Science Publishers.

Zarakovsky, G. M. 2004. The concept of theoretical evaluation of operators' performance derived from activity theory. *Theor Issues Ergon Sci* 5(4):313–37.

Zarakovsky, G. M., and V. V. Pavlov. 1987. *Laws of Functioning Man-Machine Systems*. Moscow: Soviet Radio.

Zavalova, N. D., B. F. Lomov, and V. A. Ponomarenko. 1986. *Image in Regulation of Activity*. Moscow: Science Publishers.

Zinchenko, P. I. 1961. *Involuntary Memorization*. Moscow: Pedagogy.

Zinchenko, V. P., V. M. Munipov, and G. L. Smolyan. 1974. *Ergonomic Foundation of Work Organization*. Moscow: Economic Publishers.

Zinchenko, V. P., B. M. Velichkovskij, and Vuchetich. 1980. *Functional Structure of Visual Memory*. Moscow: Moscow University Publishers.

2

The Relationship between External and Internal Aspects in Activity Theory and Its Importance in the Study of Human Work

G. Bedny, W. Karwowski, and F. Voskoboynikov

CONTENTS

2.1 Introduction

Activity theory (AT) had a long history in the former Soviet Union. It was a major psychological paradigm for the study of human work and learning. AT was initially formulated by Russian scholars Rubinshtein (1940, 1957) and Leont'ev (1978). The cultural–historical theory of the development of human mind developed by another Russian scholar, Vygotsky (1960, 1962), was also critically important for AT. In the West, Vygotsky's theory and Leont'ev's concept of activity have been reconceptualized as cultural–historical activity theory by Engenstrom et al. (1999). However, in the former Soviet Union, Vygotsky's and Leont'ev's theories were not considered unitary based on their approaches. Leont'ev and his followers for a long period considered Vygotsky's works to be a predecessor of AT. They established a good deal of distance between the concept of activity and the cultural–historical theory

(Kozulin 1986). The reason was that Vygotsky's work was under attack as being insufficiently aligned with Soviet ideology. There were also scientific reasons why Leont'ev and his colleagues separated their version of AT from Vygotsky's cultural–historical theory of development of human mind. Regardless of these contemporary political factors and scientific opinions, it is clear that Vygotsky's works had an important influence on the development of Leont'ev's version of AT.

It should also be recognized that there was another founder of AT, a well-known Soviet scientist named S. L. Rubinshtein. His work is not widely known in the English-speaking scientific community. In this chapter, we will not make an attempt to present a historical analysis of who gave more input to the foundation of AT, Rubinshtein or Leont'ev. Rather, we will simply note that both authors contributed to AT. They shared a general view on some aspects in AT but, at the same time, had fundamental differences. These differences were primarily related to the relationship between the external and internal aspects of AT, the role of social and individual object-oriented aspects of activity, understanding abilities, and considering psychic development as a process. Because Rubinshtein's works are not sufficiently recognized in the West, we will explain here some of his basic positions on the question and, particularly, the relationship between the external and internal aspects in AT. We will further attempt to trace the differences on this question between Vygotsky, Leont'ev, and Rubinshtein. We will attempt to demonstrate how the relationship between external and internal aspects can be utilized in the study of human work. We will also consider the relationship between external and internal components of activity from the viewpoint of systemic-structural activity theory (SSAT). This theory is connected to general AT and operational–psychological, or applied, AT.

We would also like to mention that SSAT significantly differs from general AT and operational–psychological AT and presents an original direction in activity study. A description of SSAT and its relation to general AT can be found in the recent work of Bedny and Karwowski (2007) and Karwowski et al. (2008). We would also like to direct interested readers to *Theoretical Issues in Ergonomics Science* (Bedny 2004). The systemic-structural concept of activity suggests a different vision of the relationship between external and internal components of human activity. One of the important specifics of SSAT is that it is not only a theoretical approach but also precisely developed practical principles and methods in the study of human work.

The principle of genetic study, according to which psychological functions are studied as they are being developed, is important not only for the cultural–historical theory of the development of human mind but also for SSAT. In Chapter 5, this method will be demonstrated as it is adapted in the frame of SSAT. Therefore, the relationship between the external and internal components of activity and human development is also the basis for different concepts in AT, including SSAT. At the same time, AT cannot be reduced to the principle of genetic study. We do not agree with some scientists who overemphasize this aspect of study in AT. AT includes many different principles and methods of analysis; this is particularly relevant for SSAT. At the same time, we focus our attention on the genetic aspects of activity study, in which the interrelationship between the external and internal components of activity becomes critically important. We attempt to turn this discussion into a more applicable direction.

2.2 Personality Principle in the Study of Human Development

The study of relationship between the external and internal components of human activity plays a leading role in understanding human development. This was central in Vygotsky's work and important in AT. An analysis of the relationship between external and internal components of human activity is specifically important for the study of social determination of individual psychic development. Consideration of the social determination of psyche requires the consideration of the role of individual or personal factors in psychic development. Therefore, to contextualize this discussion, we should examine how personal aspects have been studied in AT.

Some general philosophical and psychological principles for the study of personality in psychology in accordance with the concepts of AT were first formulated by Rubinshtein in 1934. He introduced the personality principle in psychology that integrates individual and social aspects in the study of human development. Human development, according to this principle, is the result of the interaction of material and social practice with human subjectivity. This principle eliminates the contradiction between social and intraindividual aspects of human development. In developing this principle, Rubinshtein addressed inadequacies in behaviorists' approaches and idealist psychology with mentalist orientation to the study of personality. In behaviorism, for example, the American psychologist Skinner (1974) noted that humans emerge as reactive organisms. Human instrumental reactions are considered to be the result of external influences, ignoring the subjective aspects of reactions. Subjectivity is absent in these studies. However, if the subject of activity does not exist, then neither the processes of personal development nor the features of personality exist. Behaviorism ignores those mediated functions of activity, which provide a basis for personal development. Personality is developed through a person's participation in activity, which depends on the relationship between the subject and situation and the relationship between subjects. In behaviorists' approach, external reality is portrayed as a variety of stimuli to which a person must react. In AT, the person who interacts with a situation is considered the subject. In such cases, we talk about actions and cognition, not about the stimuli to which the subject reacts.

Rubinshtein (1946) wrote that through the organization of individual practice, society shapes the content of individual consciousness. He also emphasized the dependence of activity on the subject's individual features, and this thought was formulated into his famous quotation "external acts through the internal," where the internal is a person's mental processes (Rubinshtein 1946). The same external situation determines the activity of subjects in a variety of ways. In any activity, a subject not only changes the situation but also shapes himself or herself. During activity, the subject changes the object itself, which in turn presents new requirements to the subject. Thus, we can observe changes in a product of activity, in the tools' mediating activity, and in their reciprocal influence on the subject's psyche. Rubinshtein criticized personality concepts based on mentalist approaches; he also criticized functionalism that divides human unitary cognition into separate functions such as memory, consciousness, and will. Psychology cannot be reduced to the study of person-less processes and functions. Rubinshtein (1959) stated that it was necessary to study the history of an individual human development but not his or her separate consciousness or separate psychic functions. The holistic concept of personality should be used as the basic theoretical concept of psychology because all psychic processes should be considered in connection with real persons. These critical comments still apply to some degree when we consider contemporary cognitive psychology.

Thus, Rubinshtein formulated the "personality principle," according to which the psychological existence is connected with real existence of human beings. Studies of the problem of social determination of psyche, which bypass individuals, are impossible. On this view, social aspect emerges as the real, concrete existence of individual people. The most important aspect of this principle is that it emphasizes the concept of activity of particular individuals. From this basic emphasis, the content of the personality principle, specifically the interconnection of personality and activity, is derived. Social characteristics of psychic development can be discovered not through the direct relationship with social activity but through the mediation of personality. Social aspects of experience are integrated into individual activity. The social aspect depends on the individual, just as the individual depends on the social aspect. In the same social environment, different individuals act differently, and they are impacted by the social environment in different ways. Vygotsky emphasized the role of external properties of social interactions rather than the individual's experience of these interactions. According to Rubinshtein, thinking as activity should be analyzed in terms of motives, goals, actions, and operations of thinking.

Both Rubinshtein (1959) and Leont'ev (1978) suggested that the leading role in the development of the human mind is not played by words or social interaction, but rather by the interaction of the subject and the object. Rubinshtein was one of the first to formulate ideas on subject–object and subject–subject relationships in activity and the object-related character of activity. Nature exists independently from the subject; however, object can emerge only during the interaction with the subject (Rubinshtein 1957). As with Leont'ev, Rubinshtein had not sufficiently developed the semiotic aspects of activity, which were more fully dealt with in the work of Vygotsky (1962). He attempted to overcome the separation of social and individual aspects within the consciousness of particular persons. To overcome the dualism of social and individual, he attempted to trace the historical changes in the structure of consciousness. According to Vygotsky, the historicity of consciousness is connected to the historicity of activity, which is exhibited in the way subjects utilize the tools and means of work. During activity, people utilize artificially created objects such as signs and material artifacts. Individual consciousness reproduces social structures of behavior. According to Vygotsky, social penetrates (into) consciousness and transforms its essence. This theoretical statement played a very important role in the development of AT. However, this approach is different from that of both Rubinshtein and Leont'ev. In Vygotsky's theoretical analysis, activity is considered not so much in terms of individual plane but rather in terms of general social activity. In Vygotsky's approach, individual psychological aspects of activity were not considered to a significant degree.

Rubinshtein, criticizing the cultural–historical concept of human development, stated that individual psychological features of humans are not directly derived from the social activity. The social activity, when combined with individual activity, determines its specifics, and thus, the specifics of mental development. Activity is not an adaptive system; it is transformative, constructive, and creative. That is, under the same social circumstances, individuals can act differently and develop in different ways. Rubinshtein (1958) concluded, by establishing a critically important principle, that "external influences are transformed through internal conditions." According to Vygotsky, individual experience is the result of social relations. Leont'ev stated that individual experience is the result of individual object-oriented activity, but the social activity is also considered in this concept. However, it exists under subordination to individual object-oriented activity and is interpreted as a social method of actions, which is crystallized in the way objects are utilized. Object determines fixed, unchangeable, stable ways of acting with it. Specificity of object-oriented activity depends on the specificity of social relations. Rubinshtein opposed this description of psyche as being passively dependent upon the object-oriented activity. He considered activity to be always essentially constructive, transformative, and creative. Personality is shaped according to this specificity. Rubinshtein considered the development of personality an active process where the person not only accumulates social experience but also transforms it and develops through this complicated interaction. The leading doctrine in Rubinshtein's works was that action not only changes the situation but also the subject.

Engestrom et al. (1999, pp. 26–7), in their analysis of Russian publications, wrote that after the collapse of the former Soviet Union, most authors emphasized that human activity is creative, and it has the ability to exceed or transcend the given constraints and instructions. They further noted that "perhaps this conclusion reflects the impact of perestroika on philosophy and psychology." In reality, these ideas were formulated long before by Rubinshtein (1959, 1973). After Rubinshtein's death in 1960, his ideas were suppressed by Leont'ev and his followers the same as Vygotsky's works were underestimated for a long period.

2.3 Rubinshtein's External and Internal Aspects of Human Development

As early as 1922, Rubinshtein articulated a fundamental theoretical principle of AT as "the unity of consciousness and behavior" that underpinned a general psychology and philosophy. He demonstrated that practical manipulation of material objects under direct contact conditions provides continuous control for thinking process, which regulates this manipulation. Mental processes emerge from this interaction. Rubinshtein first formulated this problem and consequently never accepted the concept of internalization. Rather, he considered the interaction of external and internal components through the prism of individual activity, which is always unique (Rubinshtein 1958). The social determination of consciousness is derived not from "outside" but from "inside" the social mode of existence of the specific individual. The consciousness of the subject is determined by his or her individual social mode of existence, which involves the reflection of both social relations with others and practical relations with objects. The major idea in the interaction of subject with object or the situation in general is that external causes are mediated by internal conditions (Brushlinsky 1979).

The external and internal aspects present themselves as a unitary developmental process. The influence of external circumstances on the individual depends not only on external factors but also on the individual's activity. Thus, activity is not a system of reactions but a system of goal-directed actions. Interaction can be understood as an uninterrupted active process. Through the transformation of the object, the subject discovers different features of the object. This particularly applies to the thinking process. Rubinshtein (1973) suggested that the thinking process primarily comprises analytic and synthetic aspects of activity. When a subject tries to understand a problem, the situation should be decomposed into elements—this is analysis. However, as the objects and their relationships do not exist in isolation, we should discover their interrelationships—this is synthesis. Analysis and synthesis are always integrated. The adequacy of the reflection of the object in this process is predetermined by the actions and operation of the subject. These operations, although dependent upon external conditions, instructions, and so on, preserve their individuality. The significance of elements and aspects of the situation for a particular subject is an important influence in this process. According to Rubinshtein, factor of significance points to how the consciousness of the individual is dependent on his or her activity. Thus, both consciousness and activity depend on each other. This shows that individual consciousness cannot be a simple replication of social consciousness and it is the result of active development. A person cognizes the world and develops his or her consciousness through activity. Through activity, people create culture and objective material world of human artifacts. Different forms of social interaction are developed in the process of human activity.

Comprehension of the world emerges for the subject as a process of cognizing the significance of elements of that world for the specific individual. This significance is connected to needs and motives. The activity of the individual emerges out of what is subjectively significant for him or her. In such case, the situation emerges as a task or problem for the subject. Normatively prescribed methods of action performance, relationships to situations, and connections with other people are always generalized and separated from actual concrete situations to some degree. Because of this, activity can only be specified and regulated at the individual psychological level.

Rubinshtein emphasized the role of independent exploration and interaction with the objective world as the source of the object world's reflection in human consciousness. Therefore, he disagreed with the concept of internalization as the driving force of mental development. He argued that a person does not simply internalize ready-made standards, rather the external world acts on the mind through the mind's own internal conditions. Activity is always independent and creative to some extent. The following important assertion regarding mental development comes out of this: "External influences on mental development always act through internal conditions" (Rubinshtein 1959). Criticizing Vygotsky's cultural–historical theory, Rubinshtein argued that psychological characteristics of the individual are not completely derived from social environment.

Regulation of activity emerges as a multifaceted phenomenon. A subject does not simply regulate his or her own actions but changes the object and situation also, and in this process, changes himself or herself. Because action comes in direct contact with objective reality, penetrating inside this reality and transforming it, action is an incredibly powerful tool in the formation of thoughts, which in turn reflects objective reality. Thought is carried into objective reality on the penetrating blade of action (Rubinshtein 1958).

According to Rubinshtein, social characteristics of cognition are not simply developed out of the direct contact with social world but are mediated through the personal characteristics. When we try to study cognition as a social phenomenon, we must also study particular individuals. We can understand consciousness only as the interaction of individual and social characteristics. In Rubinshtein's work, the category of activity emerges primarily as the activity of the individual. Out of this, the personality principle that embodied the connection of personality and activity emerged.

The connection of personality with the social world can also be understood as being mediated by activity. Rubinshtein, the same as Leont'ev, stated that the interaction of subject with object played the leading role in mental development, not the words or social interaction. However, he paid insufficient attention to the semiotic aspects of activity studied by Vygotsky.

2.4 Leont'ev's Internalization Concept

The internalization concept was not accepted by all scientists in the former Soviet Union. However, this concept is important for understanding the relationship between the external and internal processes and personal development processes. In this regard, it is necessary to consider works of Leont'ev (1978) and Gal'perin (1969). In their work, the problem of external and internal aspects of activity is considered through the lens of internalization. According to these authors, the structure of object-oriented activity determines individual consciousness to a significant degree. In relation to the latter, Leont'ev (1972, p. 104) wrote that "external and internal activity have similar structure." A major aspect of the internalization process is the transformation of external material action into internal mental actions. This approach deviated from the specificity of individual activity inasmuch as there was an emphasis on the adequacy of activity in its socially fixed way of performance. Individual-specific regulative aspects of activity were not considered. Rather, it was suggested that the structure of individual consciousness is predetermined by the structure of external object-oriented activity. Leont'ev argued that over a long period of time, external and internal activity have similar structures because external physical activity is internalized and becomes internal mental activity. However, object-related human actions cannot be considered purely external and nonpsychic (Zinchenko 2001).

Social determination of consciousness has its sources not only in "external influences," but also in "internal influences," which depend on the specificity of individual activity and methods of individual performance. Leont'ev's internalization concept continually encountered contradiction with accumulating psychological data demonstrating the impossibility of reducing the internal to the external aspect. This difference of opinion and viewpoints can be described in terms of two distinct schools of thought within psychology in the former Soviet Union. One is derived from the work of Leont'ev, and the other is derived from the work of Rubinshtein. The rise of skepticism regarding the concept of internalization has led Leront'ev to reconsider his radical view on internalization. Later, he wrote that the process of internalization is not a transformation of external aspects into internal, but rather it is a process in which this internal plane is formed (Leont'ev 1978, p. 98). This statement did not change his idea about internalization, which can be seen in the analysis of the internalization concept developed by Gal'perin (1969), who belonged to Leont'ev's school of psychology.

The process of internalization is a major feature of Gal'perin's theoretical conceptualization. This process is considered a transformation of external aspects into internal aspects. During the first stage, actions are performed in material or materialized form according to the orienting basis of actions. This is a system of referenced points that present a plan of action. The material level involves manipulation of external objects. The term "materialized form" of action means that subjects perform actions with schema, drawings, models, and so on. The mental form of actions involves internal plans of actions. Any action includes orienting, executive, and control elements within itself. In the first step, material or materialized actions are performed in a very detailed manner. In the next step, material support for actions is reduced. At the following stage, actions are performed verbally as external speech to oneself. Finally, actions are transferred into soundless utterance. Hence, external speech to oneself is gradually mastered and transferred into internal speech to oneself during the learning process. During the transformation of actions from the material form into the internal, mental plane form, these actions are abbreviated and internalized. This is a very brief description of the application of the internalization concept in the learning process. During the performance of behavioral actions, only the orienting components of action can be internalized.

Gal'perin's concept of learning, which was derived from Leont'ev's idea of internalization, was successfully applied in some situations. However, his utilization of the idea of stages and the consideration of their sequential application as a process of internalization raise some ambiguities and problems. According to the principle of self-regulation developed in SSAT, this process can be considered the active formation of actions in different forms based on the comparison of different forms of action, but not as the transformation of external into internal.

The orienting basis of actions cannot always be transformed in a ready form into a subjective feeling of having internal plane. Transformation of the orienting basis of an action into an internal plane very often requires a series of trials and analysis of errors as an important element of feedback evaluations. Not all aspects of actions can be verbalized, and the stages of action acquisition listed above were often utilized in a dogmatic way, when it was not always the case that these stages are required. The method of representation of the reference points to students was not well-developed. What Gal'perin called "reference points" were considered in the algorithmic concept of learning developed by Landa (1976) and the self-regulative concept of learning developed by Bedny and Meister (1997) as "identification features" of actions, which are required for their performance. However, some of Gal'perin's ideas are useful. For example, his idea about actions as a guiding system, which consists of orienting, executive, and evaluative components, is important. These aspects of Gal'perin's theory do not have any relation to internalization. They are more relevant to the problems of regulation or self-regulation of activity. Therefore, Gal'perin's theory (in contradiction to his own interpretation) cannot be considered to be dealing with internalization when external actions are transformed into internal mental plans. According to the principle of self-regulation of activity, internal plane of activity is a result of active formation of internal mental actions and operations based on self-regulation mechanisms. We will consider this in a more detailed manner in Section 2.7.

2.5 Vygotsky's Internalization Concept

The interrelationship between the external and internal components of activity takes a central place in Vygotsky's work. He explained the psychic development through the interrelationship of the external, social, and internal components. He connected psychic development with the historicity of activity. In Vygotsky's theory, the historicity of activity is connected with the methods of utilization of tool and means of work without which human collective labor would be impossible. Vygotsky attempted to track how the social interpenetrates individual consciousness. This idea was specifically formulated as the principle of internalization. Internalization is first of all a social process. Vygotsky wrote that "external" means "social" (Vygotsky 1971). He paid attention to the semiotic aspects of internalization. Language becomes the major mediator between social and individual functioning. External social activity is the source of internal mental activity. Therefore, semiotic mechanisms provide connection between external and internal activity. The problem of internalization in Vygotsky's work is the foundation for different ontogenetic aspects in the development of the personality. One of these problems is the zone of proximal development (ZPD). This important concept has both theoretical and practical meaning. This contrasts with Piaget's work, in which development is a more spontaneous process that depends on the stages of development. In Vygotsky's work, the development process is led by social interaction. Learning precedes development and is adequate to the ZPD. Hence, learning pulls up development. The concept of ZPD is important for the assessment of a child's cognitive ability. He argues that existing test procedures were directed toward the evaluation of past accomplishments. Therefore, test procedures should be developed in such a way that one can evaluate the child's achievements not only in the past but also their achievement in the ZPD under the supervision of more capable peers, as this indicates their potential for future development. For a more detailed analysis of the concept of internalization in Vygotsky's theory, we direct interested readers to Wertsch and Stone (1995).

The analysis of internalization in Vygotsky's work demonstrates that consciousness is connected with the historicity of activity, which depends on the way of utilization of tools and means of work. In Vygotsky's theory, consciousness is individual and activity is social. Consciousness of the individual reproduces the social structure of activity, and social penetrates into the subject's essence. Vygotsky's theoretical statements played an important role in the development of AT. However, theoretical analysis of Vygotsky's theory demonstrates that activity was considered not so much in individual plane but more in the plane of social interaction. Rubinshtein and his followers criticized Vygotsky's work on the grounds that he omitted the individual as the subject of activity. In other words, Vygotsky did not sufficiently consider the activity of particular individuals. The social aspects of activity are acquired only through particular individual activity. Activity is not purely adaptive, it is also transformative and creative. This means that in the same social environment, different individuals act and develop in different ways. The specificity of knowledge acquisition and socialization is dependent not only on social but also on the specificity of individual activity.

2.6 Activity as a Self-Regulating System

To understand the relationship between the external and internal components of activity, we should consider the systemic-structural theory of activity. According to the systemic-structural approach, activity can be considered a coherent system of internal mental processes and external behaviors and motivations that are combined and organized by mechanisms of self-regulation to achieve conscious goals. Any activity has a recursive loop-structure organization according to the principle of self-regulation (Bedny and Meister 1997). There are both automatic and conscious levels of self-regulation (Bedny and Karwowski 2007). At the automatic level, the conscious and verbalized aspects of self-regulation play a subordinate role, and this level is particularly important when imaginative and nonverbalized strategies of activity play the leading role. At the conscious level of self-regulation, verbal and logical aspects of activity are dominant. Both levels of self-regulation are interdependent and the relationship between them is dynamic. This interdependency gives rise to the formation of different strategies of activity, which are adequate to the external and internal conditions of activity. Through mechanisms of self-regulation, Rubinshtein's principle of "external acting through internal" is provided. Activity has not only preplanned but also situated components that develop during self-regulation. Learning is considered a self-regulating process during which strategies of activity are transformed. The more effective the teaching process, the fewer intermediate strategies are utilized during the learning process (Bedny and Karwowski 2006). Activity can be considered an uninterrupted process, as well as a discontinuous or discrete system. Rubinshtein considered cognition an uninterrupted process, and at the same time, object-related components of activity a discontinuous process. Rubinshtein's concept of continuity and discontinuity in human activity is not precisely described. Understanding these aspects of activity becomes possible by the analysis of self-regulation (Bedny et al. 2000) and by considering cognition as a reflective process.

One of the important principles of AT is that psychic processes perform reflective functions (Rubinshtein 1958; Platonov 1982; Leont'ev 1978). Reflection is a particular kind of interaction among phenomena, in which the reflected object preserves its topological structure within a systemic reflective medium. Therefore, the reflection process is possible only during the interaction of different objects. The most complicated type of reflection is psychic reflection. Due to psychic reflection, mental representation of reality is provided. Psychic reflection in human beings emerges through the interaction of subject and object. Sensation, perception, thinking and so on can all be considered reflective processes. Psychological reflection is not a passive mirror-like reflection; it possesses active features that imply some system of mental stages and operations. Thus, reflection can be analyzed in terms of information processing insofar as they transmit information semantically, pragmatically, and quantitatively. Semantic refers to the qualitative meanings, pragmatic refers to its utility, and quantitative refers to the density of information available. Cognitive processes can be considered distinct configurations of psychological reflections. Psychological reflection is always organized as a self-regulation process. Since this process cannot be fully determined in advance, it always contains situated elements that are developed during self-regulation process of reflection. The more complicated a person finds a task, the more important and complicated the reflective process becomes. The most complicated reflective process is thinking.

During problem solving, conditions continually change. However, the situation not only changes but also the individuals themselves change as they receive additional information about the situation. As a result, the strategies of activity continually require adjustment and correction. During the thinking process, objects or elements of the situation are transformed and get into new interrelationships with each other. Due to this, according to Rubinshtein, new features are extracted from objects as activity proceeds; the object seems to reveal new facets to the subject. He called this process "analysis through synthesis" (Rubinshtein 1958). Analysis is the extraction of some aspects of the situation or facet of objects. Synthesis is relating and connecting extracted features that are already known to the subject to the elements of the situation. It is important to note that this process is organized according to the principles of self-regulation through feedback, which provides continual correction to the process (Bedny and Karwowski 2006). At the unconscious level of self-regulation, cognition unfolds as an uninterrupted process. Automotive mental operations are not organized into cognitive actions. This can be explained by the fact that the unconscious level of self-regulation is not subordinated to conscious goals. Activity is triggered automatically and performed through unconscious automatic reflective processes. Perceptual images or other psychological phenomena can be developed as a result of these reflective processes, which are organized according to mechanisms of automatic self-regulation. Therefore, reflective process is always transformed, adapted, and provides a more precise reflection of the situation. The subject is only conscious of the results of this process.

The conscious level of self-regulation presents itself not only as a process but also as a system of logically organized actions. Each action is also organized according to mechanisms of self-regulation. Each action has a beginning and an end. At the conscious level of self-regulation, activity can be considered a hierarchically organized system of self-regulative stages of uninterrupted reflective processes. At the same time, these processes are discrete. Activity can be described as a chain of loop-structure organization units. One cycle loop transfers into another loop. Each loop has its qualitative specificity and is ended by the achievement of the goal of cognitive action. Therefore, at the conscious level, cognition is continuous and at the same time interrupted. The goal of action performs the integrative function. Understanding the principle of the self-regulation of activity (Bedny and Karwowski 2007) helps us understand the relationship between uninterrupted and interrupted aspects of self-regulation. We need to understand also that concepts such as perception, memory, and thinking are theoretical abstractions. All such psychic phenomena are interdependent and are included in the unitary process of self-regulation.

Understanding how activity is organized helps explicate the relationship between the external and internal components of human activity. During self-regulation, subjects not only change their strategies but also scope their external environment and compare external and internal components of activity. The socially determined aspects of our cognition are not only based on "external" influences, as suggested by the cultural–historical theory of development of Vygotsky. Nor do they wholly depend on object-oriented activity, as suggested by Leont'ev. Psychic activity emerges as a function of the social existence of the individual, based on which the ability for reflection develops. Psychic or cognition functions, according to principles of reflection, are organized as a self-regulation process. At the conscious level of self-regulation, this process is organized as a system of logically interdependent cognitive actions that are transformed from one into another as activity unfolds. These actions include mental operations. At the unconscious level of self-regulation, this process unfolds as automotive unconscious operations. These operations cannot be considered components of conscious actions. They function independently. For example, unconscious "gnostic dynamic" during operative thinking, which was described by Pushkin (1978) and later in our work (Bedny and Karwowski 2007), cannot be considered components of conscious actions. Based on the latter, it follows that content of activity cannot be reduced to the system of logically organized actions as it is derived from Leont'ev's concept of AT. Moreover, psychological determination does not depend on social and external factors only but also on internal influences derived from the mechanisms of self-regulation, which integrates external and internal components of human activity.

2.7 Concept of Internalization in Systemic-Structural Activity Theory

Leont'ev and Rubinshtein concentrated their efforts on the study of practical object-oriented activity as the source of mental development. Vygotsky in his works paid attention to the semiotic aspects of activity and in their connection with the problem of social interaction. According to SSAT, the aspects of activity and the development of personality are interconnected. The objective world becomes open to humans only in the unity of the object-oriented and semiotic aspects of activity and social relations. Social relations and individual practical activity exist in unity. In a hypothetical example, if a child is raised by animals and he or she is eliminated from social interaction with human beings, he or she will have an animal's intellect level when compared to the upbringing he or she would have received in the objective world created by humans. The human objective world remains closed to him or her because animals cannot communicate the knowledge of how to utilize the artifacts of human culture. During their interactions with the objective world, humans interpret situations based on the system of meanings acquired during ontogeny. A child develops into a person through the process of individuating himself or herself from the surrounding environment, which stands in opposition to the child as the object of his or her actions and cognition.

Object-oriented and social-oriented activity exists and develops in unity from the beginning of the child's life. It provides transformation of imaginative and sensory aspects of thinking into symbolical thinking and back again. The child is born into an objective world with already developed symbolical systems that correspond to this world. Human activity, from the very beginning, is both object-oriented and sign-oriented. We can talk about human activity only when there is a connection between object-oriented and symbolic activity. We can easily explain using a famous example from Vygotsky's work in which he attempts to interpret the gesture of a child, which signifies the material act into a sign. Vygotsky wrote that initially the child attempts unsuccessfully to grasp the object. When the mother realizes that the child's movement indicates something he or she wishes to grasp, she immediately helps the child to reach or grasp this object. During this social interaction with the mother, a child also interacts with the objective world.

Both aspects of activity are equally important. Due to the interaction of these two aspects of activity, the child becomes aware that his or her hand movement can be understood by others as his or her desire to reach something. Movement of hand becomes a sign. Here, we can see how object-practical action is transferred into semiotic action. Meaning is developed within subject–object and subject–subject interaction, which are totally interdependent. Signs emerge not only simply as representatives but also as mediators in human activity. In the example described above, Vygotsky underestimated the object as element of activity and overemphasized the social interaction. Both these elements of activity are interdependent.

When 1- to 2-year-old children engage in practical contact with a real object and struggle to reach it, the motor action becomes a gesture or sign not only as a result of social interaction with others but also as a result of interaction with the object. It is important to note that the individual activity of a child is creative to some degree—he discovers for himself that his gesture can transmit his desire. Intersubjective interaction may be discovered even in individual subject–object activity. Intersubjective relationships arise from the interaction with others in the past. The intersubjective features of human individual activity may be discovered through inner speech or inner dialogue (Bakhtin and Voloshinov 1973). We can summarize that social factors play an important role in the creation of human sign systems and the internal plane of activity.

During the acquisition of social experience, interaction of subjects with objects and with other subjects becomes increasingly complicated. Subjects create individual strategies for these interactions. Different abilities and interpretations emerge from these interactions depending on individuals' different experiences and strategies of interactions. Because of this, the internal world of humans cannot be portrayed as a simple result of transformation of the external into the internal. It is rather the result of the formation or construction of the internal based on the unitary process of social- and object-oriented interactions. Furthermore, it is important to note that during the development of sign systems, a child learns how to manipulate these systems in some degree independently based on individually developed strategies.

Another example of this kind of activity is mathematics. From the viewpoint of cognitive psychology, the acquisition of knowledge in mathematics is conveyed by the reorganization of the memory structure, which is a result of social and practical activities. Mentalistic explanation about memory reorganization is important but not sufficient. The teacher needs to know what kind of strategies he or she should utilize during the explanation of the required material and what kind of possible strategies students will utilize. Here, we can discover a complicated relationship between the activity strategies of the teacher and students. In such case, the specificity of individual theoretical activity (such as mathematics) and its transformative and creative character are important. The connection of consciousness with social reality is always established through the connection of particular individual with the objective and social world. Rubinshtein wrote "outside of the connection with personality, we cannot understand psychological development" (Rubinshtein 1934, p. 14). The above statement in the study of relationship between external and internal components is as relevant now as it was in 1934. The personality principle is closely connected with our understanding of the principle of self-regulation, particularly through the specifics of self-regulation as it is manifested in the unique psychic activity of particular persons in particular situations. Mechanisms of self-regulation provide the uniqueness of individual experience.

According to Leont'ev, objective meaning (meaning associated with object-practical activity) presents itself as being fixed in language and in stable methods of operation and action with objects. As a result, the structure of internal operations replicates the structure of external object-practical activity. Meaning can be studied from morphological and functional analysis perspectives. In morphological analysis, meaning is a function of standardized actions, whereas in functional analysis, meaning is the result of the self-regulation process. Meaning arises as a multidimensional phenomenon. According to morphological analysis, meaning arises as the interaction of both material and sign actions. These kinds of actions are interdependent and shape each other through the feed-forward and feedback influences.

One of the important distinguishing characteristics of meaning from the functional perspective is the connection of meaning with motivation. As the motivational state of the subject changes within the same situation or the significance of information changes, it alters the meaning of the situation for the subject. Consequently, meaning and personal sense are not a rigid, fixed structure, but a flexible and self-adjusting system. Through the process of self-regulation, objective, commonly accepted meanings are transformed into idiosyncratic sense for each individual. Moreover, objective meaning itself is also gradually formed according to the process of self-regulation. From the multiple potential semantic features that can be included in meaning, realized categories include only those features that are relevant to the goal of the person's activity, the significance of the situation for them, and their motivational state. The significance of the categorical features of an object and/or a sign can be changed through the self-regulation process. This in turn can result in changes to the strategy of categorization or interpretation, and indeed, the strategy of activity in general. The foregoing discussion demonstrates the important role of personal, individual activity in the formation of internal meaning and sense.

Conscious and unconscious strategies for gathering and interpreting the information can be triggered by cognitive and emotionally motivational mechanisms of self-regulation. Conscious strategies are important particularly for categorical or idealized meaning, which is a part of verbal categories that the subject masters during ontogeny. Categorical meaning plays a central role in the formation of social consciousness (Gordeeva and Zinchenko 1982). Categorical meaning is stable and is independent of the situation. Categorical meaning has an objective, social–historical character. This is the objective property of signs. However, during the process of individual acquisition of social experience, individual introduces elements of subjectivity into the system of objective meanings. This subjective coloring is relatively stable and less dependent on the situation than the personal sense (significance). This demonstrates that the internal never exactly replicates the external. From this perspective, the work of Pushkin is interesting (Pushkin 1978). He introduced an important notion in the study of human thinking, which he called the "situational concept of thinking." This concept is a result of the "gnostic dynamic," which he understood as the ability of a subject to change the meaning of the external situation in his or her mind when the external situation is relatively stable and unchangeable.

At the base of the gnostic dynamic lies the self-regulation process, which provides mental transformation, evaluation, and interpretation of the situation (Bedny and Karwowski 2006). Tikhomirov (1984) introduced the closely related concept of "nonverbalized situational meaning" and "nonverbalized situational sense." Situational concepts or nonverbalized meanings and senses are important building blocks in the construction of dynamic models of the situation. Therefore, they play an important role in the dynamic reflection of the situation. The gnostic dynamic activity is significantly different from external behavioral activity and, therefore, the internal plan cannot be the result of the transformation of the external into internal plan.

In the Russian language, the word *znachenie* (meaning) in its external sound form is close to the word *naznachenie* (functional purpose) of the object. This is the source of erroneous interpretation of these two concepts not only by some scientists in the former Soviet Union but also by professionals who attempt to translate data associated with the concept of meaning in AT. For example, a child's mental development is closely connected with his or her ability to manipulate correctly different objects according to the methods developed for the utilization of these objects. People live in the world when they utilize various things that usually have constant meanings for them. In contrast, for animals, things have only situated meaning. Children learn from adults to pay attention to constant meaning (*znachenie*) of objects during interaction with them. Some scientists (Nepomnayashaya 1973; and others) mixed this constant meaning with constant functional purpose (constant *naznachenie*) because of its similar sound in the Russian language. The functional purpose of objects (*naznachenie*) is one of the most important characteristics of the object, which can determine its meaning.

However, meaning of objects includes other characteristics. For example, a hammer can have different shapes and sizes, which determine the specificity of utilization of the hammer. Some objects and concepts do not have an assigned purpose at all, but they still have a meaning. For example, the concept of time in physics does not possess any assigned purpose. Acquisition of the meaning of such concepts or phenomena is much more difficult for a child and requires a long period of time. During the first stage of mental development, children acquire the meaning of different objects that have precisely assigned functional purposes; later, they gradually begin to acquire the meaning of things that do not have assigned purposes. The acquisition of meaning as an internal mental tool cannot be reduced to the manipulation with external object tool that has a particular purpose. It is not the process of transformation of external operations with external tools into mental operations with mental tools. Internal mental operations do not replicate external behavioral operations. Mental development is a result of interaction between external and internal operations based on the mechanisms of self-regulation and, therefore, always includes some unique individual components. Here, we can also see that linguistic characteristics of language sometimes influence the interpretation of scientific data. This is particularly relevant during the translation of scientific data from one language to another.

Gordeeva and Zinchenko (1982) described different stages of motor actions, which they considered "blocks." They described the blocks as "formation of motor instruction" (program formation stage), "executive block" (executive stage), and "corrective and evaluative block" (evaluative stage). The first and third are the cognitive components of motor activity. The relationship between all the blocks is constantly changing during skill acquisition. The cognitive aspects of motor actions are responsible for programming, evaluating, and correcting the motor action. Therefore, any behavioral action from psychological perspectives contains cognitive components.

According to the systemic-structural concept of activity, during the mental development of a child, with increasing complexity of motor activity, the complexity of the cognitive components of motor behavior also increases. During further mental development, a predisposition toward complication and a gradual separation of the cognitive components of activity from external motor activity is being created. Cognitive activity begins to be performed relatively independently of motor activity. For example, a program formation block, which is associated with material transformation of external object at the later stage, can be involved in the transformation of the image of the same object. At the next stage, symbolical activity is developed on the basis of the support of object-practical activity and vice versa. This genetic connection of external motor activity with internal cognitive activity influences further development of the person. At the beginning of the learning process, the learner manipulates signs and symbols by means of external objects using different instructions and schema, which facilitate the externalization of internal mental activity. We can say that mental activity is guided by external, orienting components of activity. Therefore, the external behavior and internal activity depend on each other. During feed-forward and feedback influences between external and internal components of activity, a mental plane of activity is developed. Mental components of activity through an active reconstructive process form are the basis for the development of the mental plane of activity. This leads us to conclude that mental activity cannot be considered the transformation process of external component into internal in an abbreviated form as it is described by Gal'perin and Leont'ev. The internal plane of activity can be significantly different from its external form. At this stage of performance, mental activity becomes relatively independent of external activity and can be partly developed according to its own regularities.

Speech improves the regulation of internal and external activity. As the activity is acquired, the need for verbalization decreases, and the activity becomes less conscious. In some situations, verbalization manifests itself as self-instruction. This is important in programming the actions and integrating volitional processes into performance. At the same time, it should be noted that not all components of activity can be verbalized because some have no verbal equivalents. Speech plays an important role in the transformation of self-regulation processes from the conscious to unconscious level and vice versa. What kind of self-regulation is dominant depends on the level of skill acquisition by a person. Sometimes, transformation from an unconscious to a conscious level of self-regulation can cause deautomatization and hence degradation of skills. This factor should be considered when specialists attempt to use metacognition in the study of work activity regulation. Metacognition is a person's appraisal of his or her own cognitive processes. Existing literature assume that metacognitive intervention will have a positive effect for all learners (Aron et al. 2003). In the same cases, an attempt to control a subject's thinking about how to manage skills can lead to loss of skills. There are certain relationships between realizable and unconscious components of self-regulation, which are optimal when a subject is mastering motor skills. This is often observed, particularly in the development of advanced skills in gymnastics or of complex motor skills in work environments. This fact is overlooked by some industrial/organizational (I/O) psychologists. Violation of the existing relationship between the conscious and unconscious levels of self-regulation of activity in the training process may in some cases lead to the destruction of professional skills.

The points discussed are relevant to how we interpret the process of internalization. We consider internalization a process of the mutual regulation of external and internal activities. According to this viewpoint, the terms "internalization" and "externalization" should be considered simply the terms that emphasize the interdependence of external and internal activities during personal development.

Pushkin (1978) demonstrated that the structure of the internal gnostic dynamic is different from that of external behavior. The external dynamic changes the objectively existing situation. The internal dynamic changes the situation as reflected in the mind of the subject. The internal and external dynamics of activity are interdependent, but they are not congruent to each other. Through feed-forward and feedback influences, these two dynamics influence shape and correct each other (Bedny and Karwowski 2007). Therefore, the self-regulation process is critical in this interaction. At the first stage of mental development during ontogeny, the internal dynamic is constructed or shaped, with the support of the external dynamic. Gradually, the internal dynamic changes and becomes less dependent upon external support. At the final stage of development, the internal dynamic becomes relatively independent. Therefore, we have mutual influences and development of the external and internal components of a unitary self-regulation process. In object-related activity, the external component never exists without the internal, because internal cognitive mechanisms regulate external behavior. Specifically through the unity of the external and internal components, the ability to perform internal cognitive activity develops.

Processes of transformation from the external to the internal, and vice versa, do not exist per se. Here, we preserve the term "internalization" simply conditionally to preserve a useful tradition that emphasizes the interdependence of internal and external components. However, what we understand from "internalization" is that internal activity, at the beginning of a child's development, is performed with the support of external object-oriented activity and social interaction; in further development, internal activity can be performed relatively independent of external practical and communicative activity. Object-oriented activity always includes internal cognitive components, although sometimes they exist in a much abbreviated manner. Therefore, we can say that rather than "internalization," we are referring to a process of the self-regulation and active formation and shaping of an internal plane of activity. It is not a transformation, but the interaction of the external and internal components is a major mechanism of internalization. We do not agree that internalization can be understood as a transition from intersubjective to intrasubjective; rather, these two supposedly distinct realms are in fact totally interpenetrating and inseparable. Internalization cannot be reduced to separate psychological processes and memorization in particular, and it is the process by which the formation of internal mental operations and actions based on mechanisms of self-regulation is provided. A subject acquires the ability to manipulate the internal sign system without external material activity or interaction with others. All psychological processes depending on the specificity of self-regulation process are involved in internalization.

The presented data also demonstrate that the concepts of cognitive and behavioral actions are important to the study of relationship between external and internal components. It is well known that mental imagination of motor actions is conveyed by the electric activity of particular groups of muscles. This phenomenon is called *ideomotoric actions*. Recently, neuroscientists discovered mirror neurons in the animal and human brains (Rizzolatti et al. 2001). Mirror neurons are a type of brain cells that respond equally when we perform or observe actions performed by others. Activation of these neurons is triggered automatically. Neurons can fire before the following action is started as a prediction of the following situation. Therefore, scientists discovered new predictive mechanisms involved in the formation of our expectation and forecasting during the performance of a sequence of actions. The role of mirror neurons in perception and prediction of actions can help us understand the interrelationship between the external and internal activities from the neuropsychological perspective.

2.8 Practical Application

The genesis of internal mental activity during interaction with external behavior activity promotes an important principle in the study of work activity called "unity of consciousness and behavior" or "unity of cognition and behavior." This principle did not receive much attention in the study of human work from the cognitive psychology perspectives. Based on the above-mentioned principle, the study of human work allows overcoming the criticism of cognitive psychology approach for its mentalistic orientation and ignorance of the fact that not only the internal activity regulates the external but also the external behavior influences cognition. From AT perspectives, cognitive task analysis ignores the principle of unity cognition and behavior. The interdependence of cognition and motor activity can be seen in studies of different cognitive processes. For example, in the former Soviet Union, Yarbus (1967) studied the relationship between eye movement and visual perception and interpretation of information. He summarized his studies in his famous book, which was translated into English in 1967. In his experiments with stabilized images, he discovered that without the movement of an eye in relation to the scene, after a few seconds, the scene disappears. Therefore, eye movements are involved in visual perception. Yarbus also demonstrated that eye movement strategies are far from random. He showed his subjects a picture and asked them various questions about it; he formulated various goals. He discovered that strategies of observations depend on a goal of activity, significance of the elements of the picture, and so on. Cognitive psychologists in the West overlooked Yarbus's study, which was performed in the framework of AT, and in which there were important factors such as unity of cognition and behavior, goal of activity, and motivation.

Studying sense by a touch also demonstrated clearly that touch perception is impossible without motor motions (Anan'ev et al. 1959; Turvey 1996). It was discovered that touch perception is accompanied by complicated hand and finger macro and micromovements. In other words, motor activity can be discovered during the perceptual process. Zinchenko and Vergiles (1969) studied visual thinking, which is significant in solving visual problems. They described complicated micromotions of eyes during visual problem solving. Pushkin (1978) and, later, Tikhomirov (1984) studied eye movements in the process of solving chess problems and visual games. In these studies, the following were discovered: complicated "gnostic dynamics" directed at the analysis of various elements of the situation and their interrelationships, mental transformation of the situation, and so on. Such a "dynamic" includes a system of thinking and explorative motor actions directed at the extraction of the dynamic meaning of the situation. These data were critically important in the separation of visual perception actions and visual thinking actions in applied research (Bedny et al. 2007). The difference between visual perceptual actions and visual thinking actions is that the latter is involved in discovering functional relationships between the elements of the situation rather than in perceiving the qualities such as color and form. In practical studies, scientists have to separate visual perceptual actions, which are directed at perceiving perceptual qualities of objects, from visual thinking actions, which have the purpose of discovering functional relationship between elements of the situation.

Cognitive actions can occur when motor components of an eye, hand, or body movements are reduced to a significant degree. In such situations, the activation of lips and throat muscles can be useful. Sokolov (1969) discovered that electrophysiological activation of the lips and throat muscles is heightened during problem solving. Sometimes, the activation of lip muscles is correlated with the electroencephalograms of the brain and means that verbalization is involved in the thinking process. Automatic cognitive activity can occur without external motor components. However, during the analysis of training process, one can trace some external motor activity. Internal gnostic dynamic does not replicate external activity. Internal and external components are interdependent and influence each other during mental development through a feedback. All of these demonstrate that cognitive task analysis overemphasizes the independence of cognition from external behavior. It is not only cognition that regulates behavior but also behavior that regulates and shapes cognition. It is incorrect to reduce complicated external behavior to simple psychomotor processes, as is being presented in the works of cognitive psychologists today.

Let us consider some examples. One classical study of motor behavior in cognitive psychology is the analysis of tracking tasks. From the activity viewpoint, the study of transfer functions in cognitive psychology ignores the activity components such as goal, motivation, and strategies derived from self-regulation. During the study of the tracking tasks, Zabrodin and Chernishov (1981) discovered additional micromotions that were not related to the tracking goal of the task. Investigators noticed that operator responses contained micromotions with additional harmonics, which are not anticipated by the goal of activity. From the standpoint of existing mathematical models in cognitive psychology, these additional harmonics are errors. According to SSAT, this explorative strategy includes additional micromotions generated by subjects, which produce useful information for subjects. An operator works as a self-regulative system. It was evident that the subject did not perform a linear transformation of input signals. An information processing system is a goal-directed self-regulative system in which variables such as goal, strategy, feedback, significance of information, and motivation are important.

Similar data can be discovered in pilots' activity. Experiments and analysis of accidents and pilots' activity in emergency situations demonstrated that, in cases of equipment failure, not only auxiliary eye movements take place but also explorative hand movements take place (Zavalova and Ponomarenko 1980). Both kinds of exploratory movements perform similar cognitive functions. During the performance of exploratory hand movements, pilots search for necessary information from instrumental and noninstrumental sources. During the transition from automatic to manual control in malfunctioning situations, auxiliary hand actions can be observed in 67% of cases. Pilots begin to intervene in the control of the aircraft and can perform improper motor actions that worsen the situation. From a self-regulation point of view, motor actions perform not only executive but also cognitive functions. In cognitive psychology, motor actions are considered, first of all, executive components of human performance. In contrast, in AT and ecological psychology (Turvey 1996), attention is paid to cognitive functions of external behavior. At the same time, this problem is considered from much broader perspectives in AT than in ecological psychology.

Let us analyze another example. Subjects had to perform a computer-based task. The task was to impart features to the letters so that a given arrangement could be reached. They used eye-movement analysis in this study. Traditional cumulative scan path of eye movement is utilized. The data extracted from this registration method are usually the length of a scan path as measure of search behavior, cumulative dwell time or average fixation time, and average eye movement time. These data, which are widely utilized in cognitive psychology, can be useful but are not sufficient for task analysis. SSAT suggests a new method of eye-movement analysis and interpretation. Cumulative scan path of an eye movement is divided into segments. Motor clicks are utilized as a border between segments. Eye movement during performance of each segment is extracted. As an example, one segment of a task is presented in Table 2.1.

From the table, it is obvious that it is much easier to analyze and describe the eye movements and discover the purpose of these movements when we divide cumulative scan path into segments. To discover cognitive actions during the performance of each segment of task, we developed rules of cognitive actions extraction and their classification principles (Bedny and Karwowski 2007; Bedny et al. 2007). Each saccades and gaze associated with the conscious goal is considered a cognitive action.

Actions are classified based on the following six criteria:

1. Dominance in a particular moment cognitive process

2. Analysis of the action purpose

3. Relation of gaze to visible elements on the screen

4. Purpose of the following action and particularly the motor clicks

5. Duration of the gazes and their qualitative analyses

6. Analysis of debriefing of the subjects and comparison of their reports

This example demonstrates that the analysis of motor activity of an eye and motor clicks are an important source of information for the extraction and classification of cognitive actions.

TABLE 2.1

An Example of an Eye Movement and Action Classification between Two Clicks

Eye Movement Time to Required Position (ms)	Dwell Time at Position (ms)	Total Action Time (ms)	Classification of Cognitive Actions	Scan Path Image (Eye Movements between Clicks)
150	180	330	Simultaneous perceptual actions	
150	220	370	Simultaneous perceptual actions	
180	150	330	Simultaneous perceptual actions	
180	220	400	Thinking action based on visual information	
150	190	340	Thinking action based on visual information	
210	220	430	Thinking action based on visual information	
150	330	480	Thinking action based on visual information	
150	190	340	Thinking action based on visual information	
210	190	400	Thinking action based on visual information	

The principle of unity of cognition and behavior is important in the training process. In cognitive psychology, mentalistic explanation of human learning dominates. For example, the learning process is explained as changes in the structure of memory and transformation of declarative knowledge into procedural knowledge (Anderson 1981). Learning occurs also as a result of development and modification of schemata, which is the means of information presented in a person's memory (Lindsay and Norman 1977).

This explanation is useful but not sufficient. Cognitive psychology does not pay much attention to the interaction of the subject with the external means of activity, object of activity, goals, motivation, and the relationship between cognitive, verbal, and motor actions, and so on. All these components of activity should be taken into consideration during the analysis and development of training process. According to the self-regulative concept of learning in SSAT, external practical and internal cognitive actions are interdependent. Because of the existence of feed-forward and feedback interconnections between these types of actions, the possibility of their comparison and evaluation enables to create an active formation of both external and internal components of activity. External actions can be considered a support for the performance of internal mental actions. During the acquisition of mental actions, external support for their performance is less necessary and can be performed on a mental plane. This in turn influences external motor actions. Learning is considered not as a simple reconstruction of memory structure but rather as a process of active formation of cognitive and behavioral actions and strategies of activity. Learning can be described as sequential changes in cognitive, executive, and motivational components of activity. The more complicated acquired activity is, the more intermittent strategies are utilized by the subject. This requires dynamic orientation in situation. As learning proceeds, the number of intermittent strategies is reduced. During the training, the subject utilizes cognitive, verbal, and motor actions. Relationships between these actions and content of actions gradually change. The role of feedback in training process also changes. Its influence gradually is reduced. Extrinsic feedback becomes less important. It is important that extrinsic feedback does not substitute intrinsic feedback but rather help develop it. All these data demonstrate that even in development of cognitive skills both verbal and motor actions are important. The study of the motor components of activity is deeply implicated in cognitive tasks.

2.9 Discussion and Conclusion

The analysis of the works of Rubinshtein, Vygotsky, and Leont'ev shows that each emphasized different theoretical studies and different aspects of activity and individuality. For example, Leont'ev and Rubinshtein both proceeded from the general statement that object-practical activity is the principal source of mental development. They both had some general ideas about the structure of activity. The major components of activity for them were goal, motive, action, and operation. Action was considered the basic unit of analysis of activity. The main differences in Rubinshtein's and Leont'ev's views were their different explanations of the principles of psychic development determination and the problem of internalization. Rubinshtein and his followers did not accept the idea about internalization. During his analysis of mental development, his major point was that external causes act through internal conditions. In contrast to Rubinshtein, Leont'ev did not give sufficient value to the role of internal conditions in development. The problem of determinism is considered through the lens of internalization as transformative process when external plane in abbreviated form transformed into internal plane. Rubinshtein stated that according to the internalization principle of Leont'ev, external causes act directly, bypassing internal conditions. According to Rubinshtein, the concept of internalization did not consider the creative aspects of a person's cognition. The personal principle, in which the external conditions act through the internal, is significantly omitted from Leont'ev's approach. The leading idea in Rubinshtein's concept is that human action changes not only the objects but also the subjects themselves.

Vygotsky's work was connected primarily with the study of the personality from the cultural–historical perspective. For Vygotsky, internalization was first a social process in which an important role is played by semiotic mechanisms, particularly language that mediates the interaction between social and individual functioning. Individual practical aspects of activity were underestimated. These aspects were considered in a more detailed way in the works of Rubinshtein and Leont'ev. Each of these three approaches has its strengths and weaknesses. Here, we agree that the development of the human mind cannot be reduced to the history of the development of human culture. This approach underestimates those aspects of activity that are derived from concrete, individual practical activity of the person. At the same time, this individual psychological approach does not sufficiently take into account the historicity and semiotic aspects of activity.

The weakness in Rubinshtein's conceptualization was that in the study of the external and internal components of personal development, he did not sufficiently consider the semiotic aspects of activity, nor did he discuss the mechanisms of self-regulation, which more precisely explain the interdependence of the external and internal activities.

In SSAT, the interrelationship between the external and internal components during development is considered from the perspective of self-regulation. Activity unfolds as a process with the help of feed-forward and feedback influences and mechanisms of self-regulation; activity is reconstructed, developed, and shaped. Activity can be considered the continual formation, execution, evaluation, and correction of strategies of activity. In these strategies, we can observe complex relationships between preplanned and situated components of activity. The interrelationship of these components is continually changing. The development of activity (including learning) can be considered a series of transformations from less effective to more effective strategies. During the stages of skill acquisition, a person acquires a number of different strategies of activity, each of which may be utilized in human performance. We utilize the term "internalization" as a label to designate the interdependency of external and internal planes of activity. According to the principle of self-regulation, internalization is the process of the formation of different strategies in which the internal aspect develops with the support of the external aspect. Internal activity is constructed on the basis of its interaction with the external activity. Therefore, subjects can develop different internal structures based on the same external activity. The internal and external components of activity regulate and check each other, based on mechanisms of self-regulation. Subject–object activity and subject–subject activity are interdependent and continually transform into each other. Intersubjective interaction may even be discerned in subject–object activity and may be grounded or discovered through inner speech, inner dialogue. Intersubjective relationships arise from the observation of others in the past, even without direct contact with them or from the use of socially developed informal instructions. Thus, in the study of individual object-oriented activity, intersubjective relationships must always be incorporated. Social interactions develop in the surrounding world of objects.

Similarly, interactions with objects in the world arise on the basis of social norms and standards. Thus, from the systemic-structural perspective, we can eliminate the presumptive opposition of the primacy of either the subject–object or subject–subject interrelationship, which exists between Vygotsky's cultural–historical theory on one hand and the object-oriented theories of Leont'ev and Rubinshtein on the other hand. Social interaction and material activity are interdependent elements of internalization. Internalization is not a transformation of external activity into internal; it is the process of active formation of internal activity with the support of external components.

References

Anan'ev, B. G., L. M. Vekker, B. F. Lomov, and A. V. Yarmolenko. 1959. *Sensing by Touch in Cognition and Work*. Moscow: Academy of Pedagogical Science RSFSR.

Anderson, J. R., ed. 1981. *Cognitive Skills and Their Acquisition*. Hillsdale, NJ: Lawrence Erlbaum.

Aron, M., J. Schmidt, and K. Ford. 2003. Learning within a learner control training environment: The interacting effects of goal orientation and metacognitive instructions on learning outcome. *Personnel Psychol* 56:405–29.

Bakhtin, M. M., and V. N. Voloshinov. 1973. *Marxism and Philosophy of Language*. Cambridge, MA: Harvard University Press.

Bedny, G. Z., and W. Karwowski. 2006. The self-regulation concept of motivation at work. *Theor Issues Ergon Sci* 7(4):413–36.

Bedny, G. Z., and W. Karwowski. 2007. *A Systemic-Structural Theory of Activity. Application to Human Performance and Work Design*. Boca Raton, FL: Taylor & Francis.

Bedny, G. Z., ed. 2004. Preface to special issue. *Theor Issues Ergon Sci* 5(4):249–54.

Bedny, G. Z., W. Karwowski, and T. Sengupta. 2007. Application of systemic-structural theory of activity in the development of predictive models of user performance. *Int J Hum Comput Interact* 24(3):239–74.

Bedny, G., and D. Meister. 1997. *The Russian Theory of Activity: Current Application to Design and Learning*. Mahwah, NJ: Lawrence Erlbaum Associates.

Bedny, G., M. Seglin, and D. Meister. 2000. Activity theory. History, research and application. *Theor Issues Ergon Sci* 1(2):165–206.

Brushlinsky, A. V. 1979. *Thinking and Forecasting*. Moscow: Science Publishers.

Engestrom, Y., R. Miettinen, and R.-L. Punamaki, eds. 1998. *Perspectives on Activity Theory*. Cambridge: Cambridge University Press.

Gal'perin, P. Y. 1969. Stages in the development of mental acts. In *A Handbook of Contemporary Soviet Psychology,* eds. M. Cole and I. Maltzman, 249–73. New York: Basic Books.

Gordeeva, N. D. and V. P. Zinchenko. 1982. *Functional Structure of Action*. Moscow: Moscow University Publishers.

Karwowski, W., G. Z. Bedny, and O. Y. Chebykin. 2008. Activity theory: Comparative analysis of Eastern and Western approaches. In *Ergonomics and Psychology. Development in Theory and Practice,* eds. O. Y. Chebykin, G. Z. Bedny and W. Karwowski, 221–46. Boca Raton, FL: Taylor & Francis.

Kozulin, A. 1986. The concept of activity in soviet psychology. *Am Psychol* 41(3):264–74.

Landa, L. M. 1976. *Instructional Regulation and Control: Cybernetics, Algorithmization and Heuristic in Education*. Englewood Cliffs, NJ: Educational Technology Publication.

Leont'ev, A. N. 1972. The problems of activity in psychology. *Questions of Philosophy* 9(6):104–23.

Leont'ev, A. N. 1978. *Activity, Consciousness and Personality*. Englewood Cliffs, NJ: Prentice Hall.

Lindsay, P. H. and D. A. Norman. 1977. *Human Information Processing*. New York: Harcourt.

Nepomnayashaya, N. I. 1973. Chapter 2. Mental development and teaching. In *Developmental and Pedagogical Psychology Textbook,* ed. A. V. Petrovsky, 21–37. Moscow: Educational Publisher.

Platonov, K. K. 1982. *System of Psychology and Theory of Reflection*. Moscow: Science Publishers.

Pushkin, V. V. 1978. Construction of situational concepts in activity structure. In *Problem of General and Educational Psychology,* ed. A. A. Smirnov, 106–20. Moscow: Pedagogy.

Rizzolatti, G., L. Fogassi, and V. Gallese. 2001. Neurophysiological mechanisms underlying the understanding and imitation of actions. *Nat Rev Neurosci* 2:661–70.

Rubinshtein, S. L. 1922/1986. Principles of creative activity. *Questions of Psychology* 4:101–7.

Rubinshtein, S. L. 1934. The problem of psychology in Marx's works. *Soviet Psychotechnics* 1:14.

Rubinshtein, S. L. 1940. *Problems of General Psychology*. Moscow: Academic Science.

Rubinshtein, S. L. 1946. *Foundations of General Psychology*. Moscow: Academic Pedogogical Science.

Rubinshtein, S. L. 1957. *Existence and Consciousness*. Moscow: Academy of Science.

Rubinshtein, S. L. 1958. *About Thinking and Methods of Its Development.* Moscow: Academic Science.

Rubinshtein, S. L. 1959. *Principles and Directions of Developing Psychology.* Moscow: Academic Science.

Rubinshtein, S. L. 1973. *Problems of General Psychology.* Moscow: Academic Science.

Rumelhart, D. E., and D. A. Norman. 1978. Accretion, tuning and restructuring: Three modes of learning. In *Semantic Factors in Cognition,* eds. J. W. Cotton and R. Klatzky, 328–50. Hillsdale, NJ: Lawrence Erlbaum.

Skinner, B. F. 1974. *About Behaviorism.* New York: Knopf.

Sokolov, E. N. 1969. The modeling properties of the nervous system. In *A Handbook of Contemporary Soviet Psychology,* eds. M. Cole and I. Maltzman, 671–704. New York: Basic Books.

Tikhomirov, O. K. 1984. *Psychology of Thinking.* Moscow: Moscow University.

Turvey, M. T. 1996. Dynamic touch. *Am Psychol* 51(11):1134–52.

Vygotsky, L. S. 1960. *Developing Higher Order Psychic Functions.* Moscow: Academy of Pedagogical Science RSFSR.

Vygotsky, L. S. 1962. *Thought and Language.* Cambridge, MA: MIT Press.

Vygotsky, L. S. 1971. *The Psychology of Arts.* Cambridge, MA: MIT Press.

Wertsch, J. V., and C. A. Stone. 1995. The concept of internalization in Vygotsky's account of genesis of higher mental functions. In *Culture, Communication and Cognition: Vygotskian Perspectives,* ed. J. Wertsch, 162–82. Cambridge, UK: Cambridge University Press.

Yarbus, A. L. 1967. *Eye Movement and Vision.* New York: Plenum Press. (Translation from Russian).

Zabrodin, Y. M., and A. P. Chernishov. 1981. On losing information during description of operator's transfer functions. In *Methodology of Engineering Psychology and Psychology of Work and Management,* eds. B. F. Lomov and V. F. Venda, 244–9. Moscow: Science Publishers.

Zarakovsky, G. M. 2004. The concept of theoretical evaluation of operators' performance reliability derived from activity theory. *Theoretical Issues in Ergonomics Science,* special issue ed. G. Bedny, 313–37.

Zavalova, N. D. and V. A. Ponomarenko. 1980. Structure and content of psychic images as mechanisms of regulation of actions. *Psychological Journal* 1(2):5–18.

Zinchenko, V. P. 2001. External and internal: Another comments on the issue. In *The Theory and Practice of Cultural-Historical Psychology,* ed. S. Chaiklin, 135–47. Aarhus, Denmark: Aarhus University Press.

Zinchenko, V. P., and N. Y. Vergiles. 1969. *Creation of Visual Image.* Moscow: Moscow University.

Section II

Human–Computer Interaction

3

Task Concept and Its Major Attributes in Ergonomics and Psychology

W. Karwowski and G. Bedny

CONTENTS

3.1 Introduction

Task analysis is an important branch of applied psychology and ergonomics and covers a range of independent techniques used by scientists and practitioners to describe and evaluate what should be done to achieve particular job requirements. Task analysis might be performed for a variety of purposes, such as development of more efficient job performance techniques, design of man–machine and human–computer systems, and development of training procedures. It is difficult to imagine contemporary work psychology and ergonomics without task analysis as a part of them. However, a review of recent discussions on task analysis reveals that task analysis lacks a solid theoretical background and consists of a number of independent techniques that lack sufficient scientific foundation. Task analysis is often reduced to the description of various techniques repeatedly covered in one publication after another, for example Annett (2000), but there is no discussion on the necessity to develop unified and standardized procedures that are particularly important for the ergonomic design. Poor design practices result in design failures, lower usability, and lower efficiency of performance. In addition, many problems can be encountered in designing equipment and tools for nonproduction environments, such as recreation items and home appliances.

Publications in this field do not have a clear concept of task. Definition of task in ergonomics is not based on any theoretical psychological data. Task analysis is mainly considered from two theoretical perspectives: the behavioral approach and the cognitive approach. The behavioral approach focuses on studying externally observable behavior, described as independent reactions to a variety of stimuli or as a sum of independent physical acts. In contrast, the cognitive approach focuses on studying internal mental processes required to support mental work or external behavioral acts. The behavioral approach is considered weak and inadequate to conduct a task analysis. So, the cognitive approach is dominant in contemporary task analysis. It does not totally ignore behavioral aspects of work, but rather concentrates on studying cognitive aspects of single reactions or their performance in sequence (Wickens and Hollands 2000; Kantowitz 1974).

The other aspect of studying motor components of human work includes the analysis of discrete movements such as positioning acts (Fitts 1954) or tracking movements (Crossman 1960). Cognitive psychologists collected a lot of interesting data in this field. However, human behavior cannot be reduced to these motor responses. Analysis of such motor components has certain limitations when utilized for contemporary activity analysis. Human work activity includes complicated sequences of purposeful behavioral and cognitive actions, which have a logical organization in space and time. From activity perspectives, cognitive and behavioral actions are interdependent. Therefore, methods utilized in cognitive psychology contradict with the principle of unity of cognition and behavior, first introduced to activity theory (AT) by Leont'ev (1978) and Rubinshtein (1959). Later, this principle was elaborated and adapted to the study of human work (Bedny et al. 2001; Bedny and Karwowski 2007). According to this principle, overt, observable behavior and covert cognitive processes should be studied in unity.

The definition of task and the description of its basic attributes in cognitive psychology and ergonomics are rather vague and ignore recent psychological data. Because there are multiple definitions for task and its basic components, today, task analysis is a mess (Diaper and Stanton 2004). Moreover, Diaper and Stanton thought that the main elements of task, such as goal, should be abandoned because as they are unnecessary and are one of the major sources of confusion (Diaper and Stanton 2004, p. 603). These authors insist that this is particularly relevant for the analysis of computer-based tasks and the entire future of task analysis is not clear.

On the one hand, task analysis is one of the main fields of work psychology and ergonomics; on the other, because the definition of task and its major attributes is not clear, the future of this field is now in question. The purpose of this work is to consider the concept of task and its attributes from general and systemic-structural AT (SSAT) perspectives in particular. We start our discussion with the concept of task.

3.2 Task Concept in Work Studies

In this chapter and Chapter 4 we will consider the concept of task from the viewpoint of applied and systemic-structural activity theories. "Task" is the main element of a job. "Job" is a collection of tasks that fall within the scope of a particular job title (Landy and Conte 2007). In work settings, tasks usually have some logical organization. The logically organized sequence of tasks is called production process (Bedny and Harris 2005). From technological perspectives, production process in manufacturing can be defined as a sequence of transformation of raw material into a finished product. Any production process has two main components: work process and technological process (Figure 3.1).

In automated systems, the term "operational monitoring process" is used instead of production process. An operational monitoring process is a combination of tasks essential to accomplish some automated or semiautomated system functions. One important aspect of operational monitoring process is that the operator is not simply involved in the transformation of physical material but also in the transformation of information. In operational monitoring process, the sequence of tasks does not have rigorous logical organization. This is especially relevant for human–computer interaction (HCI) tasks, in which a significant percentage of tasks are formulated by users, who decide what has to be done. The task in operational monitoring process is a problem-solving one. The operational monitoring process includes work process and control process (Figure 3.2). The structures of the production process and the operational monitoring process are similar. The only difference is that in operational monitoring process, an operator or user interacts

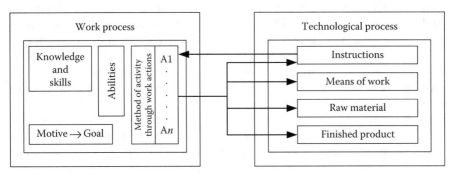

FIGURE 3.1
Structure of the production process.

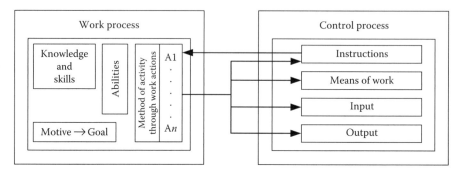

FIGURE 3.2
Structure of the operational monitoring process.

with input and output information instead of raw materials and finished product.

The HCI process is very similar to the operational monitoring process. The major difference is that in the former, a computer is the dominant means of work. This means of work produces a variety of tools on the screen that are required for a particular task and mediates HCI. Hence, we do not agree with Kaptelinin (1997), who considered computers a tool. From our point of view, this is not a precise interpretation of the role of computer in human work.

We can study any task from two main perspectives: (1) from the technological perspective and (2) from the work analysis perspective. These two approaches are interdependent but have their own specifics. In manufacturing, the task is synonymous to the production operation. A process engineer is responsible for developing the manufacturing process, which includes steps such as cutting, drilling, and reaming. All of these steps are performed in a particular sequence and sped up using special equipment, tools, computers, and so on. Ergonomists concentrate on the behavioral aspects of human work. It is obvious that the equipment configuration, manufacturing process, and computer interface influence strategies of human work activity and vice versa. Hence, changes in the equipment configuration, computer interface, and technological process will result in new strategies of human work activity in a probabilistic way. Based on this, we can distinguish three major aspects of ergonomic design: (1) design of equipment and tools, (2) design of HCIs, and (3) design of human performance. Analysis of the relationship between work process and technological process is also important for a number of other ergonomic problems, such as training and professional selection.

From the description of task demands, it is clear that task is a psychological concept. According to the *Psychology Dictionary* (Reber 1985), the task demands of a particular job are the aspects of task that require the use of certain actions, and therefore, particular patterns of perception, thinking, and feeling to accomplish the task goal. According to AT, the task is an overall goal that should be achieved in particular conditions (Leont'ev 1978;

Rubinshtein 1959). In Sections 3.3 through 3.5 we will show that the subject's goal does not exist in a ready form. Only requirements are given objectively. These requirements can be transformed into subjectively excepted goal at a later stage. Hence, there are requirements given in particular conditions and the task emerges after the transformation of these requirements into a goal. Acceptance or formulation of the task–goal is closely associated with the subjective representation of the task. The subjective or mental representation of a task is characterized by the following four features (Kozeleski 1979):

1. It depends on the objectively presented structure of the task.
2. It is a dynamic phenomenon that can change during task performance.
3. The mental representation of the task determines its performance.
4. Success in solving the task depends on its subjective representation.

Task concept will be also considered in more detail in Chapter 4.

3.3 Historic Overview of Goal Concepts in Psychology and Ergonomics

A goal is an important element of the task. The origins of the concept of goal can be traced back to German psychology (Lewin 1951). Lewin assumed that behavior is goal-driven. According to him, a goal creates tension, which remains until the goal is achieved. If the goal is not achieved, the tension associated with the goal-related behavior will remain. This was first reported by Lewin's Russian students, Zeigarnik (1927) and Ovsiankina (1928). In Zeigarnik's experiment, children were engaged in a variety of tasks. They had an opportunity to complete some tasks, but they were prevented from completing other preliminary tasks. After engaging in unrelated type of activity, they were asked to recall previous tasks. Zeigarnik discovered that the subjects better remembered uncompleted tasks. This phenomenon is presently known as the "Zeigarnik effect." Ovsiankina (1928) demonstrated that interrupted tasks have tendency to be spontaneously resumed. These studies show that a goal directs human behavior until its completion. Ovsiankina's study led to other interesting studies connected with studying substituted behavior (Sliosberg 1934). Tension created by the interruption of task performance can be eliminated by specially selected and further performed new tasks. If the interrupted tasks did not demonstrate a tendency to be performed again, then new tasks were considered substituted activity. Substituted features of new behavior depend on the similarity of tasks, their complexity, and so on.

The other important aspect of the study of human goal is associated with concepts such as the level of aspiration (Hoppe 1930). Hoppe demonstrated that human behavior cannot be explained by behaviorism using terminology such as reward and punishment. According to Hoppe, a subject, before starting his or her work, always imagines its possible result or the level of achievement. Hoppe called this imagined level of achievement the level of aspiration.

In the United States, the concept of goal is associated with the scientists McDougall (1908) and Tolman (1932). McDougall rejected the explanation of human behavior as a stimulus-triggered system. He explained human behavior as an active striving toward an anticipated goal.

Tolman developed purposeful behaviorism. According to him, goal is the end state toward which the motivated behavior is directed. Knowledge about responses and their consequences are important in both animal and human behavior. Hence, animal and human behavior is flexible in reaching its purpose. Miller et al. (1960) suggested in their model that plans are associated with goals that have motivational properties. One of the peculiarities of considered theories is that motivation and goal are not distinguished. According to AT, motivation and goal are different entities. In more recent studies, the goal concept and motivation also are not differentiated. Locke and Latham (1984) and Pervin (1989) postulated that goal has both cognitive and affective features. Pervin stated that goal can be weak or intensive, and Locke and Latham (1990) considered goal a motivational component of human behavior. Presumably, the more intense the goal is, the more one strives to reach it. Hence, the goal "pulls" the behavior or activity. Kleinback and Schmidt (1990) described the volitional process and considered the goal a source of inducing behavior (Figure 3.3).

From Figure 3.3, we can see that volitional process moves from goal to behavior. Kleinback and Schmidt also did not distinguish goal and motive.

In the field of HCI, goal is defined as a state of a system that a human, machine, or a system strives to achieve (Preece et al. 1998). Hence, human and nonhuman goals are considered similar phenomena. This is the cybernetic understanding of goal. The extraction of subgoals is based on formal, logical analysis of human performance without a real analysis of strategies of task performance and actions performed by a person. Such theories do not discuss objectively given requirements of tasks and process of goal acceptance, intermittent goals of actions, final goals of task, the difference between goals of artificial system and human goals, and so on. All these lead to the fact that scientists such as Diaper and others (Diaper 2004; Diaper and Stanton 2004) attempted to eliminate the concept of goal and some suggested to eliminate the task entirely (Nardi 1997).

Diaper wrote (2004, p. 17):

> A hierarchy of goals, as used in HTA, consists of multiple related goals, but a person can also perform an action on the basis of unrelated goals. Furthermore, unrelated goals that nonetheless motivate the same behavior cannot be simply prioritized in a list, because different goals have more or less motivational potency depending on their specific context.

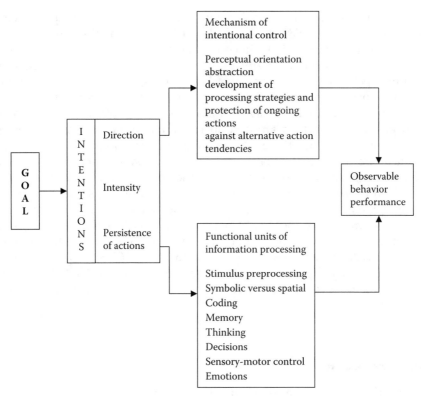

FIGURE 3.3
Scheme of the volitional process according to Kleinback and Schmidt.

For example, a chemical plant operator's unrelated goals for closing a valve might be (1) to stop the vat temperature rising; (2) to earn a salary, and (3) to avoid criticism from the plant manager. The first might concern the safety of a large numbers of people, the second is sociopsychological and might concern the operator's family responsibilities, and the third is personal and might concern the operator's self-esteem. These three goals correspond to different analysis perspectives, the sociological, the sociopsychological, and the personal psychological; and there are other possible perspectives as well. Furthermore, people might have different goals within a single perspective.

For HCI practitioners, the concept of goal is confusing. For example, what is an unrelated goal? In AT, unrelated goals do not exist. It is not Diaper's fault that he confused goal with motives; this can be explained by the fact that the concept of goal is not precisely defined in cognitive psychology and ergonomics, and goal and motives are not distinguished. Our behavior is polymotivated and can include a number of motives. According to personality and social psychology, if the goal includes various motives, a person pursues multiple goals and motives at any given time during task performance (Pervin 1989). In AT, goal is only an informational or cognitive component of human activity. In contrast, motives or motivation is an energetic, inducing aspect of activity. Therefore, in the above example, the goal of human activity is "closing a valve." In contrast, the motives, which push the operator to close the valve, may be (1) to stop the vat temperature rising, (2) to earn a salary, and (3) to avoid the criticism of the plant manager. These wishes or desires (which are described verbally) become motives during their connections with the goal. They create motivational-inducing forces to reach one single goal: "closing a valve." This example demonstrates that including motivational components into the goal of a task or actions makes it difficult to perform the task analysis in practical settings. Hence, it is not accidental that in the concluding article, Diaper and Stanton (2004) suggested eliminating the concept of goal from task analysis. The authors suggested substituting the concept of goal and the motives connected with goal with their method, called "forward scenario simulation (FSS) approach." As a result, a critically important area of psychology that is involved in studying anticipation, forecasting, formation of hypothesis, goal formation, and so on, is entirely ignored.

3.4 Anticipation from Activity Theory Perspectives

In this section, we will consider anticipation from an AT viewpoint, which helps us understand goal as one of the most important anticipatory mechanisms of activity.

The term "anticipation" was introduced by Wundt (1894), who just underlined its importance. Behavior approach dominated in psychology for quite a while, and as a result, anticipation has not been in the spotlight. However, anticipation mechanisms always played an important role in AT. Attention has been paid to the psychophysiological aspects of anticipation. This field of study is associated with the scientists Anokhin (1969), Bernshtein (1966), and Sokolov (1963). Bernshtein considered human beings an active system that is continually attempting to solve a variety of problems and particularly behavioral problems. According to him, human movements are performed based on of self-regulation process in which mechanisms of brain responsible for the creation of required neural model of the future events are very important. Similar regulation mechanisms of neural activity were suggested by Anokhin and Sokolov. According to Anokhin (1969), a neural predictive mechanism is involved in the formation of a conditioned reflex called "acceptor of action effect." This is the state of the neural system that contains the most important parameters of the desired future result of an act. Sokolov (1963) demonstrated that neural mechanisms can model the external world through specific changes in the internal structure. The set of changes produced in the nervous system by an external object is isometric with the changes to the external object. These changes comprise an internal model or image of an external event (neural model of stimulus). This model is a predictive function of the possible future event. For instance, an orienting reflex is a result of discrepancies found in the nervous system when information about the incoming signal is compared with the trace of earlier stimulus.

A goal is one of the most important anticipatory mechanisms at the psychological level of self-regulation of activity. It predicts the outcome of our activity. A goal is a psychological model of a desired future result. During an ongoing activity, this goal becomes more specific and is corrected if needed. Anticipation was considered by Lomov and Surkov (1980) from the systemic approach perspective. They described five levels of anticipation: sensory motor, perceptual, imagination, verbally logical, and subsensory. Selected levels reflect psychological processes that dominate at the corresponding levels of anticipation. This classification is a relative one. Each level of anticipation has its specific features, which should be taken into consideration when analyzing work activity. The dominant level of anticipation depends on activity goal and the task specifics. For example, if the sensory level of anticipation dominates, then the other levels perform auxiliary functions. Inadequate substitution of one level of anticipation by other levels can reduce the efficiency of task performance. Utilizing the verbally logical level of anticipation during the performance of the tracking task, for instance, can reduce the efficiency of the task. The sensorimotor level of anticipation is more effective for this task.

Let us briefly consider the above-mentioned levels of anticipation. The simplest level of anticipation is the subsensory level. This is an unconscious level of preparatory neuromuscular tone, posturing ideomotor acts that precede movement. The sensorimotor level of anticipation includes tuning up in advance to a stimulus. Such a tune-up can be beneficial for performing various reactions, forecasting emergence of a stimulus, and reacting with anticipation of a particular situation, and so on.

The perceptual level of anticipation is anticipation at the perceptual level. This level is involved in evaluating the distances between moving objects and the speed of their movements in the near future, forecasting collisions of moving objects and the possibility of avoiding such events, and developing and promoting perceptual hypotheses about possible dynamics of events in space and time. Studies demonstrate that a person often cannot verbalize developed perceptual hypothesis.

The imaginative level of anticipation means forecasting events based on manipulation of various images. Perceptual processes play a subordinate role at this level. A person not only perceives a situation but also operates with the images of events that can happen in future and promotes imaginative hypotheses, which are not completely verbalized. This level of anticipation is important for creation of a dynamic mental model of a situation. In our self-regulation model described in Chapter 9, the subblock "operative image" is a mechanism responsible for this level of anticipation.

The verbally logical level of anticipation includes anticipation sustained by complex intellectual operations and higher forms of thinking processes. At this level, external and internal speeches are particularly important when subjects promote various verbalized hypotheses and develop a verbalized plan of future cognitive and behavioral actions. In our model of self-regulation (see Chapter 9), there is a "situation awareness" (SA) block, that is responsible for creation of such models. We borrowed the term "situation awareness" from Endsley (1995). In our work, this mechanism has some similarities with SA, but it is considered in SSAT from a different perspective. The SA subblock of self-regulation can reflect not just current information but also near future and past events. This is a verbally logical anticipatory mechanism that provides broader and deeper forecasting hypotheses.

Another important kind of anticipation is the creation of the goal of someone's own activity. Usually, such anticipation is a result of a complex interaction of the verbally logical and subsensory levels of anticipation. These levels of anticipation will be considered in Section 3.5.

An analysis of the presented material demonstrates that there are qualitatively different levels of anticipation. Knowledge about these levels of anticipation is important for the analysis of human performance. Anticipation includes the formation of verbalized and nonverbalized hypotheses. The basis for promoting such hypotheses is various strategies of gnostic explorative activity. Exploration can be a combination of internal or cognitive and external or behavioral actions. The ratio of these two kinds of actions in the activity as a whole can vary. Sometimes, explorative activity can be purely mental. Based on the feedback, a person can evaluate the promoted hypothesis. Hence, exploration is an example of a self-regulative process (Bedny and Meister 1997).

From a functional analysis perspective, not only cognitive but also emotionally evaluative and motivational aspects of activity play an important role in the promotion of hypothesis and in anticipation processes in general (Bedny and Karwowski 2007).

There are two blocks in the self-regulation model of orienting activity presented in this book (see Section 9.4) that depict emotionally evaluative and motivational aspects of self-regulation. One block is called "assessment of sense (significance)" and the other is "formation of level of motivation." Both cognitive and emotionally evaluative and motivational mechanisms determine what elements of situation will be selected for formation and promotion of hypotheses. Hence, not only cognitive but also emotionally motivational mechanisms are involved in the selection and interpretation of information. These aspects of human activity are not sufficiently studied in cognitive task analysis.

The unconscious level of self-regulation is also involved in the formation and selection of hypotheses, and "set" is a very important mechanism for the promotion of these hypotheses. A hypothesis can be considered a probabilistic model of possible solution. It can include potential goals of activity and mental representation of possible development of events. Such hypotheses are called orienting hypotheses. There are also instrumental hypotheses that are responsible for forecasting methods of goal attainment.

3.5 Concept of Goal in Activity Theory

Our behavior can be regulated at two levels (Reykovski 1974). The first level reflects the performance of a variety of reactions in response to presented stimuli. This is reactive behavior. The second level of human behavior is the actualization of purposeful, consciously and voluntarily performed actions directed to achieve the goal of actions. This kind of behavior is goal-directed human activity. Human activity includes both levels. However, goal-directed activity dominates. The simplest kind of human goal-directed actions resemble reactive behavior to some degree, but still include voluntary and conscious aspects of action regulation. Therefore, involuntary reactions and voluntary, goal-directed actions are basic components of human work. Highly automated actions are similar to reactive behavior. For example, an alarm sounds at a nuclear control station when a certain parameter exceeds the limit. This is a signal for the operator to take a specific highly automated action. However, this is still a meaningful and purposeful action (not a reaction), because it has a corresponding specific goal or future desired result. Elimination of the task's goal and goals of separate actions during task performance reduces human work activity to a chain of reactions or responses. Such behavior is passive and can be triggered by outside stimulation. Each new stimulus initiates new reaction. Generally, such behavior entirely depends on external, environmental stimulation. AT psychologists who study human work cannot agree with such interpretation of human work behavior.

In this light, it is interesting to consider some publications in the *Handbook of Task Analysis*, edited by Diaper and Stanton (2004). Analysis of this book makes it obvious that in HCI there is no clear understanding of the concept of goal. This can be explained by the fact that this concept is not clearly defined in cognitive psychology and ergonomics. Diaper and Stanton attempted to substitute the concept of goal with FSS processes (Diaper 2004; Diaper and Stanton 2004). As we already discussed in Section 3.2, Diaper mixes goal with motives because goal in cognitive psychology is not distinguished from motives. In his theoretical substantiation of the concept of goal by FSS process, he wrote that this approach is multitheological versus existing single-theological approach that utilizes a single desired future state of the system. Hence, this author mixes the anticipatory stage of activity, when a subject can formulate multiple potential goals and hypotheses about the state of the system, with the final stage of the goal-formation process when a subject chooses one goal of the task from a number of potential goals. The goal-formation process is followed by the development of a number of instrumental hypotheses and selection of one of them. Sometimes, the goal is formulated in ambiguous terms. For example, "What happens if I perform this particular course of action?" A subject can check his or her instrumental hypothesis mentally or practically, abandon the hypothesis, and formulate a new one. If the subject formulates a new goal of task, it leads to a completely new task. Moreover, the user can formulate not only a final goal of task but also multiple intermittent goals that depend on a selected strategy of task performance, which is subgalling process of breaking down the overall goal of the task into smaller goal-directed steps (Pushkin 1978). This process, to some degree, is similar to the means–ends analysis that was introduced by Newell and Simon (1972).

In SSAT, the process is treated as a goal-directed self-regulation process. In a computer-based task, the final goal of a task is often formulated by a subject, and the goals and strategies of task performance are not defined in detail in advance. Subjects progress from the anticipatory stage of activity to the executive stage of activity, while the overall goal of the task gradually becomes clearer. The anticipatory and executive stages of activity include complex exploration of possible strategies of task performance. The user can mentally operate with various elements of a task presented on the screen, compare, and combine these data with the information in memory. As a result, the same external situation is constantly changing in the user's mind. This phenomenon is called "gnostic dynamic" (Pushkin 1978). The self-regulation process, with its mental transformation, evaluation, and correction of the situation, is the basis of gnostic dynamics (Bedny and Karwowski 2007). This process utilizes not only conscious mental actions but also unconscious mental operations. There is a complex relationship between these conscious and unconscious components of activity that can sometimes be transformed into each other (Bedny and Meister 1997) when a user eventually selects subjectively accepted goal and possible strategies of its achievement. Hence, Diaper's (2004, p. 17) "unrelated goals" are simply hypothetical motivational factors.

There are five psychological approaches to the understanding of the concept of goal (Tikhomirov 1984; Bedny and Karwowski 2007):

1. The goal is not a scientific notion. For example, Skinner (1974) described a person's behavior with the following terms: stimulus, reaction, and reinforcement. Here, the goal is not considered a psychological concept.

2. The goal is considered the physical location of an object or a formal description of the final situation or state, which can be achieved through the functioning of technical or biological system.

3. The goal is the end state toward which the motivated behavior is directed and by which it is completed. This is an interpretation given by Tolman (1932), the founder of purposeful behaviorism.

4. The goal is a cognitively motivational factor that pulls human behavior in a certain direction until the desired outcome is achieved.

5. The goal is considered a conscious mental representation of the desired future result of a subject's own activity that is connected with motives. This connection creates a vector "motives → goal," which directs activity until a desired goal is achieved. This is the understanding of goal in AT.

From the above approaches, one can see that a goal in AT is a conscious image or logical representation of the desired future result. An image of future result when a subject is not directly involved in achieving this result is not a goal of activity. An imaginative, future result emerges as a goal only when it is consequence of the subject's own activity. For example, a student may know that an excellent grade requires 4 to 5 hours of preparation. Such knowledge does not create a goal unless the student is motivated "to achieve the desired grade." Hence, only when a conscious image of desired future result joins with the motivation and the student actively prepares for the examination, this future result is transferred into the goal of activity. In AT, the goal is connected with motives and creates the vector "motives → goal" that lends activity a goal-directed character. We also need to distinguish between the "overall or terminal goal of task" and "partial or intermittent goals of actions and subgoals of a task."

In AT, there are energetic and informational (cognitive) components of activity that are interdependent but still different. A goal is a cognitive component of activity. Motives or motivation is an energetic component of activity. A goal cannot be presented to the subject in a ready form but rather must be an objective requirement of the task. However, these requirements should be conscious and interpreted by the subject. At the next stage, these requirements should be compared with the past experience and the motivational state that leads to the goal acceptance process. The subjectively accepted goal does not always match the objectively presented goal (requirements). Moreover, subjects can often formulate the goal independently. As can be seen, the goal always assumes some stage of activity, which requires interpretation and acceptance of the goal. So, we can conclude that the goal does not exist in a ready form for the subject and cannot be considered simply an end state to which the human behavior is directed.

3.6 Goal and Motivation in Task Analysis

Freud (1916) was the first to introduce a concept of energy into psychological studies. According to Freud, people are complex energetic systems. Energy is necessary for the functioning of mental processes. The idea that energy is an important component of mental functioning is borrowed from biological sciences and physics. At the same time, Freud did not describe the interaction between energy and informational aspects of human functioning in detail. Today, concepts such as energy and information are clearly defined in psychology. Information processing is a cognitive function. The energy concept, which is derived from neural system's electrophysiological function, has been utilized in the studies of emotionally motivational aspects of behavior or activity. Although both these concepts are distinguishable phenomena in AT, they are considered closely interconnected. There are several types of information–energetic interconnections in activity (Vekker and Paley 1971, Vekker 1974). The first type of interconnection was shown in psychophysical studies that discovered that an increase in the intensity of external stimuli increases the sensory qualities. The second type of interconnection is related to the reticular activating system of the brain, which plays an important role in controlling the state of arousal and awareness. For example, Kahneman's (1973) view on attention includes the concept of energy. This model demonstrates that success of cognitive processes depends not only on physical characteristics of information presented to the subject but also on the level of activation of neural system. The third type of interconnection is linked to the emotional–motivational components of activity and to the specificity of information processing. These groups are interdependent. Currently, the cognitive approach dominates in ergonomics, in which a person is considered a pure information-processing system that picks up information from external situation and memory because the relationship between cognition and energetic aspects of activity is not sufficiently studied. However, the interpretation of information depends on emotionally motivational stage of a person, and moreover, information cannot be transmitted without energy.

Motivation is an intentional or inducing component of activity, which is closely connected with human needs. A need is an internal state of individuals that is less than satisfactory and produces a feeling of a desire for something (Carver and Scheier 1996). Some needs are biological in nature, and other needs such as achievement and power are secondary or psychological needs. Human needs are not just a result of biological evolution but also the result of an experience acquired through human culture.

Needs become motives if they are connected with the goal of an activity. Motives are an inducing force that catalyzes a person's desire to achieve the goal of the activity. Thus, needs operate through another construct, called motives. Motives derive from needs, but they are closer to our activity or behavior. For example, a need for food as physiological state can be transferred into motivational state called "hunger." This state is experienced cognitively and affectively. Motives are also influenced by external events, which are sometimes called "press" (Murray 1938). These external stimuli create a desire to obtain or avoid something. If somebody received recognition, this can trigger his or her own motives for further recognition. Hence, internal

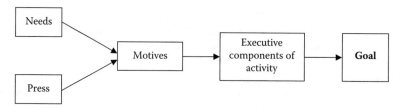

FIGURE 3.4
Relationship between needs, press, motives, and goal in the activity structure.

needs and external press can activate motives to engage in a particular kind of activity to achieve a conscious goal. The relationship between internal needs, external press, motives, and goal is presented in Figure 3.4.

Motivation is another important construct that includes diverse motives that have a hierarchical organization. Some motives can be more important than others; some can be conscious, and others can be semiconscious or even unconscious. The relationship between motives is typically dynamic and can be modified during human activity. The goal can often be induced not by one but by several motives. Activity results can coincide with the goal or deviate from it, and a subject should correct her or his actions or activity strategy. Deviation of the result from the goal can be considered an error if this deviation is outside of the subjective criteria of success, which is not always the same as the objective one. Sometimes, the activity result that deviates from its goal can be useful for the subject when it is a desired accessory result.

There is another aspect of motivation that is associated with the role of emotions in the motivational process. According to Zarakovsky and Pavlov (1987), emotions reflect the relationship between our needs and real or possible success in their satisfaction. The need-information theory suggested by Simonov (1982) is very interesting from this perspective. Simonov stated that emotions are reflection of the actual needs (its quality and level) and the probability of their satisfaction in our brain. The relationship between these variables is demonstrated by the following formula:

$$E = f[N(U_{ir} - U_{iex})]$$

where N is intensity and quality of an actual need, U_{ir} is information about means, resources, and time presumably required to satisfy a particular need, U_{iex} is information about means, resources, and time the subject has to satisfy a particular need.

The difference $U_{ir} - U_{iex}$ is assessment of possibility to satisfy the need. The more this difference is, the less is the subjective confidence in satisfaction of this particular need. According to Simonov, emotion is a specific "currency" of the brain or universal measure of utility. Our feelings (happiness, anger, outrage, etc.) present a measure of quality and level of our need in relation to a possibility to satisfying it. Of course, it is not a rigorous mathematical formula. The last equation simply shows the relationship between information and needs related (energetic) aspects of activity.

The analysis of the relationship between goals and motives demonstrates that actions during task performance can be either successful and unsuccessful. Each action can be evaluated in relation to the goal of the action and in relation to the goal of the task. Evaluation of actions or activity always includes the emotional components or subjective significance of the obtained result. Nardi (1997, p. 242) wrote that the notion of task does not suggest motive or directive force, and, therefore, this concept should be abandoned. However, this remark is true only for cognitive psychology. From the AT perspective, the task always includes a motivational component and there is no such a thing as an unmotivated task.

3.7 Analysis of the Concept of Goal in Human– Computer Interaction Task Analysis

We now concentrate on the concept of goal in the context of users interacting with the interface, which was not an object of the previous analysis. The hypothetical task and interface were developed by Bedny et al. (2008). The interface is divided into three areas: goal, object, and tool (Figure 3.5). These areas of the screen correspond to three basic elements of activity (except activity elements such as subject and result), which is depicted by the following schema of activity.

$$\text{Subject} \to \text{Tool} \to \text{Object} \to \text{Goal} \to \text{Result}$$

The goal area stays the same until the task is completed. It represents the desired arrangement of elements (result) that should be achieved by manipulating elements in the object area. More accurately, this is an objectively given task requirement that is transformed into a subjectively accepted goal of task only after being accepted by the subject in particular conditions. The object area consists of the elements whose state should be manipulated to achieve the final arrangement that is given in the goal area. The tool area includes elements utilized to impart the desired features and required arrangement in the object area. These are externally given tools that interact with mental tools that exist in the subject's mind.

The subject's attention to specific areas of the screen corresponds to her or his strategies of task performance that have been outlined by the basic theoretical concepts of AT. Dividing the screen into these areas is very helpful for understanding the activity structure during the skill acquisition process. Hence, when interface usability issues exist, study of the eye and mouse

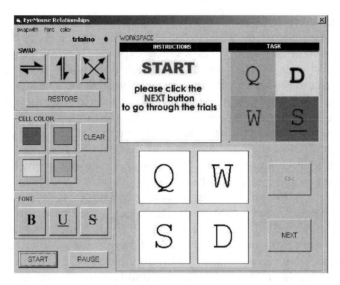

FIGURE 3.5
Task interface.

movements in these areas can provide an understanding of the difficulties encountered by the user during task performance and helps discover more efficient versions of the interface or better methods of computer-based task performance.

In the described task, we have an external goal. When a goal is not given on the screen in a ready form, any elements of the screen, with which the user interacts during period of time spend to clarify the task's goal, can be considered the goal area. There are also externally given tools that interact with internal mental tools such as words, images, and instructions, which the subject keeps in her or his memory. As per AT terminology, the main focus of the task is to alter the object's features using the available external and internal tools to reach the externally presented or internally formulated goal.

The interface of the task under consideration is given in Figure 3.5. This interface has been designed in such a way that certain sequence of actions or steps can be inadequate so the users have some constraints in the possible strategies of task performance. These constraints cannot be discovered just by looking at the interface but rather uncovered only by trail and error. Hence, there is a possibility to explore different strategies of task performance until one of the subjectively preferable strategies of task performance has been selected. In each trail, a new version of goal and object areas is presented. Participants cannot achieve the task's goal all at once. They need to break the tasks into subtasks. Each subtask has its own subgoal. Moreover, cognitive and behavioral actions that are a part of the task execution also have their goals. The goals of individual actions are usually kept in working memory during a short period of time. Feedback provided by the obtained result helps users discover the constraints of the task performance and to select the sequence of cognitive and behavioral actions that permit them to complete the task. Hence, task performance involves explorative activity and therefore thinking and decision-making processes. A preliminary experiment demonstrated that there was a variety of possible strategies of task performance that can be limited to three main groups. Variations in strategies inside each group can be neglected because they are insignificant. Two strategies considered were equivalent in their complexity and had a probability of 0.8. We have selected one of them for further analysis.

In our study, we have utilized methods such as eye movement analysis, mouse movement registration, chronometric analysis, debriefing, discussion, and observation. Extraction of cognitive and behavioral actions and analysis of the task performance strategies has been performed based on theoretical principles and methods developed in SSAT (Bedny and Karwowski 2007, 2008). One important new method is combining eye movement and motor action (mouse movements and clicks) registration. This method derives from the principle of "unity of cognition and behavior" (Bedny et al. 2001). Traditional methods of eye movement interpretation use a cumulative scan of the entire task (Figure 3.6) when scientists extract data such as scan path length (in pixels), cumulative dwell time or average fixation time, number of fixations, or number of saccades (Viviani 1990). Such data are useful but not sufficient and can be related to the parametric method of study.

In systemic-structural analysis, we divide a task into a number of relatively independent fragments. Usually, each fragment traces eye movement between two clicks (Bedny and Karwowski 2008). Each fragment is associated

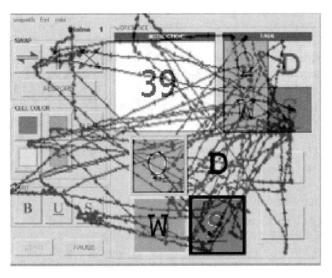

FIGURE 3.6
Cumulative scan path of the human–computer interface task.

with one or several images of eye movement. These images allowed us to trace eye movement from one element of task to the next with high precision and to compare it with the mouse movement. Based on such data, rules of action, extractions, and their standardized classification, we can accurately describe the activity structure during task performance. In Figure 3.7, the first two images of eye movement out of 12 developed during the task analysis are depicted.

To better understand our approach to eye movement analysis, the cumulative scan path of eye movement for the entire task is presented in Figure 3.6. As can be seen, the cumulative scan path used by traditional eye movement analysis is less informative than the fragments utilized in our method. The two images presented in Figure 3.7 are more readable. It is much easier to trace the eye movements, their dwell time, or their average fixation time in Figure 3.7 than in Figure 3.6.

Eye movements should be considered exteriorized elements of mental activity. For example, from Figure 3.7, we can see how a subject develops a mental model of situation that reflects the first stage of task performance. The subject shifts her or his sight from elements in the goal area to the elements in the object area multiple times. These actions are not simply perceptual ones. The subject explores the relationship between the elements of the situation at hand. Therefore, this fragment also includes thinking actions. Elements of a situation are structurally organized. The subject selects element D in the object area and clicks on it so that she or he formulates the goal to change this element's position after intensive explorative activity as has been discovered through eye movement analysis. We have observed three stages of activity: creation of a mental picture of the situation, selection of a desired element (Figure 3.7), and selection of a necessary tool and shifting of the letter D up after developing a new mental model of the situation.

Not only the overall goal of the task but also intermittent goals are critical elements of task performance. The subgoaling process regulates all stages of task performance. These subgoals have a hierarchical organization. For

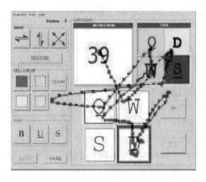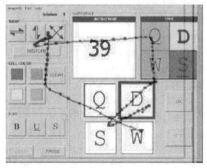

FIGURE 3.7
First two images of eye movement.

example, during performance of the first stage of the task (Figure 3.7), the subject formulates intermittent goals to find out the goal area, the object area, the relationship between elements in the goal and object areas, or "I need to select an element in the object area," and so on. Some of these goals that flash across the mind can be quickly forgotten. However, mental representation such as "what I wish to achieve" (the cognitive aspect) makes our behavior purposeful and goal directed, not reactive. However, achieving the required result is also connected with various levels of intense desires or wishes to complete the task successfully. It is the emotional-motivational factor. In the experiment, this factor was determined primarily by the fact that subjects recognized that their actions were observed and recorded during the experiment.

Our method helps extract standardized cognitive actions performed by the user and compare them with the motor actions. We will not go into the details of eye movement analysis here and only briefly consider the two images depicted in Figure 3.7 that reflect the goal acceptance stage and process of formatting the mental model of the task at the beginning of task performance. These are important anticipatory mechanisms that influence the strategies of task performance.

The following is the sequence of actions extracted from the eye movement path based on the SSAT method: The eye moves from the start position to the element Q in the goal area (G_Q). The action performance time was 330 milliseconds (movement took 150 milliseconds and dwell time was 180 milliseconds). According to the existing SSAT principle of action classification, these are simultaneous perceptual actions. In further discussion, the performance time of analyzed actions will not be mentioned, but we would like to stress that action performance time is an important information source for their classification and analysis. By performing the first action, the user just wanted to identify the task's goal. In the user's visual field ($\alpha \approx 10°$), he or she can simultaneously perceive four elements in the goal area. From Figure 3.7, we can see that the user moves her or his eyes from G_Q to O_Q (from element Q in the goal area to element Q in the object area). This is the first shift of the gaze into the object area. Then, the user did one short shift into the tool area to check the tools position because he or she wants to receive general information about a new version of the task. These two actions, developed according to the SSAT principle of action classification, are simultaneous perceptual actions. At the next stage, the eyes move from the tool area to the goal area again. The purpose of further eye movement is not simply to receive information (perception), because here the user is striving to figure out the relationship between the elements of the situation and the task goal, which are not perceptual features of the task. This requires the involvement of the thinking process performed based on the visual information. According to the SSAT classification system of actions, users at this stage of the task perform simple thinking actions based on visual information (Bedny and Karwowski 2007). Eye movements during this period demonstrate this fact. Based on explorative thinking actions, the user creates a mental model of the situation and makes a decision about which element should be activated. Figure 3.7 shows that element O_D was activated, and later, this element was shifted up and element O_W was moved down. Hence, mental actions that choose activating element D in the object area and clicking a vertical position tool can be classified as decision-making actions based on visual information (the user decides which element should be activated before it can be moved and which tool should be used). According to the SSAT classification, they are the simplest decision-making actions at the verbally thinking level. Figure 3.7 demonstrates that the user utilizes various complex combinations of conscious and unconscious processes. At the final stage, we observe the conscious, voluntary goal-formation process to activate element O_D and shift element "D" up using a corresponding tool (Figure 3.7). This is the final goal of these stages of task performance that are associated with the given images of the eye movements. At the same time, it is an intermittent goal in relation to the overall goal of task. Moreover, during performance of cognitive actions, a user also formulates various goals of separate actions, and the user forgets the majority of them.

It is interesting that the letters initially chosen as elements for the object area can influence the sequence of manipulations of the elements in this area. This sequence does not derive from the given instructions. For example, if the top row in the object area contains letters A and B and the bottom row contains letters C and D, it affects the sequence of manipulation with these letters because the subject starts manipulating with them utilizing their alphabetic order in spite of no instructions to do so. Hence, the sequence of letters impacts a particular sequence or plan of subgoaling, which is not given in the instructions. This example demonstrates that the goal-formation process depends not only on objectively given requirements or instructions but also on the past experience of the subject, and more specifically, it depends on a given situation, subjectively preferable, and not always the conscious strategies of the information interpretation. Thus, the letters chosen for the experiment are Q, W, S, and D (Figure 3.5).

Let us consider the goal-formation process. A user received verbal instructions to reorganize the object area based on the externally presented goal. The user had performed other versions of this task before and therefore has a preliminary developed goal of what needs to be done. This goal is formulated and accepted by the user in a very general manner based on verbal instruction and past experience. It should be specified in each particular trail. The user needs to know the conditions (givens) in which this goal is presented. Without specifying the goal and conditions of the task, the user cannot develop a mental model of the situation and the program of task performance. This can be clearly observed during the eye movement analysis. At the first stage of task performance, explorative activity includes perceptual, simple thinking, and decision-making actions, which are associated with some motivational state that can be labeled as a desire to reach the goal of the task. This motivational state is clearly observed during the first trails when users report emotional tension during the task performance.

In general, the purpose of this stage of activity is to specify what kind of goal of the task is presented, find out the initial state of the object area, compare the goal area with the object area, and create a dynamic model of situation. A dynamic model includes not just what exists in the present situation but also in the near future if the user follows the selected strategy. Only then does the user formulate the program of performance. This demonstrates that a new method of eye movement analysis helps us find out how gnostic activity is performed and the relationship between explorative and executive activities. The user formulates a number of potential goals that can be quickly forgotten. Such goals are associated with gnostic hypotheses, which very often are not verbalized. However, some of such hypotheses can be verbalized and become conscious at the further stages of activity. Eye movement helps observe unconscious aspects of activity or the aspects of activity that were conscious during very short periods of time and quickly forgotten.

The overall goal of the task is not precisely defined at the beginning of its performance. For instance, when playing chess, the goal of the subject is to win the game. The final goal can be imagined only in some general form.

3.8 Conclusion

Task analysis has arisen out of work psychology and has been utilized in ergonomics and later in software engineering. It studies how people perform their work and how it should be performed. The major concept of task analysis is the task, but this concept is not clearly defined in work psychology, ergonomics, or the HCI field. This makes it difficult to conduct an efficient task analysis. Task analysis provides a better understanding of the relationship between human behavior or activity and technology or interface. Tasks can be studied from technological and/or behavioral perspectives. In work psychology and ergonomics, the behavioral aspects of task analysis prevail. The structure of human activity during task performance depends on the equipment configuration or the computer interface in a probabilistic manner. If we change the computer interface, this changes the strategies of activity performance in a probabilistic manner. Hence, through analysis of the activity structure during task performance, the efficiency of design equipment, technology, or interface can be evaluated.

Training and professional selection are another aspect of task analysis. The same equipment or interface can be utilized by various strategies of task performance. Training is required to develop the most efficient strategies of task performance. The task is first of all a psychological concept. However, there are aspects of task analysis that are associated mostly with the technological aspects of task performance. For example, in manufacturing processes, an engineer can develop a technological process specific to a particular task (production operation). However, his or her knowledge is not sufficient to take the behavioral aspects of work, and he or she should collaborate with an ergonomist.

The concept of task can be understood only when we understand the general principles of work activity regulation and the major attributes of the task. Anticipation, goal, motivation, and cognitive and behavioral actions are critical for understanding the task.

Historical analyses of the concept of goal demonstrate that it is not sufficiently developed in cognitive psychology, in which task analysis usually does not include motivational aspects. However, goal and motivation greatly influence the specificity of information processing. Some practitioners mix motives with the goals of the task because human behavior is usually poly-motivated. They conclude that the same task can have multiple overall goals. This contradicts with AT, where one task has only one final goal. If the final goal of the task totally changes, a subject formulates a new task. The overall goal of a task should be separated from the goals of individual actions.

Another problem arises from the integration of goals with motives. Including motivational aspects of activity into the goal makes it intensive, but according to AT, a goal cannot be intensive. It is just an anticipatory, cognitive mechanism of activity of the subject's own activity. The goal can be precise or imprecise, can include verbally logical and imaginative components, and can be defined preliminarily and later specified more precisely. If the goal is not sufficiently defined at the beginning of task performance, it can be specified or modified during task performance. The goal is an informational or cognitive component of activity and should be separated from emotionally motivational or energetic aspects of activity. The goal and the motive are interdependent, just like information and energy, but they are different phenomena.

In production processes, the task and its goal are usually presented to a worker in a ready form. However, very often, in operationally monitoring processes and in computer-based tasks in particular, the overall goal of the task can be formulated by the user independently. Those are self-initiated or self-formulated tasks that can be considered a task problem. For example, in explorative activity, which is particularly important in computer-based tasks, the subject often formulates her or his own goal of task. In an ambiguous situation when the goal is not very clear, a trial-and-error strategy is used intentionally to see the outcome. Sometimes, these trials and errors are just mental activity. As a result of such explorative activity, the subject can formulate a more precise goal of activity. Therefore, a goal can be corrected and reformulated based on the feedback due to the activity of the self-regulation process.

There is another aspect of the goal in task analysis. Some specialists considered the goal an end state of the object or system toward which human activity is directed. This is a cybernetic understanding of the goal that is popular among HCI specialists. The shortcoming of such an interpretation is that the goal of a technical system is not distinguished from a human goal. The goal cannot be considered an end state of the system and cannot be presented to the subject in a ready form. There are only objectively given requirements of the task, which need to be transferred into a subjectively excepted goal. The analysis of the users' performance demonstrates that even in situations when a task's goal is clearly defined, a person has to interpret, clarify, reformulate, and accept it. Then, the person formulates intermittent goals, the achievement of which brings him or her closer to the overall goal of the task. So, there is one final goal of the task and multiple intermittent goals of task achievement. Any task performance includes anticipatory and executive mechanisms. The goal is just one anticipatory mechanism of activity. The more complex the task, the more anticipatory mechanisms it includes.

In cognitive psychology, emotionally motivational aspects of activity are not sufficiently considered, whereas in AT, motivational aspects of activity are considered to be of importance because the goal of a task is associated with motives and creates the vector "motives → goal." The presented material demonstrates that in AT, a task has both motivational and cognitive components.

The subjective significance of the goal depends on its relationship to emotionally evaluative and inducing aspects of motivation. If there are contradictory motives, the formation of the goal emerges in a contradictory motivational state. At this stage, volitional processes are critical. The emotionally motivational state of activity influences the information-processing state of activity. So, task analysis should not ignore energetic aspects of activity.

The task includes the initial situation, the accepted or formulated goal that is associated with motives or motivational state as the inducing force, the cognitive and behavioral actions required for achievement of the overall task's goal, and the elements of external environments. Every task has requirements and conditions. When requirements are interpreted and accepted by a person, they become the goal of the task. This transformation process is one of the major means by which the goal and therefore the task are formed. Including anticipatory mechanisms (including the goal), motivational components, and problem-solving aspects of activity in the task concept shows that understanding of task in AT is significantly different from its understanding in cognitive psychology.

References

Annett, J. 2000. Theoretical and pragmatic influences on task analysis methods. In *Cognitive Task Analysis*, ed. J. M. Schraagen, S. F. Chipman, and V. L. Shalin, 25–40. Mahwah, NJ: Lawrence Erlbaum Associates.

Anokhin, P. K. 1969. Cybernetic and the integrative activity of the brain. In *A Handbook of Contemporary Soviet Psychology*, ed. M. Cole and I. Maltzman, 830–57. New York: Basic Books.

Bedny, G., and W. Karwowski. 2007. *A Systemic-Structural Theory of Activity. Application to Human Performance and Work Design*. Boca Raton, FL: Taylor & Francis.

Bedny, G., W. Karwowski, and M. Bedny. 2001. The principle of unity of cognition and behavior: Implications of AT for the study of human work. *Int J Cogn Ergon* 5(4):401–20.

Bedny, G., and D. Meister. 1997. *The Russian Theory of Activity: Current Application to Design and Learning.* Mahwah, NJ: Lawrence Erlbaum Associates.

Bedny, G. Z., and S. Harris. 2005. The systemic-structural theory of activity: Application to the study of human work. *Mind Cult Act* 12(2):128–47.

Bedny, G. Z., W. Karwowski, and T. Sengupta. 2008. Application of systemic-structural theory of activity in the development of predictive models of user performance. *Int J Hum Comput Interact* 24(3):239–75.

Bernshtein, N. A. 1966. *The Physiology of Movement and Activity.* Moscow: Medical Publishers.

Carver, C. S., and M. F. Scheier. 1996. *Perspectives on Personality.* London: Allyn and Bacon.

Crossman, E. R. F. W. 1960. The information capacity of the human motor system in pursuit tracking. *Q J Exp Psychol* 12:1–7.

Diaper, D. 2004. Understanding the task analysis for human-computer interaction. In *The Handbook of Task Analysis for Human-Computer Interaction*, ed. D. Diaper and N. Stanton, 5–49. Mahwah, NJ: Lawrence Erlbaum Associates.

Diaper, D., and N. Stanton. 2004. Wishing on a sTAr: The future of task analysis. In *The Handbook of Task Analysis for Human-Computer Interaction*, ed. D. Diaper and N. Stanton, 603–19. Mahwah, NJ: Lawrence Erlbaum Associates.

Endsley, M. R. 2000. Theoretical underpinnings of situation awareness: A critical review. In *Situation Awareness Analysis and Measurement*, ed. M. Endsley and D. J. Garland. Mahwah, NJ: Erlbaum.

Fitts, P. M. 1954. The information capacity of the human motor system in controlling the amplitude of movement. *J Exp Psychol* 47:381–91.

Freud, S. 1916/1917. *A General Introduction to Psychoanalysis.* New York: Pocket Books.

Hoppe, F. 1930. Success and miszerfold. *Psychol Res* 14:1–62.

Kahneman, D. 1973. *Attention and Effort.* Englewood Cliffs, NJ: Prentice Hall.

Kaptelinin, V. 1997. Computer-mediated activity: Functional organs in social and developmental contexts. In *Context and Consciousness. AT and Human-Computer Interaction*, ed. B. Nardi. Cambridge, MA: The MIT Press.

Kantowitz, B. H. 1974. Double stimulation. In *Human Information Processing*, ed. B. H. Kantowitz, 457–460. Hillsdale, NJ: Lawrence Erlbaum Associates.

Kleinback, U., and K. H. Schmidt. 1990. The translation of work motivation into performance. In *Work Motivation*, ed. V. Kleinback, H.-H. Quast, H. Thierry, and H. Hacker, 27–40. Hillsdale, NJ: Lawrence Erlbaum Associates.

Kozeleski, J. 1979. *Psychological Theory of Decision Making.* Moscow: Progress (translation from Polish).

Landy, F. J., and J. M. Conte. 2007. *Work in the 21st Century. An Introduction to Industrial and Organizational Psychology.* Malden, MA: Blackwell Publishing.

Leont'ev, A. N. 1978. *Activity, Consciousness and Personality.* Englewood Cliffs, NJ: Prentice Hall.

Lewin, K. 1951. Intention, will, and need. In *Organization and Pathology of Thought*, ed. Rapaport (Translator), 95–153. New York: Columbia University Press.

Locke, E. A., and G. P. Latham. 1984. *Goal Setting: A Motivational Techniques That Works.* Englewood Cliffs, NJ: Prentice Hall.

Locke, E. A., and G. P. Latham. 1990. Work motivation: The high performance cycle. In *Work Motivation*, ed. V. Kleinback et al., 3–26. Hillsdale, NJ: Lawrence Erlbaum Associates.

Lomov, B. F., and E. N. Surkov. 1980. *Anticipation in Structure of Activity.* Moscow: Science Publisher.

McDougall, W. 1908. *An Introduction in Social Psychology.* London: Methuen.

Miller, G. A., E. Galanter, and K. H. Pribram. 1960. *Plans and the Structure of Behavior.* New York: Holt, Rinehart, and Winston.

Murray, E. J. 1938. *Exploration in Personality.* New York: Oxford University Press.

Nardi, A. 1997. Same reflection on the application of AT. In *Context and Consciousness: AT and Human-Computer Interaction*, ed. B. A. Nardi, 17–44. Cambridge, MA: The MIT Press.

Newell, A., and H. A. Simon. 1972. *Human Problem Solving.* Englewood Cliffs, NJ: Prentice Hall.

Ovsiankina, M. 1928. The resumption of interruptive goals. *Psychol Res* 11:302–79 (in German).

Pervin, L. A. 1989. Goal concepts, themes, issues and questions. In *Goal Concepts in Personality and Social Psychology*, ed. L. A. Pervin, 173–80. Hillsdale, NJ: Lawrence Erlbaum Associates.

Preece, J., Y. Rogers, H. Sharp, D. Benyon, S. Holland, and T. Carey. 1998. *Human-Computer Interaction.* Wokingham, UK: Addison-Wesley.

Pushkin, V. V. 1978. Construction of situational concepts in activity structure. In *Problem of General and Educational Psychology*, ed. A. A. Smirnov, 106–20. Moscow: Pedagogy.

Reber, A. S. 1985. *Dictionary of Psychology.* New York: Penguin Books.

Reykovski, J. 1974. *Experimental Psychology of Emotion.* Moscow: Progress (translation from Polish).

Rubinshtein, S. L. 1959. *Principles and Directions of Developing Psychology.* Moscow: Academic Science.

Simonov, P. V. 1982. Need-motivational theory of emotions. *Quest Psychol* 6:44–56.

Skinner, B. F. 1974. *About Behaviorism.* New York: Knopf.

Sliosberg, S. 1934. Dynamic of substituted behavior in games and real dangerous situations. *Psychol Res* 19:122–81.

Sokolov, E. N. 1963. *Perception and Conditioned Reflex.* New York: Macmillan.

Tikhomirov, O. K. 1984. *Psychology of Thinking.* Moscow: Moscow University.

Tolman, E. C. 1932. *Purposive Behavior in Animals and Men.* New York: Century.

Vekker, L. M., and I. M. Paley. 1971. Information and energy in psychological reflection. In *Experimental Psychology*, ed. B. G. Ana'ev, 3:61–6. Leningrad: Leningrad University.

Viviani, P. 1990. Chapter 8. In *Eye Movements and Their Role in Visual and Cognitive Processes*, ed. E. Kowler, 352–71. Amsterdam: Elsevier Science.

Wickens, C. D., and J. G. Hollands. 2000. *Engineering Psychology and Human Performance.* Upper Saddle River, NJ: Prentice Hall.

Wundt, W. M. 1894. *Lectures on Human and Animals Psychology* (Translation from the 2nd German edition by J. E. Creighton and E. B. Titchner). New York: Macmillan.

Zarakovsky, G. M., and V. V. Pavlov. 1987. *Laws of Functioning Man-Machine Systems.* Moscow: Soviet Radio.

Zeigarnik, B. 1927. The undischarged reserve and pending negotiations. *Psychol Forsch* 9:1–85.

4

Task Concept in Production and Nonproduction Environments

G. Bedny and W. Karwowski

CONTENTS

4.1 Introduction

The interaction of humans with technology imposes certain demands on operators' behavior. Task analysis helps us reveal the extent of these demands. Thus, analysis of human–technology interaction usually begins with a description of the tasks. The concept of task and the techniques of task analysis are the most potent tools for the analysis of the human–technology relationship. They determine the steps by which humans interact with technology. From the activity theory (AT) perspective, task analysis can be defined as the study of work activity strategies in terms of logical organization of cognitive and behavioral actions to achieve the goal of the task that derives from the demands of the system. According to this definition, the goal of the task and the goal of the system are not the same. The overall goal of a task should be accepted or formulated by the performer based on the requirements of the system. Furthermore, there is never just one optimal way of task performance. Possible strategies of task performance should be developed based on constraint-based principles that derive from the requirements of the system. This understanding of task analysis eliminates contradictions between the so-called normative and constraint-based principles of task analysis, which exists in the ecological approach to task analysis (see, for example, Vicente 1999). However, the nature of work constantly changes. Recently, the main change in human work is its computerization. This evokes new issues in the study of human work associated with informational technology. These issues can be addressed by task analysis of computer-based systems.

One weakness common to cognitive approaches is the disregard for the emotional-motivational aspects of human work activity. Usually, in ergonomics, this problem is reduced to the study of the emotional or psychic tension of an operator in a stressful situation. Psychic tension is a state of the operator that arises in difficult activity conditions. In the West, this is known as emotional stress and is important for reliability analysis. Nayenko (1976) distinguishes between operational and emotional tension. Operational tension is determined by a combination of task difficulty and time limitations in performing the task. Emotional tension is determined by the personal significance of the task. These two kinds of tension are closely interrelated. Under certain conditions, one causes the other.

This brief analysis shows that the emotional-motivational aspects of task analysis are reduced to a relatively narrow problem of task performance in stressful conditions. This approach does not consider positive emotions in task performance, the relationship between emotions and motivation, and so on. In industrial or organizational psychology, motivation is considered separately from the human information-processing approach developed in cognitive psychology. Usually, motivation is described from personality, group dynamics, and productivity perspectives. Another approach to studying motivation in industrial/organizational psychology involves analyzing the goal of human behavior (Pervin 1989). The goal is considered to be a standard according to which behavior is evaluated. The goal and motives or motivation are not clearly distinguished. According to some authors, the goal includes motivational factors. Presumably, the more intensive the goal, the more it affects human behavior (see, for example, Kleinback and Schmidt 1990). Because human behavior is polymotivated, the same task can involve multiple goals. This makes it impossible to utilize the concept of goal in task analysis.

The functional analysis of activity (Bedny and Karwowski 2007) clearly demonstrates that cognitive functions are closely connected with emotional-motivational factors of activity. People can process the same information differently depending on its subjective significance (positive or negative) for a person. Hence, systemic-structural activity theory (SSAT) considers activity to be a goal-directed self-regulated system that integrates not only cognitive and behavioral components but also emotional-motivational components. When activity is considered to be a self-regulative system and major units of analysis are function blocks, according to SSAT, this is a functional analysis of activity.

The emotional-motivational aspects of activity should be considered in unity with cognition. Today, computer-based technology opens a wide new area of nontraditional ergonomic design that includes recreational and nonproductive design. Computer technology is now increasingly used for nonproductive purposes. There is a trend to design human–computer systems for recreational activity, education, games, and so on (Karat et al. 2004). In order to improve the design of such systems, we need a good understanding of task, goal, emotional-motivational aspects of human performance, and so on. Any vague definitions of such basic concepts in cognitive psychology is a source of confusion, erroneous terminology, and so on.

In ergonomics, the concept of affective design has recently emerged (Helander 2001). This concept of design attempts to introduce pleasure-based principles as an important field in ergonomics. Concepts such as emotion, motivation, or aesthetic requirements are particularly important in this design. However, in AT, the emotional-motivational aspects of human performance are always important. Human information processing cannot be separated from the emotional-motivational aspects of activity. An individual is not a computer and cannot be considered to be simply a logical device. He or she always emotionally relates to the presented information. Hence, emotional-motivational processes always interact with cognition. The interpretation of information depends on the emotional-motivational aspects of activity. The pleasure-based design is only one aspect of the implication of the principle of unity of cognition, behavior, and the emotional-motivational components of activity. In the traditional design, this principle was also critical. The conceptualization of the principle of the interdependency of cognition, behavior, and motivation can be found in SSAT. In the pleasure-based design, we need to know how a user emotionally evaluates a product, an interface, or a system. The application of emotional-motivational aspects of affective design requires the development of adequate concepts, terminology, and a theory that would consider their relevance to the design process. Design principles and concepts developed in SSAT can be very useful for this purpose. In this chapter, we consider the concept of task and its main attributes from the SSAT perspective.

4.2 Meaning and Sense in Task Performance

In this section, meaning and sense are two interrelated concepts. In Russian, meaning is *znachenie* and sense is *smysl*. Comprehension and interpretation of information and the content of the task are critical aspects of task analysis. In human–computer interface (HCI), task performance largely depends on the subjects' ability to operate with the sign system. So, greater emphasis should be placed on the analysis of the semantics of the work domain (Rasmussen and Goodstein 1988). The concepts of meaning and sense are the theoretical foundation of AT. The evolution of human culture has always depended on humans' ability to use a sign system. From the functional analysis perspective, the interpretation of sign systems depends on various functional blocks of self-regulation and particularly on such functional blocks as meaning and sense (Bedny and Karwowski 2007; see also Chapter 9 of this book). Anything that is significant for the subject is deemed to have a personal sense. During the process of mental development, an individual internalizes various sign systems and uses them as internal mental tools for thinking (Vygotsky 1978). There are two kinds of signs: one originates in the external world and the other is in the subject's mind. The signs that reside in the subject's mind fulfill the role of internal psychological tool. An understanding of the interaction of internal and external sign systems is important for the analysis of how the interpretation of meaning takes place during task performance. Manipulation with the signs in their material form is not sufficient for the interpretation of the sign system. Understanding the meaning of a sign is the most important aspect in the interpretation of information.

Currently, the study of the meaning of signs focuses on the relationship between the signs and the object or its denotation. However, from the AT perspective, the signs should be also studied in relation to the human activity in which this sign system exists and from which it acquires its meaning and personal sense (Shchedrovitsky 1995). A subject can perform cognitive actions not only with the sign in its material form but also with its meaning. During such activity, the subject extracts and fixates relevant aspects of an object and its phenomena. Hence, certain aspects of the object in its sign form determine a system of actions and operations that correspond to that particular sign form. The sign functions as a tool of cognition because cognition is a system of mental actions that operate with the sign's meaning (Bedny and Karwowski 2004a). The meaning and the function of an object are closely interconnected during task performance. A symbol is a sign only because people relate to it as such. However, this does not mean that the interpretation of a sign is a purely subjective process. The meaning of a sign has an objective character in that it is the result of a sociocultural development. This sociocultural development gives a sign standardized methods of its interpretation.

There is also a subjective interpretation of the meaning of an object, sign, or word (a verbal sign), along with its objective meaning. The interpretation of the objective meaning is its transformation into subjective sense. The sense is more personal and depends on the goal and the emotional-motivational state of the subject. Meaning and cognitive aspects of sense often overlap. They can diverge when there is a possibility of multiple interpretations of the same facts and data. Emotional evaluative aspects of the sense (see block 7 in Figure 9.1, "significance") influence a subjective interpretation of the meaning. This understanding of meaning and sense derives from the earlier philosophical concept of meaning and sense developed by Frege (1948/1892). He considered denotative meaning to be an objective characteristic of an object that should be distinguished from its idiosyncratic interpretation. Sense can be viewed as a structural organization with various images and representations of the same object. Meaning determines stable nonsituational features and characteristics of the object or situation. Sense is a dynamic entity, which is formed based on specific strategies of extracting subjectively important features and attributes of the same object. Of the multitude of potential semantic features that can be chosen for meaning, their subjective interpretation includes only those features that are relevant to the goal of activity and significant for the person in her or his motivational state (Bedny and Karwowski 2004c).

The algorithm of interpretation and categorization of meaning depends on the strategies of activity. Hence, the interpretation of information cannot be separated from emotional-motivational aspects of activity as it is done in cognitive psychology. Emotional-motivational factors are important not only in aesthetic design but also in traditional ergonomic design. From the functional analysis perspective, human information processing always includes emotional-motivational aspects (Bedny and Karwowski 2007). Comprehension and understanding of a sign system can be viewed as an information flow in the form of a system of meanings and senses that reflects the given information and motivational state of the subject. Emotional-motivational aspects of information processing also involve the mechanism responsible for evaluating task difficulty. From the functional analysis or self-regulation perspective, we need to distinguish between the objective complexity of the task and the subjective evaluation of task difficulty. Complexity is the objective characteristic of the task. Task difficulty is the subjective evaluation of task complexity. The more complex a task, greater the probability that task will be evaluated as a difficult one. This evaluation process depends on the specificity of the task, the individual features of the subject, and his or her past experience. We consider this mechanism in more detail in Chapter 9. This cognitive evaluative mechanism is important not only in the process of cognitive regulation of activity but also in emotional-motivational regulation.

Figure 4.1 illustrates the concepts of meaning, sense, and complexity in task performance from the functional analysis perspective. Meaning has two important characteristics: the level of informativeness and the level of complexity. The emotionally evaluative aspects of sense also have two characteristics: positive and negative significances.

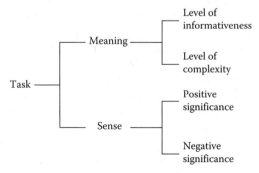

FIGURE 4.1
Relationship between meaning and sense during task performance.

The level of informativeness demonstrates to what extent the meaning of the task or situation is informative for a person. The level of complexity demonstrates how comprehensible is the meaning of the task to a person and how much mental effort is required to comprehend it. Positive significance reflects a subjective value to reach a desired result of a task, with positive emotions accompanying the achievement of a task's goal. Negative significance is accompanied by a negative emotional state associated with a fear of failure, danger, obstacles, feelings of boredom, and so on. Figure 4.1 and the models of self-regulation of activity (see Figure 9.1) allow the correct interpretation of some aspects of task performance. For example, if the existing information is not well organized and meaningful, its interpretation is difficult, and the subject can misinterpret the task. If a given task is objectively very complex and the subject underestimates its complexity, he or she will fail to achieve the goal of the task.

The functional mechanism of self-regulation, such as "sense (significance)," interacts with the functional mechanism of the "goal" of the task. If the goal of the task has only positive value for the subject, then the functional mechanism or block "sense" has a homogeneous structure and has only positive significance. On the other hand, if the goal of the task possesses attributes that might have positive and negative personal value, then the functional block "sense" has a heterogeneous structure and includes positive and negative significances. The ratio of these two types of significance determines the integrative character of the evaluation of task significance. For example, in gambling, the goal is to win money in a risky situation. Achievement of the goal is conveyed also with the possibility of losing money. The relationship of these two kinds of significance in the functional block "sense" determines the subject's involvement in gambling. Moreover, there is also a possibility that a risk-taking addicted person chooses a task that includes a high probability of failure. For such a person, it is important not only to obtain something with positive significance but also to experience the danger. Elimination of the danger immediately eliminates the feelings of positive significance in gambling for such a person. Risk-taking addicted people are usually involved in gambling based on some ratio of positive and negative significances associated with the mental representation of a probability to lose or win money. The feeling of some level of risk (to lose money) increases the feeling of positive significance of such task. This in turn triggers inducing components of motivation (another functional mechanism of self-regulation). Risk-taking addicted people, after the success of their risky behavior very often get involved in even riskier tasks until they fail. Hence, there are individual differences between risk-taking addicted and non-risk-taking addicted people in the selection of activity strategies in risky situations. Some individuals always attempt to be involved in tasks that have some negatively significant factors. They usually get involved in tasks that have different proportions of positive and negative significances. Moreover, this proportion can have different values for different individuals. For some individuals, this relationship is dynamic and depends on their history of successes and failures. These aspects of motivation are important for the development of computer-based gambling tasks and tasks in nonproduction environments in general.

The relationship between positive and negative significances is also important in safety analysis. This relationship can produce conflict of motives during the achievement of the task's goal. It is well known that safety requirements very often contradict with productivity. The factor of productivity is significant for workers because it influences their wages and social status. Therefore, workers will have a high aspiration to reach the goal (i.e., increase the productivity). However, safety rules often are restrictive. As a result, workers face an ambiguous situation. The achievement of high productivity can contradict with the safety requirements. This produces some conflict between positive and negative significances in task performance. In this case, the worker's strategy depends on what kind of significance dominates. Often, workers ignore high risk with low subjective probability in favor of more probable and valuable productivity.

4.3 Thought Process during the Performance of Computer-Based Tasks

Thought process plays a leading role in the regulation of activity. This mental process plays an important role in such functional mechanisms or blocks as interpretation of meaning (block 1), formation of a goal (block 2), subjectively relevant task conditions (block 9; they are responsible for the creation of a dynamic model of a situation), decision making, and program formation (block 10). All these blocks can be found in Figure 9.1.

Newell and Simon (1972) provided a concept for understanding human thinking and problem solving that has its foundation in artificial intelligence and computer simulation. However, the human mind does not work like a computer. Thus, studying the mechanisms of the thinking process is critical for understanding the nature of computer-based tasks and strategies for performing those tasks. Here, we present new data that derive from applied AT and SSAT.

Operative thinking is an important concept in applied AT and SSAT. It was introduced to AT by Pushkin (1965). Under operative thinking, one understands such a problem-solving process when a subject develops a dynamic mental model of the situation and formulates a plan of her or his task performance using this model, which facilitates the achievement of the task's goal usually in time-restricted and stressful conditions. Operative thinking includes the identification of the problem, formation of a hypothesis, identification of intermittent goals and the task's goal, and the development of a performance plan and a plan of mental and behavioral actions necessary to solve the problem at hand.

The four most important features of operative thinking are as follows:

1. Operative thinking is directed toward obtaining a solution to a practical problem.
2. Practical actions are formed and immediately analyzed through operative thinking. The performance of these practical actions, in turn, allows the subject to correct and change thinking processes on the fly.
3. Operative thinking is often performed under time constraints and might be accompanied by stress. Consequently, operative thinking is studied in connection with the emotional-motivational aspects of activity.
4. Operative thinking performs diagnostic, planning, control, and regulative functions.

The following are the most important components of operative thinking: (1) meaningful interpretation of a situation; (2) structuring of a dynamic model of the situation (the development of meaningful units of operative thinking and their structuring based on the connection between various elements of the situation); (3) dynamic recognition of the final state of a situation without considering the intermediate steps to achieve it; (4) anticipation of the effect of the final result based on an already-developed mental structure; (5) formation of an algorithm of situation transformation (developing principles and rules of task solution); and (6) development of an action plan (develop and determine the sequence of actions in a particular situation).

The content of verbalized or nonverbalized thinking actions and operations is important for the study of operative thinking. Actions are the conscious elements of the thinking process, whereas operations are its unconscious elements. In operative thinking, problem solving and task performance are based on visual information, and thus, eye movement is an important component of the thinking process. Consequently, eye movement registration is widely used in the study of operative thinking. AT distinguishes between perceptual eye movements and thinking eye movements. Eye movements, which are involved in the thinking process, are explorative in nature and are used to extract the meaning of the situation in order to mentally reconstruct it.

The differences between perceptual and thinking eye movements can be understood based on the analysis of their purpose. The purpose of perceptual eye movement is to receive visual information such as the color, shape, size, and position of the stimulus in space. The purpose of thinking eye movement is to discover the functional relationship between the elements of the situation. The relationship between elements of the situation and the task's goal are not perceptual features of the task and require the thinking process to be performed based on the visual information. Hence, the thinking process is a dominant one during the performance of such actions. According to the action classification principles (Bedny and Karwowski 2007), cognitive actions can be classified based on the cognitive process dominant during its performance. Usually, the duration of thinking actions is longer than that of perceptual actions except when perceptual actions are involved in analyzing a complex object or perception of an object from a distance, and so on. In such cases, simple thinking actions, which are performed based on visual information, can be shorter. Hence, the time of fixation and qualitative analysis of data also can be used as one of the criteria for the classification of action. For a more detailed analysis of this issue, we direct the readers to Bedny and Karwowski (2007). Thinking is an uninterrupted process of moving from one mental action or operation to the next (Rubinshtein 1973; Brushlinsky 1979). In creative processes, the sequence of such actions and operations cannot be predetermined. As a result of the thinking process, there is a transition from one meaning of a situation to another, which leads to a deeper comprehension of the situation.

Thinking actions perform such functions as analysis, synthesis, comparison, and generalization. Analysis is the extraction of various facets, elements, properties, and relationships of the object or situation. Analysis breaks up the object or situation into various components and extracts those aspects of the object that are subjectively most significant at a given moment. The operations and actions of the analysis are always related to the operations and actions of synthesis. Synthesis connects the extracted content of reality with previous knowledge stored in memory. Rubinshtein (1973) termed the interrelationship of analysis and synthesis as "analysis through synthesis."

Pushkin (1965) called the practical or applied tasks that involve operative thinking "operative tasks." They can emerge in the course of performance of work, play, and/or learning. One specific characteristic of such tasks is their strong connection between the perception of a quickly changing situation, its meaningful interpretation, and formation of the strategies of task performance. In such tasks, we can extract static and dynamic elements. For example, while playing chess, we can consider the squares of the chessboard as a static element and the pieces, which are mentally manipulated by subjects, as dynamic elements. Another example is the panel of a railroad dispatcher that also contains static elements such as the train tracks and dynamic elements such as the moving train. Of course, some pieces during the chess game or some trains at a particular time can be also considered static elements in relation to others that are considered dynamic ones. This depends on the mental representation of the situation. The purpose of emerged operative task is to find the way to transfer the initial position of the dynamic elements into a position required according to the developed subgoal of the subject. Sometimes, it also requires understanding and forecasting of the possible manipulation of the situation by the opponent or forecasting dynamic development of the situation over time.

While performing an operative task, an operator often has to integrate independent elements into a holistic mental picture of the situation and extract the most significant element from it. Such elements are dynamic components of the situation. The following components of the thinking process are important: structuring the situation (the development of meaningful units of thinking based on connections between various elements of the situation), dynamic recognition of the final result based on the initial state, anticipation of the effect of the final result, formation of the solution program (development of principles and rules of task performance), and deciding on the sequence of cognitive and behavioral actions.

Operative tasks are often formulated by a person independently during the performance of her or his duties. An ability to quickly recognize and formulate the operative task in a particular situation is a critically important feature of operative thinking. It is also important to formulate quickly a system of hypotheses, evaluate them, select the most significant one, and formulate adequate subgoals of the task. Subgoals of the task are considered conscious elements of task performance that can be quickly forgotten during task performance. Each hypothesis is compared with the situation at hand and evaluated mentally and practically. Hence, thinking is a self-regulative process responsible for the creation of strategies of task performance.

A significant part of the thinking process can be unconscious. Unconscious thoughts are an important aspect of the operative task performance, and verbalization is not always possible during task performance. Hence, eye movement registration techniques are important in such situations. Eye movement can be considered an exteriorization of some aspects of the thinking process. We have developed a new method of eye movement interpretation for the analysis of computer-based tasks (Bedny and Karwowski 2007). Using subsequent eye fixations, the subject can extract various distinct essential characteristics from the same situation that are germane to the solution of the operative task. These features are not always available to consciousness or verbalization. According to Tikhomirov (1984), such "thinking" eye movements facilitate the extraction of "nonverbalized operational meaning." The notion of "nonverbalized operational meaning" is close to the "situational concept of thinking" developed by Pushkin (1978). There is another important concept of operative thinking called "gnostic dynamic" (Pushkin 1978), which reflects the fact that due to operative thinking the same external situation can constantly change in the subject's mind. Self-regulation process lies at the base of gnostic dynamic, which provides the mental transformation, evaluation, and correction of the situation (Bedny and Meister 1997; Bedny and Karwowski 2007; Bedny and Harris 2005). Gnostic dynamic reflects the complex relationship between conscious and unconscious processes. The thinking process can be presented as the interaction of conscious and unconscious strategies and their mutual transformation. However, only some parts of the unconscious dynamic can be transferred to the conscious level.

It is beneficial for analysis of gnostic dynamic to study eye movements in combination with hand movements, so the external and internal components of activity are studied in unity (Bedny, Karwowski and Bedny 2001). This principle is important especially for the study of computer-based tasks (Bedny, Karwowski, and Sengupta 2008).

Our study also demonstrates that operative thinking includes imaginative components and intuition. In operative thinking, images are dynamic; that is, the subject's mind manipulates not just words and other sign systems but images as well. Therefore, it is not always possible to verbalize the thinking process. Conscious and unconscious components of activity interact and influence each other. However, the subject might be unaware of this relationship. For example, a subject who has memory issues can unconsciously select strategies of thinking that compensate for the shortage of memory. This is called an individual style of activity (Bedny and Seglin 1999) or intellectual style of thinking (Hogan et al. 2008). A subject is conscious of his or her own thoughts. This self-awareness is a fundamental aspect of human thinking. Despite the importance of intuitive mode of thinking, conscious processes play a leading role in the thinking process. Unconscious thinking processes require at least some periodic guidance from the conscious thinking mode. Consciousness provides a frame for the subject's intuition. At the same time, intuition can override the border of conscious thoughts, and thinking process starts to evolve in the other direction.

The analysis of operative thinking in computer-based tasks shows that they can be categorized as operative tasks. Computer-based tasks are often accompanied by emotional stress and performed in time-restricted conditions. Data are presented on the screen in visual form; interface includes verbal, graphic, and imaginative elements. The information has a discrete dynamic structure and can be modified by a user. Computer tasks are often formulated by users. Solving such tasks requires a gradual modification of the data and a step-by-step approach to the final goal of the task and includes the promotion of hypothesis, formulation of final and intermittent goals, and so on. It also requires gradual modification and clarification of the task's goal during task performance. Hence, the concept of operative thinking and operative task are applicable to computer-based tasks. Involvement in computerized games requires a similar understanding of the task. The same theoretical approach can be applied to the study of entertaining computerized tasks.

The analysis of operative thinking helps us develop a simplified functional model of operative thinking that is useful in the study of computer-based tasks in which the main units of analysis are functional mechanisms or functional blocks. Hence, not only activity but each cognitive process can be described as a self-regulative system. In SSAT there are morphological and functional analyses. As mentioned, morphological analysis utilizes actions as units of analysis. Functional units or function blocks describe functionally specialized stages of information processing (Bedny and Karwowski 2007). In cognitive psychology similar terminology has different meaning. Morphological units depict the construct of physiological and anatomical mechanisms responsible for the specialized stage of information processing.

In cognitive psychology, some scientists also describe cognition using various qualitative stages or modules of information processing (see, for example, Sternberg 2008a, 2008b). Sternberg (2008a, p. 113) clearly distinguishes between functionally specialized processors and processes: "The distinction between processes and processors is sometimes overlooked. Processes occur over time; their arrangement is described by a flowchart. In contrast, processors are part of a physical or biological device (such as brain); their arrangement can often be described by a circuit diagram." There are some similarities and differences between the concept of modules and functional blocks. The latter is a result of multiple experimental studies and their further qualitative interpretations. Modules are more specific and small units, which are usually derived from carefully developed specialized experimental procedures. These two approaches can be considered to be macroanalysis and microanalysis. Instead of the term "modules," functional analysis uses the term "functional mechanism" or "functional block." When describing some independent stages of information processing, we use "functional mechanism." If we develop a functional model of self-regulation of activity and uncover the relationship between functional mechanisms, in addition to this term, we also use the term "functional block." In such a model, each block performs specific functions. The content of a functional block depends on the specificity of the task performed by a subject. The relative importance of various functional blocks is contingent on the task at hand.

The other important difference between the analysis of modular processes in cognitive psychology and SSAT is that functional analysis describes cognition as a goal-directed self-regulative system, in which a goal is a conscious desired result of activity. Such a system has a complex structure. Conscious voluntary aspects of self-regulation of human activity or behavior are important for such models. These models describe the real strategies of task performance. A subject can voluntarily vary the strategies of her or his task performance widely despite the fact that he or she has similar mental modules or mechanisms of information processing or the same biological device (brain). Cognitive psychology has developed a set of models that describe the inner architecture of human information-processing system. These models are useful but not sufficient for the analysis of human performance. They can efficiently describe the internal mental structure but cannot describe consciously regulated mental strategies of activity during task performance. This important voluntary aspect of human activity regulation is often overlooked in cognitive psychology.

We present a thinking process as a self-regulative system that is comprised of various stages, which can be presented as interdependent functional blocks (Figure 4.2). At this stage, we use the hierarchical principle of functional analysis of activity. According to this principle, we can develop a self-regulation model of activity, a self-regulation model of separate cognitive process, and a self-regulation model of motor or cognitive action. We present

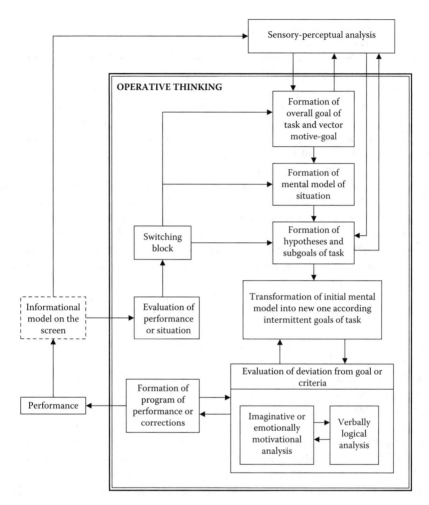

FIGURE 4.2
Model of the thinking process.

an approximate model in Figure 4.2 that is derived from multiple studies of thinking process in cognitive psychology and AT. When a user perceives information on a computer screen, the most essential data about the task on that screen can be considered an information model. The user selects information that is subjectively important for him or her from the screen.

An overall goal of the task is essential in determining the significance of the information on the screen. Therefore, the internal mental state of the user is a key factor in the selection of information, and not just externally given information on the screen. The selection of information also depends on emotionally motivational aspects of activity. The motivational state and the goal of activity create the vector "motives → goal." Goal is considered a cognitive component and motive is an informational component of activity. Hence, the vector "motives → goal" is a filter that determines the selection of information from the screen and from memory. This process is described as an interaction of two subblocks: "sensory-perceptual analysis" and "formation of overall goal of the task and vector motive → goal." The user develops a mental model of the situation based on her or his past experience, visual information on the screen, and the goal of the task. However, this initial mental model of the task often cannot facilitate the achievement of the desired result. So, there are two other blocks called "formation of hypotheses and subgoals of task" and "transformation of initial mental model into a new one based on intermittent goals of task." The last block depicts the transformation of the initial model of task into a new one according to a new hypothesis and a subgoal of the task in a previous block. The obtained result is evaluated in the next block "evaluation of deviation from the goal" based on its correspondence to the goal or subgoal of the task and a subjective criterion of success, which can deviate from the objective goal. A person can evaluate her or his own result as a successful one even if it is lower than required. For example, blue-collar workers can reduce the quality in order to increase quantity.

A goal is formulated in advance, but subjective criteria of success can change during task performance. For example, if tired, a worker can decide to reduce productivity and stop working if she or he thinks that she or he has done enough for the shift. The development of subjective criteria of success can be based on logical-verbal analysis of deviation from the goal or imaginative and emotional-motivational analysis of such deviation. The emotional-motivational analysis is not always conscious. At the final stage of task performance, the goal and subjective criteria of success often can coincide. The function block "evaluation of deviation from goal or criteria" contains two sub-blocks. The left-hand sub-block utilizes imprecise and not sufficiently verbalized and sometimes even unconscious criteria. The right-hand sub-block utilizes consciously verbalized criteria.

Based on this evaluation, the user can develop her or his program of performance and correct it. The block "formation of performance program and correction" influences the "performance" block, and the user modifies the informational model on the screen accordingly. This stage can interact with the preliminary stage based on feed-forward and feedback corrections. Information on the screen interacts with the block "sensory-perceptual analysis" again. At the same time, the created mental model is evaluated. This is facilitated by the functional block "evaluation of performance or situation." Due to the interaction of these stages, the informational model on the screen is constantly modified according to the goal of the task, and the program of performance changes to reflect it.

The block "evaluation of performance or situation" can selectively influence blocks such as "formation of hypothesis and subgoals of task" (small circle of self-regulation) or the upper blocks (big circle of self-regulation) through the "switching block." The big circle of self-regulation, if successful, reflects the completion of the task. When the small circle of self-regulation is active, the "sensory-perceptual" block interacts directly with the block "formation of hypotheses and subgoals of the task." The self-regulative process that provides constant reformulation of the task and a gradual approach to the final goal of the task are the basis for solving a computer-based task.

4.4 Analysis of Task Performance in the Nonproduction Environment

There are two main types of activity: object-oriented and subject-oriented activity. The first type is performed by a subject using tools or material object. The simplest scheme of such activity can be presented as follows:

$$\text{Subject} \rightarrow \text{Tools} \rightarrow \text{Object}$$

With mental and external tools, the object is modified in accordance with the required goal. Objects may be either concrete or abstract (mental signs, symbols, images, and so on). The next type of activity refers to what is commonly called social interaction and can be presented as follows:

$$\text{Subject} \leftrightarrow \text{Tool} \leftrightarrow \text{Subject}$$

Like an object-oriented activity, social interaction begins with subjects' goal formation, orientation in situation, and so on. However, this kind of activity includes understanding of partners' personal features and prediction of their goals and strategies of activity. Social interaction includes three aspects: exchange of information, personal interaction, and mutual understanding (Bedny and Karwowski 2007). Intersubjective aspects of activity can be found even in subject–object activity when there is no direct contact with others. These data are presented in the works of the famous philosopher and literature theorist Bakhtin (1982). Social interaction between subject and object can be uncovered through "inner dialogue."

There is another classification of activity: play, learning, and work. All three kinds of activity have similar structure and include goals, motives, cognitive, and behavior actions. Games play an important role in the study of HCI in a nonproduction environment. For our analysis, we consider the role of playing in mental development. Vygotsky's (1978) work gave the most wide-ranging account of the psychological characteristics of game and their role in mental development. When a child plays, this activity fulfills two functions: formation of a child's needs and formation of her or his cognitive functions. The purpose of playing is not in achieving some useful result of activity but rather the activity process by itself. However, this fact does not eliminate the goal formation and motivational aspects of activity. The actions of the child are purposeful and goal directed. For example, an adult demonstrates how to feed a doll. Children have their own experience of being fed. Despite the fact that a child cannot really feed the doll, she or he still performs conscious goal-directed actions. Moreover, imaginative aspects of play become critical when children operate with a variety of objects, which have a particular purpose and are associated with the objects' meaning. The child at play operates with meanings that are often detached from the corresponding objects (Vygotsky 1978). For example, the child can take a stick, say that it is a horse, and start performing meaningful actions that are similar to a rider's actions. Play is associated with pleasure and it develops motivational components of activity. Gradually, play becomes more and more important and a child develops rules for her or his play. Subordination to rules makes playing more complex, and play is transferred into a game. The imposed rules force children to learn how to suppress involuntary impulses. Hence, childhood games precede real adult activity.

In games, the same as in work, tasks can be formulated in advance by others or formulated by the subject independently. Independently formulated tasks exist even in jobs that have rigorous rules and requirements. For example, during a flight from New York to Moscow, a pilot performs not only the prescribed tasks but also multiple tasks formulated independently based on the weather, information from the air controllers, and so on. Similarly, there are multiple self-initiated tasks in performing computer-based tasks in productive and nonproductive environments. In this type of task, goal formation and motivational aspects of activity are particularly important. Analysis of games and other types of activity demonstrates that goal is one of the central concepts in psychology, but in cognitive psychology, it is mixed with motivation, which is one of the main confusing factors in applied fields. For example, Diaper and Stanton (2004, p. 611) wrote:

> The basic idea is that there is some sort of psychological energy that can flow, be blocked, diverted, and so forth. Goals as motivators of behavior would seem to be a part of this type of psychological hydraulics. Given that there is no empirical evidence of any physical substrate that could function in such a hydraulic fashion, perhaps we do not need the concept of goals as behavior motivator.

In AT, a goal is a cognitive component that interacts with motives and creates a vector "motives → goal," or more specifically "motivation → goal." In this regard, Klochko (1978) studied the contradictions between the task elements (conditions) and emotional tension during task performance. Subjects did not know about contradictions in task conditions and could not report them verbally. Galvanic skin response was found to increase when a subject read the text of the task description related to contradictions. Electric resistance of the skin is an indicator of emotional tension. Hence, emotional reactions emerged as an indicator of contradictions in problem situation without its awareness. This is just an example that demonstrates a complicated relationship between emotional-motivational and cognitive components of activity. An analysis of relationship between cognitive and emotional-motivational aspects of task performance can be found in the study by Bedny and Karwowski (2004b).

With the development of the computer industry, games became important even for adults. Motivational factors play a particular role in a game. Thus, we will consider the stages of motivational process in human activity. According to the concept of motivation developed by Bedny and Karwowski (2006), there are five stages of motivation: (1) the preconscious stage, (2) the goal-related stage, (3) the task-evaluative stage, (4) the executive or process-related stage, and (5) the result-related motivational stage. According to the principle of self-regulation, these stages are organized as a loop structure, and depending on the specificity of the task, some stages can be more important than the others. Depending on task specificity, scientists should pay more attention to some of these stages and their relationship.

The preconscious stage of motivation predetermines motivational tendencies. This stage is not associated with a conscious goal but rather with an unconscious set that can be later transferred into a conscious goal and vice versa. The goal-related motivational stage is important for goal formation and acceptance. This stage can be developed in two ways: by bypassing preconscious stage of motivation or through the transformation of an unconscious set into a conscious goal. When the current task is interrupted and attention is shifted to a new goal, the previous goal of the task does not disappear, but is transformed into a preconscious set. It helps a subject to return to an interrupted task, if necessary, through the transition of a set into a conscious goal. The third motivational stage, the task-evaluative stage, is related to the evaluation of task difficulty and significance, which has been discussed in Section 4.2. The fourth stage, the executive or process-related motivational stage, is associated with executive aspects of task performance. Goal formation, task evaluation (evaluation of task difficulty and its significance and their relationship), and process-related stages of motivation are particularly important for understanding risky tasks, games, and the development of recreational computer-based tasks. The result-related motivational stage depicts the evaluation of activity result (completion of task). All stages of motivation can be in agreement or in conflict.

Let us consider some examples. The relationship between process-related and result-related stages of motivation is important for production environment. In some cases, the work process itself does not produce a positive emotional-motivational state. This can be observed during the performance of a boring job when the work process–related stage of motivation is negative. In order to sustain a positive motivation during such task performance, commitment to the goal-related (stage 2) and result-related motivational stages (stage 5) should be positive to offset it.

In computer-based games, the process-related stage (stage 4) is critical and should be associated with the positive emotional-motivational state. The result-related stage (stage 5) should vary when positive results are combined with negative results, producing a mixture of positive and negative emotional-motivational states. At the same time, if the results of the computer-based games are always positive, this can reduce interest in such games. A simple game without a risk of losing can reduce the positive aspects of the process-related stage of motivation. Hence, the complexity of the task should be regulated depending on the previously obtained results. If the game is designed for children, the possibility to obtain a positive result should be significantly increased. Even in gambling, when people can lose their money, some relationship between success and failure is important. The strength of positive and negative emotions during different stages of motivation is also important. This is particularly relevant for risk-taking addicted people, since manipulation with process- and result-related stages of motivation is critical. Of course, other stages of motivation also should be taken into consideration. In nonproductive tasks, the simplicity or difficulty to obtain a desired result is an important factor. Understanding motivation as a sequence of interdependent motivational stages helps us create a desired motivational state in the production and nonproduction environment.

Sometimes, the goal of a task or game is not precisely defined, and at the beginning the goal is presented in a very general form. Only at the final stage of the game does the goal become clear and specific. This is not new. For example, when playing chess, the goal can be formulated only in a very general form "to win" or "to tie the game." A chess player also formulates in advance some hypotheses about his or her possible strategies that are closely connected with the goal of the task. For this purpose, a player uses some algorithmic and heuristic rules that he or she stores in his or her memory. When a chess player selects a possible strategy, he or she starts to formulate multiple intermittent goals that correspond to his or her strategy. The selected strategy can be corrected or totally abandoned depending on the strategies of the opponent. A clear and specific understanding of an overall goal is possible only at the final stage of the game, just before a checkmate. Even when the goal of a task is externally given in a very precise form, the subject can reach this goal by using a variety of strategies and various intermittent goals. Therefore, a goal cannot be considered an end state of the system that the human or machine wishes to achieve, as has been stated by Preece et al. (1998). A goal of a system and a goal of a person are two totally different concepts.

Let us consider an example of one possible task from everyday life. Suppose one of the authors of this chapter wants to get ready for a formal meeting. He needs to select a tie that matches his suit. First of all, he formulates the goal "I need to select the most suitable tie." Then, he opens his closet door and looks at his ties, compares their colors with his suit, asks his wife if he made the right choice, and so on. The goal of this task is not precise at the beginning of task performance and only at the final stage of the task does the subject identify what tie should be selected. However, the goal of the task exists in a general form. If somebody needs to find a tie, he does not look for shoes.

This understanding of goal is radically different from its description as a clearly defined end state of the system as it is considered by some usability engineers. Karat et al. (2004) stated that a task in a production environment has a clearly intended purpose or goal. According to these authors, the HCI field shifts its focus from a production environment with its clearly defined tasks and goals to the nonproduction field, where the major purpose is communication, engaging, education, game, and so on.

> HCI professionals might say that people use technology because they have the goal of reaching a pleasurable state, but this is awkward and has not proven useful as a guiding approach in design. This is partly because of the difficulty in objectively defining the goal state, and without this there is not much the field can say about the path to the goal.
>
> **(Karat et al. 2004, p. 587)**

Here, the authors mix the goal of the task with the motive and insist that there are no tasks and goals in nonproduction environments and particularly in games. However, in our above examples, we have shown that the goal is to reach the desired future result of the game. The motive is to obtain a pleasurable state during a game and some satisfaction after the game. The goal of the game cannot be precise at the beginning of task performance in nonproduction and production environment. In designing a task in any environment, we should find out how the initially formed goal is gradually clarified and specified during further task performance. The above-described data in combination with data obtained in AT (Bedny and Karwowski 2007; Pushkin 1978; Tikhomirov 1984; Rubinshtein 1973) and cognitive psychology (Newell and Simon 1972) can be used for analysis of goal acceptance and the goal formation process. In cognitive psychology, intermittent goals are known as subgoals. The subgoaling process is performed based on means–ends analysis, in which the desired subgoal is considered the end state of the step. This end state is compared with the present state of knowledge. From the AT perspective, these data require some additional interpretation because subgoals are mental representations of desired future results. Hence, a subgoal is a cognitive and conscious entity. There is also a need for a general motivational state that creates an inducing force to produce this subgoaling process. The comparison of a future hypothetical end state with an existing state is provided by feedback, which is performed in the mental plane. This demonstrates that thinking works as a self-regulative process. Moreover, there are well-defined and ill-defined problems. Well-defined problems are those that have a clearly stated goal. The performance of ill-defined problems in more complex situations begins with searching for and formation of the goal. A subject promotes hypotheses, formulates hypothetical goals, and evaluates them mentally or practically. Only after that can he or she formulate hypotheses to achieve a defined goal. Hypotheses formation process can include conscious and unconscious components. Unconscious hypotheses are not verbalized, and they can be performed, for example, in the visual plane (Pushkin 1978; Tikhomirov 1984). Some of these hypotheses can be later transferred into the verbalized, conscious plane. There are also hypotheses that are conscious during a short time, and then they are forgotten. People are goal-directed systems.

According to Karat et al. (2004) and Diaper and Stanton (2004), the advent of technology in the home environment and the everyday life of people eliminates task-oriented activity. Moreover, Karat et al. (2004, p. 588) wrote that "the science of enjoyment is not capable to define a goal-directed approach." First, we emphasize the fact that there is no such science as "science of enjoyment." In psychology, this term simply refers to a certain emotional state. Then, we note that the concept of "task" is very important even in entertainment. Our activity or behavior strives toward anticipated goals in production and nonproduction environments. The analysis of a variety of tasks that people perform in everyday life demonstrates that they attempt to break down the flow of activity into smaller segments or tasks, which are often self-initiated. Users' everyday life activity cannot be understood without referring to motivational and goal-formation processes. In contrast to cognitive psychology, in AT a task always has its desired final goal and motivational forces. Similarly, social interaction, learning, playing, and games are always, as any other type of human activity, motivated and goal directed.

Karat et al. (2004) substitute complicated concepts such as motivation by the term "value," which simply has a common meaning in his discussion. Hence, the classification of HCI systems as communication-driven, content-driven, and so on is questionable. The authors came up with a new "science of enjoyment," which, in their words, is not a goal-directed approach. It is hard to agree with such an interpretation of enjoyment. People can enjoy drugs, alcohol, work, sports, and so on, depending on the motivational factors. Hence, the study of motivation should be associated with enjoyment. Is there a need for a new "science of enjoyment" when there is psychology and motivation as its branch?

Any technology is just a means or tool for human work or entertainment activity. Hence, we need to study the specifics of utilizing such tools or means of work in various kinds of activity. For instance, in order to design a computer-based system of person-to-person communication, such a system should be adapted for social interaction activity, providing means for understanding the partners, prediction of their goal and motivational state, ability to formulate the general goal of communication, ability to emotionally interact with each other, and so on. Usability engineers should work together with psychologists to improve the design of computer-based tasks in various environments.

4.5 Classification of Tasks and Their Basic Characteristics

We will now describe the major characteristics of tasks and present task taxonomy from the SSAT perspective. The concept of task is not just utilized for studying human work, but the task itself is a basic component of activity. Our lives can be conceptualized as a continuing chain of various tasks. For example, people run a variety of tasks to maintain their houses, such as cleaning, washing dishes, and so on. These kinds of tasks can be an object of ergonomic studies when working on designing or improving household devices. The design principles for the development of machines, computers, and kitchen appliances have similar principles. We can evaluate the adequacy of equipment, computer interface, or home appliances only by assessing them in the context of task performance.

The whole diversity of tasks has two main types: skill-based tasks and problem-solving tasks. These two types of tasks require different levels of automaticity with which they are performed. Skill-based tasks require standardized methods of performance. An example of such a task would be a simple deterministic production operation. These tasks very often are repetitive and have high level of task performance. The more complex problem-solving tasks require discovering the unknown based on what we already know. These are purely creative tasks that seldom can be encountered during the performance of production processes. They sometimes can be related to ill-defined task requirements.

The elements of a task include means of work, input information, required actions, and the goal of the task, which organizes all task elements in activity as a whole. There is a general agreement in AT that a task can be described as a logically organized system of cognitive and behavioral actions that contribute to objectives to reach the goal of a task. Such organization includes some problem-solving aspects, because task-performing activity is constructed based on the mechanisms of self-regulation (Bedny and Meister 1997; Bedny and Karwowski 2007).

The basic characteristics of task in production and nonproduction environment are listed next. Any task includes objective requirements (goal) of tasks and conditions (givens). Anything that is presented to a person and should be considered during performance is a task condition. What has to be achieved including finding a solution is known as task requirements. When the task requirements are given and accepted by an individual, they become a personal goal. Hence, goal recognition, goal interpretation, formulation, and goal acceptance are means by which the task's goal is formed. Thus, the objectively given goal (requirements) and the subjectively accepted goal are not the same, which itself is an important aspect of task analysis. If the task includes a problem-solving aspect, the task conditions should be considered keeping in mind the following components (Tikhomirov 1984): (1) habitual and nonhabitual situations that determine a person's ability to reach the goal of the task by using an existing method; (2) specificity of the representation of the situation (verbal description, visual presentation, and so on); and (3) the degree to which key elements for solving the problem can be distinguished from the less significant ones.

Task performance involves an initial situation, a transformed situation, and a final situation. Elements of the situation possess certain meaning that can change along with the situation. Any transformation can be evaluated in relation to the final result. The overall goal of a task can be often presented in a very ambiguous form at the beginning of task performance, which gives the activity just a general direction. Through the formulation of intermittent goals and their achievement, a person can approach the final goal and make it more precise and specific. Therefore, the goal of a task cannot be considered a ready-made standard that a person wishes to achieve. The concept of task in cognitive psychology lacks the problem-solving aspect, or, if it does include it, considers it only in the context of a problem-solving situation. In AT, the notion of task is inherently a problem-solving endeavor and emphasizes strategies and self-regulation of activity in task performance.

Actions are one of the main units of task analysis. They can be cognitive and behavioral. The performance of actions leads to achievement of the goal. The performance of all actions required by a task leads to the achievement of the overall goal of the task. Cognition is not just storage for images, concepts, or propositions but also a system of mental actions and operations carried out with and upon them. Cognitive and behavioral actions are classified and described according to standardized principles (see Chapter 1 of this book). Material presented in this chapter and in Chapters 1–3 allows us to draw a schema of task formation in the activity structure (Figure 4.3). The structure of the task influences the method of task performance. From Figure 4.3, we can see that conditions that are related to motives form the goal. The relationship between the goal and the conditions determines task performance.

In cognitive psychology, task taxonomy is now based on Rasmussen's work (1983), in which there are skills, rules, and knowledge-based tasks. Each category of taxonomy has its own level of cognitive control of activity. This taxonomy is well-known in literature. Figure 4.4 depicts the task taxonomy that is derived from SSAT (Bedny and Karwowski 2007). We distinguish two initial classes of tasks: skill-based tasks and problem-solving tasks. Problem-solving tasks in turn can be divided into two basic groups: algorithmic and nonalgorithmic tasks.

Algorithmic tasks are performed according to some logic and rules and can be divided into deterministic and probabilistic tasks. In deterministic

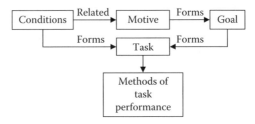

FIGURE 4.3
Scheme of task formation in the structure of activity.

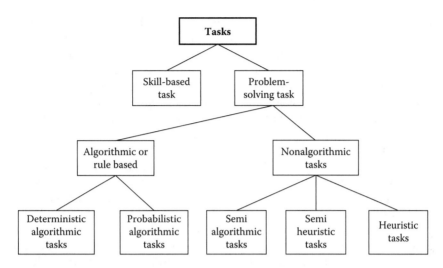

FIGURE 4.4
Task's taxonomy in systemic-structural activity theory.

algorithmic tasks, workers make simple "if-then" decisions based on familiar perceptual signals and rules. Each decision has only two outputs. For example, "if the red bulb is lit, then perform action A; if the green bulb is lit, then perform action B." Algorithmic tasks completely define the rules and logic of actions to be performed and guarantee successful performance if the subject follows the prescribed instructions. According to Rasmussen terminology, deterministic algorithmic tasks can be compared with rule-based tasks.

Probabilistic algorithmic tasks involve logical conditions with the possibility of three and more outputs, each of which possesses a different probability of occurrence. This probabilistic element significantly increases the operator's memory workload and the complexity of task performance in general. Probabilistic algorithmic tasks can be very complex and can present a significant workload not only for memory but also for thinking.

Due to the inability to remember all possible rules for the performance of probabilistic algorithmic task and insufficient familiarity with probabilistic characteristics of the task, this task very often becomes semialgorithmic and even semiheuristic. As per Rasmussen (1983), probabilistic algorithmic tasks and semialgorithmic tasks can be considered knowledge-based tasks.

More complex tasks are nonalgorithmic. According to Landa (1976), this category of tasks can be divided into three subgroups: semialgorithmic, semiheuristic, and heuristic. The distinction between these types of tasks is relative, not absolute. We recommend the following main criteria for the classification of tasks: (1) undetermined initial data and a goal of the task (when the characteristics of a goal are not precisely defined and the goal should be specified during task performance); (2) the goal is not defined and should be formulated independently by a subject; (3) the existence of redundant and unnecessary data for task performance; (4) there are contradictions in task conditions, the specificity of the relationship between elements of the situation, and the complexity of the task; (5) there are time restrictions in task performance; (6) the provided instructions inadequately describe task requirements and/or possible strategies of performance; (7) the subject's past experience is adequate in relation to the task requirements. Very often, these criteria are presented in various combinations.

A task is semialgorithmic if the instructions contain some uncertainty resulting from vague criteria that determine the logical sequence of actions and therefore require the subject not only to perform actions based on the prescribed rules but also to choose his or her own independent cognitive actions, known from past experience.

If uncertainty is even greater and includes some independent solutions without precise criteria for the selection of the subject's own actions, the tasks are semiheuristic. This class of tasks does not fully determine the required actions and also asks for explorative actions for the analysis and comprehension of the situation for task execution. Semiheuristic problems may include algorithmic and semialgorithmic subproblems. Computer-based tasks are often probabilistic algorithmic tasks, semialgorithmic, or semiheuristic tasks. The major criteria for categorizing a task as *heuristic* are as follows: indeterminacy of goal of the task, indeterminacy of initial data, and undefined field of solution. Semiheuristic and heuristic tasks are creative tasks.

The past experience of a subject is critical for task classification. If the subject does not possess the required knowledge, even an algorithmic task can become a creative one. The suggested classification of tasks was first developed for production environment. However, the classification of tasks for nonproduction environment is the same. The selection of tasks in nonproduction environments depends on their purpose in this environment. For tasks that provide social interaction for the wide audience, the task should be relatively easy. They can be deterministic algorithmic or even skill-based. In case of games, there is a trend to increase the task complexity or to allow for regulating the task complexity based on the type of audience. Depending on the purpose of the game, it can be a probabilistic algorithmic task, semialgorithmic task, and so on.

The basic characteristics of a task are its structure, the complexity and difficulty associated with it, and the degree of required physical effort. The task structure is the spaciotemporal organization of task elements and logic of the organization of actions in task performance. The critical aspect of these characteristics of the task is the relationship between the externally given structure of the task and its internal or mental representation (mental model of task). The structure of the task influences the strategies of its performance.

Task complexity depends on a number of its static and dynamic components and the specificity of their relationship. The degree of uncertainty or unpredictability of a task is also an important component of its complexity.

A number of interactions between task components, specificity of given instructions, indetermination of the task, concentration of attention, and excessive cognitive requirements are some characteristics of the task complexity. The more complex the task, the more mental efforts are required for its performance. Complexity is an objective characteristic of a task.

Difficulty is a subjective characteristic of a task. The more complex the task, the higher is the probability that it will be difficult for the subject. Difficulty depends on the relationship between past experience and individual differences and on the objective complexity of the task. Hence, a task of the same complexity can be perceived by a subject as less or more difficult. Task complexity and difficulty can be studied from morphological and functional analysis perspectives (Bedny and Karwowski 2007).

Morphological analysis involves the use of quantitative methods to assess the task complexity. This information is important to predict the mental effort required during task performance and determine the critical characteristics of the task. Functional analysis considers the relationship between complexity and difficulty from the perspective of activity self-regulation. For example, functional analysis is important in the study of work motivation when the task is not objectively very complex but is subjectively perceived as difficult and not highly significant for the user, then the motivation is low.

Another example is that an entertainment task can be objectively sufficiently complex. However, it may be perceived as a task with low difficulty by a specific individual. This may reduce the subjective significance of the task, and the subject will not be motivated to perform such entertainment tasks.

The heaviness of the task, which determines the degree of physical efforts, is an important characteristic of manual tasks. Physical efforts are easier to measure objectively. At the same time, physical efforts also have a subjective component as a feeling of physical stress, which depends on the physical conditions of a performer. Physical and mental efforts can affect each other.

4.6 Conclusion

The concept of task is important in ergonomics and work psychology because an accurate task analysis is impossible without its clear understanding. Currently, there is no consensus on the understanding of this concept and the main task attributes. Task analysis is a multitude of independent techniques, sometimes not sufficiently grounded from the theoretical perspective. There is a reasonable opinion that at this point task analysis is a mess. Moreover, some practitioners raise questions about the future of task analysis. There are even suggestions to eliminate the concept of task because it ignores motivational forces or to eliminate this concept just for entertainment systems. Other professionals insist on eliminating the concept of goal in task analysis.

From the SSAT perspective, any task has means of work, materials and mental tools, work processes, and technological or control process. Hence, a task can be studied from behavioral or technical perspectives. Of course, these two aspects of the study are interdependent. In ergonomics and work psychology, behavioral aspects of task analysis are more important. At the same time, these two aspects can be considered relatively independent. For instance, a process engineer can develop technological process that does not consider behavioral aspects of task performance. However, this independency is not absolute. Means of work, including information technology, often significantly influence human work activity. One aspect of task analysis in the design process is to find out the relationship between the physical configuration of the equipment or the software configuration and the structure of activity during task performance. Task analysis is important for the training process, safety, and development of selected methods for very demanding professions such as pilots, flight operators, medical technicians, and so on. Therefore, task analysis is important for the study of human work from the behavioral point of view.

Understanding the concept of task and its main attributes is critical for the study of human work. Cognitive task analysis, which concentrates mainly on the cognitive aspects of human work, does not pay enough attention to the interaction of cognitive, behavioral, and motivational aspects of human work. Work motivation in industrial or organizational psychology is considered separate from ergonomic design. However, human information processing inherently interacts with emotional-motivational aspects of activity. The more complex and significant the task is for a person, the more mental efforts it takes and the higher is the required level of motivation. Hence, we should consider these aspects of activity not only in pleasure-based design but also in traditional design, and not just for pleasure but also for a more efficient interaction with technology in its broader sense. All of the above are particularly relevant for the study of HCI.

In this chapter, we have considered some basic characteristics of task for production and nonproduction environments. We have demonstrated that this concept is important for the study of any kind of human activity, including games and entertainment in general. In contrast to cognitive psychology, a task in AT includes motivational forces and a goal as its main attributes. We disagree with attempts to eliminate the goal as an attribute of the task. The goal is not well-developed in cognitive psychology, and it is often not considered at all or is considered to be a combination of cognitive components with a number of different motives. As a result, some practitioners mix goal with motives, and as a result the same task has multiple goals. In contrast, in AT, the goal is a cognitive component and motives are the energetic component of activity. "Motives → goal" is an inducing vector around which all elements of task are organized. The goal is not the end state of the system. From a psychological viewpoint, it is a mental representation of a desired future result, which can be clarified and specified during task performance. If a subject totally changes his or her goal, he or she changes the task itself.

Attempts to eliminate the concept of task and its main attributes, such as the goal, are a source of confusion and strange terminology such as "nontask system" and "the science of enjoyment."

This chapter demonstrates that the functional analysis of activity helps one to understand the role of motivational factors in task performance. Functional analysis views activity as a self-regulative system. In such systems, motivation is a result of the interaction of various functional blocks or mechanisms of self-regulation. Motivation is presented as a chain of interacting stages of motivational process, which play an important role in the activity regulation. From the SSAT perspective, concepts such as task, motives, goal, and interdependence of cognitive, behavioral, and energetic aspects of activity are important in the study of human work.

References

Bakhtin, M. M. 1982. *The Dialogic Imagination*. Austin, TX: University of Texas Press.

Bedny, G. Z., and S. R. Harris. 2005. The systemic-structural theory of activity: Application to the study of human work. *Mind Cult Activ Int J* 12(2):128–47.

Bedny, G. Z., and W. Karwowski. 2004a. A functional model of the human orienting activity. *Theor Issues Ergon Sci* 5(4):255–75.

Bedny, G. Z., and W. Karwowski. 2004b. The situational reflection of reality in activity theory and the concept of situation awareness in cognitive psychology. *Theor Issues Ergon Sci* 5(4):275–96.

Bedny, G. Z., and W. Karwowski. 2004c. Meaning and sense in activity theory and their role in study of human performance. *Ergonomia* 26(2):121–40.

Bedny, G. Z., and W. Karwowski. 2006. The self-regulation concept of motivation at work. *Theor Issues Ergon Sci* 7(4):413–36.

Bedny, G. Z., W. Karwowski, and T. Sengupta. 2008. Application of systemic-structural theory of activity in the development of predictive models of user performance. *International Journal of Human-Computer Interaction* 24(3):239–74.

Bedny, G., W. Karwowski, and M. Bedny. 2001. The principle of unity cognition and behavior: Implication of activity theory to study human work. *Int J Cogn Ergon* 5(4):401–20.

Bedny, G., and W. Karwowski. 2007. *Systemic A-Structural Theory of Activity. Application to Human Performance and Work Design*. Boca Raton, FL: Taylor & Francis.

Bedny, G., and D. Meister. 1997. *The Russian Theory of Activity: Current Application to Design and Learning*. Mahwah, NJ: Lawrence Erlbaum Associates.

Bedny, G. and M. Seglin. 1999. Individual style of activity and adaptation to standard performance requirements. *Human Performance* 12(1):59–78.

Brushlinsky, A. V. 1979. *Thinking and Forecasting*. Moscow: Thinking Press.

Diaper, D., and N. Stanton. 2004. Wishing on a sTAr: The future of task analysis. In *The Handbook of Task Analysis for Human-Computer Interaction*, eds. D. Diaper, and N. Stanton, 611. Mahwah, NJ: Lawrence Erlbaum Associates.

Frege, G. (1948/1892). Sense and reference. *Philos Rev* 57:207–30.

Helander, M. G. 2001. Theories and methods in affective human factors design. In *Usability Evaluation and Interface Design. V. I of the Proceedings of HCI 2001*, eds. M. J. Smith, G. Salvendy, D. Harris, and R. J. Koubeck, 357–61. Mahwah, NJ: Lawrence Erlbaum Associates.

Hogan, R., J. Hogan, and P. Barrett. 2008. Good Judgment: The intersection of intelligence and personality. In *Ergonomics and Psychology. Development in Theory and Practice,* eds. O. Y. Chebykin, G. Bedny, and W. Karwowski, 357–76. Boca Raton, FL: Taylor & Francis.

Karat, J., C.-M. Karat, and J. Vergo. 2004. Experiences people value: The new frontier for task analysis. In *The Handbook of Task Analysis for Human-Computer Interaction,* eds. D. Diaper, and N. Stanton. Mahwah, NJ: Lawrence Erlbaum Associates.

Kleinback, V., and K. H. Schmidt. 1990. The translation of work motivation into performance. In *Work Motivation*, ed. V. Kleinback, H. H. Quast, H. Thierry, and H. Hacher, 27–40. Hillsdale, NJ: Lawrence Erlbaum Associates.

Klochko, V. E. 1978. *Goal-Formation Dynamic of Emotionally-Motivational Regulation of Task Performance*, Ph.D. Diss. Moscow: Moscow University.

Landa, L. M. 1976. *Instructional Regulation and Control: Cybernetics, Algorithmization and Heuristic in Education*. Englewood Cliffs, NJ: Educational Technology Publication. (English translation).

Nayenko, N. I. 1976. *Psychic Tension*. Moscow: Moscow University.

Newell, A., and H. H. A. Simon. 1972. *Human Problem Solving*. Englewood Cliffs, NJ: Prentice Hall.

Pervin, L. A. 1989. Goal concept in personality and social psychology: A historical introduction. In *Goal Concept in Personality and Social Psychology*, ed. L. A. Pervin, 1–17. Hillsdale, NJ: Lawrence Erlbaum Associates.

Pushkin, V. V. 1965. *Operative Thinking in Large Systems*. Moscow: Energy Publishers.

Pushkin, V. V. 1978. Construction of situational concepts in activity structure. In *Problem of General and Educational Psychology*, ed. A. A. Smirnov, 106–20. Moscow: Pedagogy.

Rasmussen, J. 1983. Skills, rules and knowledge; signals, signs, and symbols, and other distinctions in human performance models. *IEEE Trans Syst Man Cybern* SMC-13:257–66.

Rasmussen, J., and L. P. Goodstein. 1988. Information technology and work. In *Handbook of Human-Computer Interaction* ed. M. Helander, 175–201. Amsterdam: Elsevier.

Rubinshtein, S. L. 1973. *Problems of General Psychology*. Moscow: Academic Science.

Shchedrovitsky, G. P. 1995. *Selective Philosophical Works*. Moscow: Cultural Publisher.

Sternberg, S. 2008a. Identification of mental modules. In *Ergonomics and Psychology, Development in Theory and Practice,* eds. O. Y. Chebykin, G. Z. Bedny, and W. Karwowski, 111–34. Boca Raton, FL: Taylor & Francis.

Sternberg, S. 2008b. Identification of neural modules. In *Ergonomics and Psychology, Development in Theory and Practice,* eds. O. Y. Chebykin, G. Z. Bedny, and W. Karwowski, 135–66. Boca Raton, FL: Taylor & Francis.

Tikhomirov, O. K. 1984. *Psychology of Thinking*. Moscow: Moscow University.

Vicente, K. J. 1999. *Cognitive Work Analysis. Toward Safe, Productive, and Healthy Computer-Based Work*. Mahwah, NJ: Lawrence Erlbaum Associates.

Vygotsky, L. S. 1978. *Mind in Society. The Development of Higher Psychological Processes*. Cambridge, MA: Harvard University Press.

5

Microgenetic Principles in the Study of Computer-Based Tasks

T. Sengupta, I. Bedny, and G. Bedny

CONTENTS

5.1 Introduction

One important principle in activity study is the genetic method; the essence of this principle is to study various psychological phenomena in the process of their development. Scientists study psychic phenomenon based on the analysis of the sequential stages of its formation, reconstruction, and dialectic genesis. This principle of study in activity theory has been introduced by Vygotsky (1960), Rubinshtein (1973), Leont'ev (1972), and others. Here, we apply the microgenetic method of study that derives from this principle. In systemic-structural activity theory (SSAT), the microgenetic method of study is understood as an analysis of process of activity structure formation during a relatively short time period (Bedny and Karwowski 2007; Bedny and Harris 2005; Bedny et al. 2000). This method is based on the analysis of skill acquisition during experimental study. Such a method is necessary because if the regularities of the formation of the activity structure and the specificity of the acquired activity structure are known, we can evaluate the efficiency and reliability of human performance more accurately and uncover the interrelationship between various equipment or software configurations and the efficiency of human performance.

When an activity is already acquired and stabilized, the users become less sensitive to changes in the equipment or software design. The qualitative differences can be discovered later under unusual circumstances or stressful conditions. A complex task that requires a long and complicated process of skill acquisition can be performed as efficiently as the much simpler one once the skill has been fully acquired. At the same time, the activity that has a more complicated acquisition process is more sensitive to the disturbances and stressful work conditions and proves to be less reliable. Therefore, analysis of the acquisition process or microgenesis of activity can predict the reliability and efficiency of task performance in adverse work conditions. Thus, the study of acquisition processes might be very useful.

Formative (teaching) experiments can be considered a version of the microgenetic method of study. The essence of such experiments is that the researcher actively interferes with the subject's activity during the experiment and guides the task acquisition process. Usually, this method is utilized by pedagogical psychologists, but it can also be applied in ergonomic psychology and work psychology.

To use the genetic principle of study in an experiment, we had to develop methods of registration and analysis of the subject's activity data during the experiment. One such method is the development of the time structure of the formatting and already formed or acquired work activity (Bedny and Karwowski 2007). Another method of study is the development of learning curves that depict the activity acquisition during the experiment. There are also other methods of registration of the activity acquisition process in various conditions. For example, we can use bar charts that demonstrate how activity characteristics change during the acquisition process. Based on the obtained data, we can analyze various strategies for activity acquisition during the task performance. Certain features of tasks, equipment and software configuration, and so on are considered independent variables, whereas the specific characteristics of the activity acquisition process are considered to be the dependent variables. Learning curves are also used in the genetic method of study. In contrast with the analysis of the training process, in which attention is concentrated on the correlation between the teaching methods and the learning curves, in the genetic method, special attention is paid to the specificity of the task performance, the equipment and software design that is used in the work activity, and the acquisition process. For instance, subjects in the experiment perform the same task using different versions of equipment or software for further comparison of the acquisition processes and efficiency of the design.

The genetic principle assumes that the training process is not closely supervised by the trainer and is rather a process of independent skill acquisition. The experimenter interferes in the acquisition process sporadically, only when it is important or is necessary according to the carefully developed plan. There are supervised learning and self-learning or independent learning in which a subject uses explorative strategies. Self-learning is particularly important for computer-based tasks. Learnability is an important feature of a user interface. Analysis of the learnability of a task can give important insights to the difficulty of task performance and hence the usability of the interface. The ability to support learning or self-learning is an important aspect of usability of the human–computer interface.

The relationship between practice and improvement in performance is of fundamental concern not only in the study of learning and training but also in the analysis of task complexity, the efficiency of equipment, the usability of software, the reliability of human performance, and so on (Bedny et al. 2006; Bedny and Sengupta 2005). Tasks with various levels of complexity can be performed with the same efficiency at the final stage of the acquisition. The performance time can be identical, but more complex tasks require longer acquisition processes. This can be explained by the fact that a worker uses more intermediate strategies of performance during learning process (Bedny and Karwowski 2007). Such tasks are less reliable in extreme and stressful conditions.

We have concentrated our efforts not on the training methods, but rather on the acquisition of activity in various working conditions. Thus, we use the term "acquisition curve" instead of the term "learning curve." There are various methods for developing activity acquisition curves. In some cases, the acquisition curve can depict the process of the activity acquisition for a group of subjects. Then, the acquisition curves are developed for individual subjects. At the next step, the curves developed for the individual subjects can be compared with those developed for the group of subjects. It is also very useful to develop an activity acquisition curve that depicts not just the acquisition for the complete task but also for the separate elements of the task (Bedny and Meister 1997). This method also allows us to analyze skills reconstruction and skills flexibility. The experimenter can include or exclude some task elements, change their sequence, and develop the acquisition curves for individual elements of activity and for the activity as a whole.

The development of a training process based on the analysis of learning or acquisition curves is widely used in the study of traditional works. However, there are no clearly developed methods for creating such curves for computer-based tasks. Therefore, in order to use the genetic principle of study for the analysis of computer-based tasks, we needed to create procedures for developing different kinds of learning or acquisition curves. The utilization of such curves and their comparison for the analysis of activity acquisition is an important tool for applying genetic principles to analyze computer-based tasks. Thus, the main purpose of this work is to demonstrate the creation of such curves for the study of computer-based tasks.

In this chapter, we use the method of modeling a computer-based task. The tasks are divided into four groups based on their level of complexity. The versions of task varied from trial to trial within each group of task complexity. We conduct an analysis of the activity formation process during task performance for each group of tasks.

The task we have chosen for our study is a model of computerized task that requires a relatively short skill stabilization period to achieve an average skill acquisition level. Despite this fact, as we show later, the tasks presented to the subject were sufficiently complex. The variation in the level of complexity was achieved by manipulating various features of tasks and particularly by changing the location of the tools on the screen in relation to the other features of task and the sequence of actions. Since all the data obtained in this study cannot be presented in this chapter, we describe just the most illustrative ones for the basic principles of genetic analysis.

5.2 Systemic-Structural Activity Theory as a Framework for the Study of Computer-Based Tasks

SSAT describes activity structure as a system. This approach suggests that activity during the performance of computer-based tasks can be considered a complicated structure that has hierarchical and logical organization. The

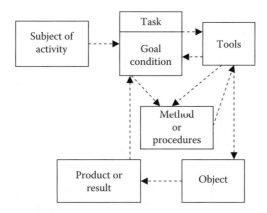

FIGURE 5.1
Main elements of activity.

systemic-structural approach considers activity to be a multidimensional system and describes it by using diverse experimental methods and formalized descriptions. Therefore, activity cannot be described completely by just one "best" method. Interdependent methods of study that are organized into stages and levels of analysis should be used for the description of the same activity.

From the activity theory perspective, a task may be defined as a logically organized system of mental and behavioral actions. Activity during the task performance is our object of study, in which actions are the main units of analysis in morphological study and functional blocks are the units of the functional analysis of activity when activity is considered to be a self-regulative system. Activity as a multidimensional system includes a number of main elements that are presented in Figure 5.1.

The *subject* of activity is understood as a socially constituted individual who performs the task in accordance with the conscious goal of activity. The *task* is the logically organized system of cognitive and behavior actions directed to achieve the goal of the task given under particular conditions. In activity theory, the task is a major object of study. The *goal* is a mental conscious representation of the future desired result of activity. We should distinguish between the goal of separate actions and the goal of the task. A goal is always associated with motives and creates the vector "motives → goal." A motive is an energetic component of activity, whereas a goal is a cognitive component.

The *tool* of an activity can be external or internal. Internal tools are internalized or acquired signs and symbol systems that are used by mental activity, and they can be used for the transformation of an idealized object of activity according to the goal of the task. External tools are used for the transformation of a material object of activity.

The *product* or *result* is achieved by a subject during activity performance. The product can be material or nonmaterial. It can match the goal of activity or deviate from the goal as it is defined according to the established criteria. Based on the feedback, subjects can evaluate these deviations and correct their activity.

The *object* of activity is something that is transformed by the subject according to the goal of the activity. It can be an idealized or material entity. An entity can be considered an object only during its interaction with the subject. The *method* is the socially established requirements and procedures for the activity or task performance. As can be seen in Figure 5.1, the goal and object are treated as distinct components of activity. Understanding of these basic elements of activity is essential for further analysis and discussion.

In this chapter, we also use the principle of unity of cognition and behavior introduced by Rubinshtein (1973). This principle highlights the importance of considering external behavior and cognition and their interaction in the study of computer-based tasks and work activity in general (Bedny et al. 2001). Many internal actions and operations originate as a result of the interaction of external and internal elements of activity. External motor actions interact with internal cognitive actions.

Internal activity is shaped during its interaction with external behavior. At the first stage, mental activity is supported by external behavior or activity. During the final stages of activity acquisition, internal activity can be performed independently (Bedny and Karwowski 2007). Yarbus (1967) invented an ingenious technique for the analysis of eye movements during visual task performance. The main purpose of such a method of study is to discover the relationship between cognitive and motor activity. Micromotions of the eyes are required to perceive the visual information. Bedny et al. (2001) described the principle of the unity of cognition and behavior and the utilization of this principle during the study of human work. Motor activity is an important component in receiving the information using sense by touch (Anan'ev et al. 1959). Fingers were found to perform multiple micromotions while exploring an object. These are gnostic micromotions that are involved in the exploration of the object. Fingers also perform various functions during the exploration of the object. For example, the thumb is stabilized in relation to the object and is used as a "points of reference" for the other fingers that perform explorative micromotions. Turvey (1996) obtained some interesting data in this area, which is known in the West as the study of dynamic touch. Speech mechanisms and motor components of verbalization are involved in the thinking process (Sokolov 1969). Hand and eye movement coordination has been studied by Zinchenko and Ruzskaya (1966). Eye movements, micromotions during the sense by touch, electrophysiological activation of lip muscles during thought process, and so on, demonstrate that motor and cognitive activity are closely interdependent.

These and other studies are the theoretical basis for the utilization of the principle of unity of cognition and behavior in applied research (Bedny et al. 2001). In our study, this principle was realized by the registration of eye movements and mouse movements and the analysis of the collected data in the study of the computer-based task performance. Our analysis and interpretation of the obtained data differ from the methods that exist in cognitive psychology. Some activity principles and their interpretation will be considered in Sections 5.5 and 5.6.

5.3 Experimental Design and Data Collection

We chose a laboratory experiment because it gave us an opportunity to carefully control the variables of the task performance. To conduct the experiment, a special computer-based model of the task was developed. To adequately design the computer-based model of the task, we needed to know the context of the task performance. Some tasks are designed for users who would work with a few screens for years. Other tasks are intended for the ever-changing workforce where the workers are hired for a 4- to 6-week peak season. Some tasks can be performed upon request or under stress, time limit, or accompanied by various interruptions. There are also computerized tasks that are used by consumers, who might perform it once or just a few times. The study of task acquisition should depend on the conditions in which the task is going to be performed. For example, in the first case, the allowed acquisition time can be much longer than in the next three cases. In the second case, training cannot exceed half a day. In the third case, related to the stressful work conditions, the concentration should be on simplifying the task performance by reducing the workload on working memory, simplifying the decision making, and so on. If the software is intended for the consumers, it should be self-explanatory because its usage is not preceded by any training at all.

In this study, we analyzed how task sequence requirements and display structure affect the usability of a graphic user interface. For this purpose, we have used the genetic principle of study. The usability of a graphic user interface refers to the ease of use, efficiency, effectiveness, and satisfaction while interacting with the system through the interactive elements on the interface display. Display structure is the arrangement of elements (icons, menus, and so on) on the interface screen, which influences the understanding of the system and strategies of interaction with the display. The quality of the graphic user interface can affect the efficiency of performance of the users and thereby the usability of the interface.

The task was to change the position of the letter arrangement and impart the features to these letters according to the presented goal on the screen, which varied from trial to trial. Any sequence of actions could be possible by the user, but the user (subject) was only instructed to reach the final arrangement. There was a limitation on the sequence of tools selection. The subject could not complete the tasks unless she or he understood that task performance sequence depends on tool arrangement characteristics of the task. The subject could complete the task by using various sequences of actions. The software allowed the subject to use the tools in order corresponding to each version of compatibility. The sequence of actions could not contradict the tool arrangement, which constrained the task performance. Violation of existing constraints made it impossible to change the features of the objects. This feedback informed the subject about erroneous actions, and he or she could correct the sequence of actions. Hence, these task acquisition processes included self-learning components, which can be observed during the performance of computer-based tasks. From trial to trial, subjects reduced the possible errors until they achieved a strategy that was perceived by the subject as the final one. Once the subject had learned the possible strategies of performance, she or he achieved the plateau on the learning curve, which signified that the skill acquisition was completed. Each group of subjects was involved in performing a particular type of tasks. After completing one version of the task, they were involved in performing a new version of the task.

The main goal of the study was to observe how subjects developed preferable strategies to achieve the task requirements. The setting was similar to a real situation in which users attempt to perform relatively new tasks. The usability of such tasks can be evaluated based on the analysis of the self-learning process. The limitations imposed by the program and the necessity to discover possible strategies made these tasks sufficiently complex.

The three features of the task that were manipulated by the subjects during the trials were as follows:

1. Position: the location of the letters with respect to each other
2. Color: the color of the cell containing the letters
3. Format: the format of the letters

These features resulted in three functional groups (based on the interface guideline of functional grouping) in the interface, results that were commonly observed in a variety of software. Almost all interfaces include functional groups for the users to understand the general functions of a tool group. For example, in Microsoft Word, format and alignment groups were observed (Figure 5.2).

The tools designed for manipulating these features and their functional groupings are depicted in Table 5.1. Their functional grouping and manifestation in the interface are given in Table 5.2. Therefore, the main focus of the task is to alter the features of the objects with available tools. Initially, the letters had no special features.

FIGURE 5.2
Format and alignment group in Microsoft Word.

TABLE 5.1

Tools for Tasks Designed as Icons on the Interface with Intended Functional Grouping

	Position		Color		Format
⇌	Swap horizontal position	■	Red	B	Bold
		■	Green		
↕	Swap vertical position	☐	Yellow	U	Underline
✕	Swap diagonal position	■	Blue	S̶	Strikethrough

TABLE 5.2

Functionally Grouped Structure of the Tools in the Interface

Position	Color	Format

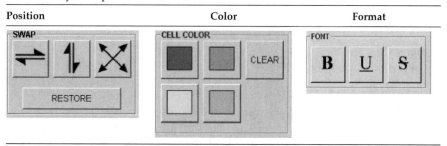

The interface for task performance is given in Figure 5.3. According to the existing basic elements of the activity (Figure 5.1), we can extract three functionally relevant areas on the screen: the tool area, object area, and goal area. The tool area consists of the tools used for transformation of the objects. The object area consists of the letters to be manipulated according to the goal of the task, and the goal area shows the arrangement that must be achieved as a result of the transformation of the object area. Analysis of subjects' activity strategies in these areas corresponds to the method of study related to the functional analysis of activity (Bedny and Karwowski 2007). The object and tool areas of the task are generally seen on the interface. In most cases, the goal area is not presented. In self-initiated tasks, the goal is formulated by the subjects independently. In such cases, any area of the screen associated with the stage of activity, the purpose of which is formulation of the goal of activity, can be considered the goal area of the screen. If it is impossible to define the stage of the activity associated with the analysis or formulation of the goal, only the object and tool areas are used during task analysis.

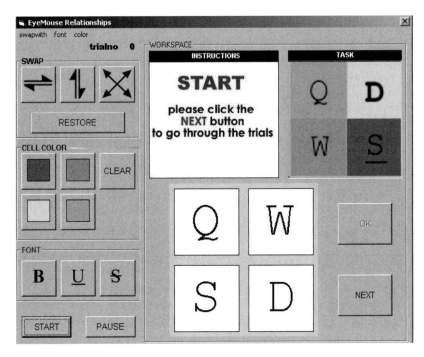

FIGURE 5.3
The experimental interface.

The functional purpose of various elements of the screen depends on the goal of the task or goal of actions. In the first situation, the goal of the task is fixed and therefore the object and tool areas on the screen also have fixed meaning. The goal of actions is not the same and depends on the purpose of actions. In such situations the object and tool areas of the screen are not fixed.

Task performance begins by pressing the START button and finished by pressing the OK button. Performance of the next task (trial) begins by pressing the NEXT button. During the trials, only one out of four squares inside the object area can be activated at a time. For this purpose, the subject clicks the corresponding square. As a result, the borderline of the corresponding square is highlighted by a bold line. After that, the features of the square can be changed.

According to the principle of unity of cognition and behavior eye movement and mouse movement registration were used. The software for the interface was additionally coded to capture the mouse movement and eye movement data. Eye movement was registered using an ISCAN eye tracking system. Although the eye movement point-of-regard coordinates were recorded, there were inherent difficulties with the analysis of the point-of-regard coordinates due to equipment restrictions. As a result, the eye movement data were obtained through the analysis of the video of the point of regard.

5.4 Classification of Tasks According to Their Compatibility, Tool Arrangements, and Design of the Experiment

Before explaining the various interfaces that influence the task complexity, a definition of the utilized terminology is required.

Compatibility: The functional relationship between the display structure and the embedded task sequence requirements. If the display structure or layout (top to bottom or bottom to top) supports a particular sequence, then it is compatible, if not, then it is incompatible.

Tool arrangement: The display structure or the arrangement of the interface elements on the screen, which supports a particular operation. It can be top to bottom or left to right or any other spatial sequence. By manipulating the tool arrangement and compatibility, we can change the complexity of the task. In the pilot study, the most preferred sequence of manipulation of various features was the positions of letters followed by the color and then the format of the letters. There were only two versions of tool arrangements and four embedded task sequences associated with them. The first version of tool arrangement was position → color → format. The second version of tool arrangement was color → format → position. The combination of tool arrangements and four embedded task sequences are as follows:

Compatible from the top: Compatible from the top interface uses a tool arrangement, which is consistent with the embedded task sequence (Figure 5.4). In this case, the tool arrangement is from top to bottom with the positioning tools at the top, color tools in the middle, and the format tools at the bottom. The embedded task sequence is also in the same order as that of the layout of the display, that is, the positioning tools being the first, then the color tool, and then the format tools.

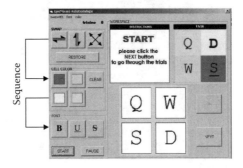

FIGURE 5.4
Compatible from the top (embedded sequence: position → color → format: display from the top: position → color → format).

Compatible from the bottom: In this case, the tool arrangement is from the bottom to top with positioning tools being the lowest tool group, then the format tools, and then the color tools. The task sequence embedded with the interface is also based on the same order, which is counting from the bottom position first, the format, and then the color (Figure 5.5).

Incompatible from the top: Incompatible from the top interface used the same display as that of the compatible from the top but a different embedded sequence in the interface, which was not congruent with the display structure of the tools. In this case, the embedded task sequence was positioning, format, and then color (Figure 5.6).

Incompatible from the bottom: The display used in this case was the same as that used for the compatible from the bottom display, but a different task sequence was used, which was not congruent with the display order from the bottom. In this case, the sequence used was positioning, color, and then format, which was not congruent with the display layout from the bottom (Figure 5.7).

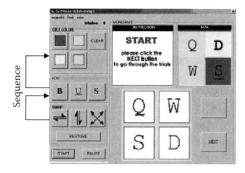

FIGURE 5.5
Compatible from the bottom (embedded sequence: position → format → color: display from the bottom: position → format → color).

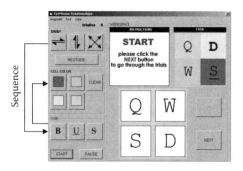

FIGURE 5.6
Incompatible from the top (embedded sequence: position → format → color: display from the top: position → color → format).

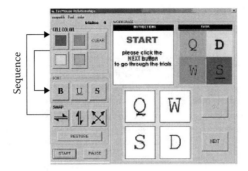

FIGURE 5.7
Incompatible from the bottom (embedded sequence: position → color → format: display from the bottom: position → format → color).

Hence, in this experiment, two display layouts had compatible task sequences, whereas the other two were incompatible due to lack of matching of the tool arrangement and the task sequence. This resulted in four groups with different levels of compatibility of task sequence requirements with the tool display (tool arrangement). The design principles were deliberately violated to observe the difference between the groups with respect to the task complexity and dependent variables. The purpose of using two displays (from top and from bottom) was to observe any effect of the visual scanning—top down or bottom up on the performance. In general, the simplest task was the compatible from the top task and the most complex task was the incompatible from the bottom task. Compatibility and tool arrangement and their combination were the major factors of task complexity in this experiment.

The tasks forced the users to explore different strategies of task performance using the trial-and-error strategy. The users could not complete the tasks unless they understood the rules and limitations in the sequence of task performance through self-learning. Once the users learned the rules, they performed the tasks using the stable strategy that they chose as the most preferable.

The roles of independent variables in the objectives are mentioned, and the dependent variables are described according to their basis and method of derivation. The final experimental design is shown in Table 5.3. The four interfaces described above are analyzed in Section 5.5.

The within-group effect in the learning phase on the subjects' performance will be considered further based on the acquisition curves of the groups, and the model for analysis of variance (ANOVA) will be formulated on the basis of the final design addressing both the between-group analysis for factors of compatibility and tool arrangement and the within-group analysis for the factor of phase of learning.

TABLE 5.3

Factorial Design for the Experiment

	Tool Arrangement 1	Tool Arrangement 2
Compatibility 1	From top—position, color, format Sequence (compatible)—position, color, format Subjects: 8	From bottom—position, format, color Sequence (compatible)—position, format, color Subjects: 8
Compatibility 2	From top—position, color, format Sequence (incompatible)—position, format, color Subjects: 8	From bottom—position, format, color Sequence (incompatible)—position, color, format Subjects: 8

The summary of the between- and within-subject analyses is presented below:

- *Between-subject analysis*: (1) The main effect of compatibility (C); (2) the main effect of tool arrangement (T); (3) the interconnection of compatibility and tool arrangement ($C \times T$)
- *Within-subject analysis*: (1) The phase of learning (exploratory and post); (2) the interaction of phase and compatibility ($P \times C$); (3) the interaction of phase and tool arrangement ($P \times T$); (4) the interaction of phase of learning, compatibility, and tool arrangement ($P \times C \times T$)

In the preliminary study, 16 subjects were involved (four subjects in each group). In the experimental study, 32 subjects took part in the major experiment. They were divided into four groups (eight subjects in each group). All subjects (15 female and 17 male subjects) had at least several years experience working with computers.

5.5 Performance Measures during Skill Acquisition

5.5.1 Measures of Motor Performance

We have employed time performance data, mouse movement data, eye movement data, and error analysis for the analysis of the acquisition process. We now present new methods of analysis for the acquisition process. The measures of motor performance are described below.

5.5.2 Measures Based on the Mouse Movement Data

5.5.2.1 Time per Click

Time per click (TC) was the time between two clicks. This time shortens during the skill acquisition.

5.5.2.2 Motor Efficiency (Click Efficiency)

A minimum number of actions can be used to accomplish each task. However, the users used either the minimum number of actions or more actions depending on the number of errors they made and the strategy they chose. The motor efficiency (E) was calculated using the formula:

$$E = \frac{\text{Number of actions required } (n_{\text{ideal}})}{\text{Number of actions used } (n_{\text{actual}})}$$

which is expressed as ratio or percentage.

Real efficiency E_{actual} is less than 1. The better the motor efficiency, the more E_{actual} approaches to 1.

Efficiency in this case is affected by the following four factors:

1. Errors due to misapplication of tools
2. Errors due to embedded task sequence
3. Errors due to omission of a feature
4. Excess clicks due to inefficient strategy

The efficiency basically reflects errors due to the compatibility of the task sequence with the tool arrangement and hence is the indicator of efficiency of the subjects' performance. The lower the efficiency, the higher the number of actions used to complete the task, and hence the less the usability of the interface. This is because subjects make mistakes due to the inconsistency of the task interface relationship. This results in a lower usability of the interface, caused by the incongruence of the task and interface features.

5.5.2.3 Distance Mouse Traversed per Time—Mouse Traversal Rate

Here, movement in terms of distance traverse is considered.

$$MT_i = \frac{M_i}{t_i} \text{ (pixels/second)}$$

where M_i is the mouse traversal in task i, t_i is the time to complete the task i. This measure reflects the speed at which the users are performing the task as well as how efficient they are in accessing the tools. The larger traversal indicates more efficient movement and performance. Mouse movements depend upon various factors. They mostly corroborate either with the events during the task performance or with indecision on the part of the users when they are unable to find the icon needed to perform the task or the subtask. The lower the mouse traversal per unit time, the more time is required by the users to perform a certain action. Users may also click more in less time.

5.5.3 Measures Based on Eye Movement Data

Search efficiency during the task performance, which is defined as the number of visits required and also the processing time per visit to the various areas of the screen, can be obtained by studying the eye movement data. The total number of fixations along with the fixation duration is taken as the gaze at the particular area of interest. Shorter fixations should be considered in the later stages of the experiment. The shorter duration threshold for eye fixation was taken as 100 milliseconds, as has been recommended by Yarbus (1967).

In this study, the eye movement data have been analyzed; the classification for each area on the screen is presented in Figure 5.1. Defining areas of interest based on already preidentified tool, object, and goal area is an important aspect in the eye movement study. Tools, goal, and objects are the main elements of the task in the functional analysis of the subject's performance. Using tool, object, and goal areas sets, a general paradigm for identifying strategies of task performance was used by different users. For example, the total number of visits to the tool area by the eye will be calculated based on the number of transitions the eye made to the tool area during the task performance.

5.5.3.1 Visual Fixation Time in Different Areas of the Screen during Skill Acquisition

These data represent the amount of confusion that existed at the initial stage of the task performance. During skill acquisition, the duration of visual fixations on various areas of the screen should decrease. Moreover, this duration can decrease unevenly for different areas depending on the specificity of a particular task.

Hence, if VTg is visual fixation time in the goal area, VTt is visual fixation time in the tool area, and VTo is visual fixation time in the object area for the ith trial, then the total visual fixation time for the ith trial is given by

$$VTg + VTt + VTo = VT_i$$

5.5.3.2 Total Number of Eye Visits to Different Areas per Click

This measure also represents the amount of confusions that existed at the initial stage of the task performance. During skill acquisition, the number of eye visits to different areas of the screen should decrease. The total number of visits (VN) here is calculated as follows:

[Number of visits to the goal area + number of visits to the tool area + number of visits to the object area for a particular trial]

Hence, if VNg is the number of visits to the goal area, VNt is the number of visits to the tool area, and VNo is the number of visits to the object area for a particular trial i, then the total number of visits for the trial i is given by

$$VNg + VNt + VNo = VN_i$$

5.5.3.3 Average Processing Time per Visit

If the total time required to complete the trial i is T and the number of visits during the trial i is V, then the average processing time per visit (tv) is given by $tv = V/T$. Because the amount of time spent on each visit within the particular trial may not be the same and hence average time is used in this case.

5.5.3.4 Ratio of Eye Visits to the Object Area to the Number of Eye Visits to the Tool Area

This is calculated using the below formula:

$$ETO = \frac{\text{Number of eye visits to the object area}}{\text{Number of eye visits to the tool area}}$$

This ratio represents the difficulty the user experiences in executing actions in the tool area. If ETO is >1, then it represents difficulty for the user to execute the actions in the tool area. The minimum value of ETO depends upon the familiarity of the user with the interface. A ratio of 1 is optimal since the user is visiting the area for carrying out an action by mouse click. A higher ratio indicates that the user is facing difficulty in associating the task with the tools and thereby moving back and forth between the tool area and other areas of interest related to the task.

We also use measures such as total time of task performance and number of errors during the skill acquisition process. Some other measures can be developed based on the described principles. The obtained measures can be calculated as average data for all groups and as measures related to a particular subject. Therefore, we can develop a learning curve or a curve of acquisition process for a group or for an individual. This helps us study individual strategies used by the subjects during skill acquisition. In Section 5.6 we will demonstrate how some of these measures can be used for task analysis by using the genetic principle of study.

5.6 Results

5.6.1 Preliminary Set of Experimental Study

An analysis of the users' performance demonstrates that they have used explorative strategies. The users do have one fixed goal of the task in mind, which is the final arrangement. However, during trials and errors, the users have different subgoals that they can change during the trials based on the feedback about the task performance. Each action gave them feedback based on which the users understand better what kinds of restrictions are imposed by the interface. In most cases, the users would change their subgoals when they encounter the limitations or the rules of the task. If the output of the actions was not what the users desired, they would change the subgoal and therefore the strategy of their task performance. Once they were accustomed to all the rules, the performance became more stable and the errors and changes in strategies of performance became infrequent.

Based on theoretical assumptions, we developed learning curves that demonstrated the process of skill acquisition. The subjects gained their expertise through task performance. In the preliminary study, the complete sequence was introduced, that is, all subjects performed every task during each trial. This was done to find the number of trials required to achieve the stable level of performance. TC was used for these curves. In this work, we present only two curves. One for the task called "compatible from top" and the other one for the task, which is more complicated, called "incompatible from bottom." For all the groups, the logarithmic curves of the TC were tested to observe whether the skill has reached the steady state before beginning the main experimental set.

Figures 5.8 and 5.9 show the power curves for the TC data for the overall sets of tasks. Tasks 1–12 consist of the first learning set, tasks 13–24 consist of the second learning set, and tasks 25–40 consist of the experimental set. Two power curves (Figures 5.8 and 5.9) show a consistent correlation with empirical data according to F-criterion. For the group of tasks compatible from the top (Figure 5.8), $F = 70.66$, $p < .001$, $R^2 = 0.65$, and for the group of tasks incompatible from the bottom (Figure 5.9), $F = 76.36$, $p < .001$, $R^2 = 0.67$. We can see that all curves reached a steady stage of performance before trial number 25 (task number 25).

The curve demonstrates that a relatively stable stage of learning can be achieved after about 25 trials. In each trial, subjects performed different versions of the same type of task. Here, we did not discuss the statistical analysis for the differences between the stages of learning within groups and between groups. We will discuss this briefly during the analysis of the basic experiment in the next section. Here, we can see that TC can be used in the development of learning curve.

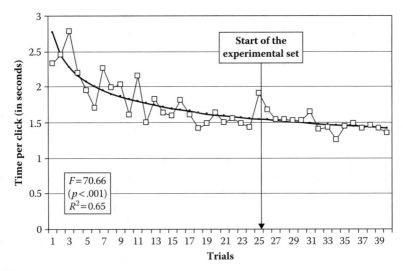

FIGURE 5.8
Acquisition curve of performance for the compatible from the top group of tasks.

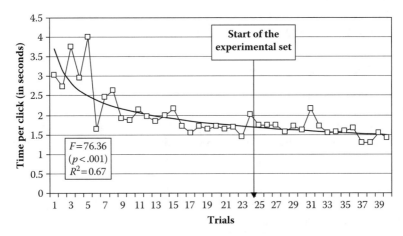

FIGURE 5.9
Acquisition curve of performance for the incompatible from the bottom group of tasks.

5.6.2 Main Set of the Experimental Study of the Acquisition Process Based on the Motor Activity Analysis

In this study, we used an acquainted set of experiments with different subjects. At the first acquainted set of experiment, subjects manipulated a single feature of the task during 12 trials. The second acquainted set required manipulation of just two features of the tasks. Therefore, the subjects were sufficiently familiar with the strategies of performance. This reduced the effect of learning to some degree. Nevertheless, if we can discover the effect of learning even in this situation, then our methods are sufficiently sensitive. In general, the acquainted stage depends on the specifics of the task. For example, in our case, it is sufficient to have three to five trials in both first and second acquainted sets. This requires expert analysis in each case to determine the number of trials for the acquainted set or for the preliminary experimental study. We should consider not only the cognitive factors but also the emotional-motivational adaptation factors. These factors influence the skill acquisition process. In the main set of the experiments, subjects performed the whole task. In this set of experiments, all subjects performed 16 trials. The first set of five tasks belongs to the exploratory learning stage, and the final set of five tasks is considered to be the postlearning stage (Carroll 1987).

There were four groups of tasks with various levels of compatibility and tool arrangement. The between–within ANOVA model was used for the statistical analysis of data obtained in these groups. The compatibility of the task sequence and tool arrangement was the between-subjects factor and the learning phase was the within-subjects factor. Only two levels of within-group factors were considered, and the variables for each analysis were the same. Since only two blocks were used in this case (exploratory and postlearning), the sphericity test was not necessary.

5.6.2.1 Total Task Completion Time

The total task performance time was the simplest measure useful for the performance comparison. These data are presented in Table 5.4, where N is the number of subjects in each group. From Table 5.4, we can see that the more complex the task, the more time is required for its performance. However,

TABLE 5.4

Mean and Standard Deviation of the Total Task Completion Time for the Exploratory and the Postlearning Stages[a]

Level of Compatibility	Level of Tool Arrangement	N	Exploratory		Postlearning	
			Mean	Standard Deviation	Mean	Standard Deviation
1	1	8	125.73	16.51	98.49	18.67
1	2	8	126.37	28.99	108.98	24.16
2	1	8	130.20	37.46	103.70	20.87
2	2	8	139.42	30.84	114.75	28.16

[a] Statistical data are selected and presented in such a manner that it permits the reader to understand the basic principles of its utilization. For example, statistical data were collected for other experimental results in similar way as presented in Table 5.4. However, we showed such data only one time for demonstration.

the difference was not statistically significant, and we consider it a tendency. There were also differences in standard deviation. The more complex the task, the more variations were present in the task performance.

Statistical analysis revealed that compatibility had no major effect on the task performance. Compatibly and tool arrangement had no effect on the total task performance time. Within the groups, there was a significant effect of the stage of learning [$F(1, 28) = 11.2, p < .01$]. All of the other null hypotheses were accepted. The total task performance time can be used only for the preliminary data analysis. Sometimes, tasks have approximately the same total performance time at the final stage of skill acquisition, but the tasks' learnability can be different.

Moreover, the task complexity cannot be evaluated based on the total performance time, because they have a different amount of elements to be manipulated during their performance. Hence, a task that requires more time can be easier than the shorter one. Learnability is a more sensitive criterion for the evaluation of task complexity (Bedny and Meister 1997). In Section 5.7, we will consider some other measures for the comparison of the task performance.

$$\text{Motor efficiency } (E) = \frac{\text{Number of actions required } (n_{\text{ideal}})}{\text{Number of actions used } (n_{\text{actual}})}$$

The acquisition curve was developed based on 16 main trials. Figure 5.10 shows the average efficiency of the groups across different tasks. An initial observation suggested that for the incompatible tasks at the exploratory learning

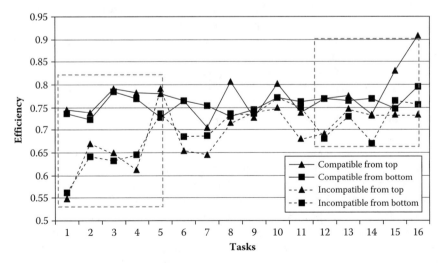

FIGURE 5.10
Average efficiency across different types of tasks for four groups of subjects.

stage of performance the efficiency was quite low. In the graph, the dotted area represents the exploratory learning stage and the postlearning stage (Figure 5.10). In all the graphs showing average measures of different groups across the tasks, the dotted area on the left indicates the exploratory learning stage and that on the right indicates the postlearning stage (Figure 5.10).

Qualitative analysis of these curves demonstrated that incompatible groups have lower efficiency than the compatible ones. During the trials, the efficiency of all the groups of subjects, which performed different types of tasks, increased. At the end of the trials, the three groups of subjects that performed various types of tasks, excluding those compatible from the top, demonstrated similar efficiency.

The difference in motor efficiency between the groups that performed compatible tasks and incompatible tasks was more significant in the beginning of the trials than at the end. Therefore, the more complicated the tasks, the more significant efficiency improvement can be observed. The variation in average efficiency of the groups can be observed in Figure 5.10. Let us consider ANOVA for the between-group effects. There was a significant main effect for the factor of compatibility [$F(1, 28) = 11.1, p < .01$]. There was no significant interaction effect. The tool arrangement did not have any significant effect on the performance efficiency.

The learning stage has a significant effect on the performance efficiency [$F(1, 28) = 4.21, p < .05$]. The between–within interaction of stage and compatibility had a major effect on efficiency [$F(1, 28) = 18.32, p < .001$], and no other between–within interactions were observed.

For the compatible from the top task group, the efficiency remained more or less at the same level with a mean of 78% (standard deviation [SD] = 5%) for the exploratory stage and a mean of 76% (SD = 7%) for the postlearning stage. However, incompatible task groups showed a drastic improvement in performance; for tasks from top tool arrangement, the efficiency increased from a mean of 61% (SD = 9%) to 77% (SD = 10%) in the postlearning stage, and for tasks from the bottom tool arrangement, the efficiency increased from a mean of 64% (SD = 8%) to 72% (SD = 5%).

The between-group hypothesis (H$_1$: $\mu_{Compatible} \neq \mu_{Incompatible}$) for efficiency was accepted at the 0.05% significance level (data statistically significant). However, due to the interaction effect, this should be considered only for the exploratory learning stage. The null hypothesis (H$_0$: $\mu_{From\ top} = \mu_{From\ bottom}$) for tool arrangement was rejected. In the postlearning stage, null hypotheses for both the compatibility and tool arrangement were rejected.

Incompatible task groups showed a significant improvement in the efficiency. For tasks from the top tool arrangement, the mean was 61.5% and SD was 9% at the exploratory stage. In the postlearning stage, the efficiency increased and was 77.7% with SD of 10.3% ($p < .005$). For tasks from the bottom tool arrangement in the exploratory stage, efficiency was 64.3% with SD of 8.1%. In the postlearning stage, efficiency was 72.4% with SD of 5.5% ($p < .005$). From the comparison of exploratory and postlearning stages, it can be seen that efficiency also increased for the compatible task groups. However, these changes were not statistically significant.

For the compatible task groups, the null hypotheses for differences in efficiency in the exploratory and the postlearning stages were rejected. However, for the incompatible task groups, the null hypotheses for the increased efficiency in the postlearning stage (H_0: $\mu_{Exploratory}$ = $\mu_{Postlearning}$) versus in the exploratory learning stage was accepted at a significance level of $\alpha = 0.05$.

5.6.2.2 Time per Click

The TC data were initially used to estimate when the subjects had reached a steady state of performance and to define the preliminary number of trials. Here, we will describe how this measure has been used in the main experiment. The mean TC of the different groups across tasks is given in Figure 5.11.

Let us consider some data as an example. For the group performing the compatible from the top task, the TC is reduced from a mean of 1.51 seconds (SD = 0.18 seconds) in the exploratory stage to a mean of 1.30 seconds (SD = 0.12 seconds) in the postlearning stage. The group that performed the compatible from the bottom task reduced the TC from a mean of 1.44 seconds (SD = 0.22 seconds) in the exploratory stage to a mean of 1.32 seconds (SD = 0.23 seconds) in the postlearning stage. The group that performed the incompatible tasks showed a similar drastic improvement in performance.

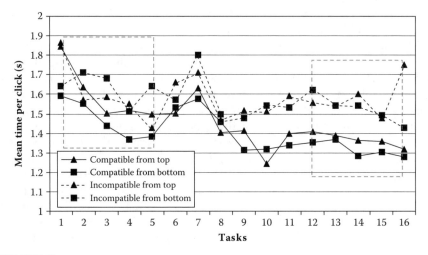

FIGURE 5.11
Time per click across tasks for different groups during comparative analysis of the acquisition curves based on the diverse measures.

The ANOVA for the between-groups effects demonstrated that there was no significant interaction effect. There was a significant main effect for compatibility [$F(1, 28) = 6.3$, $p < .01$]. The tool arrangement did not have any significant effect on the TC measure. The learning stage had a significant effect on the TC [$F(1, 28) = 21.6$, $p < .0001$]. No interaction effects for the between–within factor were observed. We did not consider any other statistics related to this measure.

In general, the results suggest that there was a significant effect of compatibility on subjects' performance as far as the time spent in between clicks was concerned, which means that incompatible interfaces are time consuming. However, the TC is reduced as has been expected due to the increased speed of performance, as the subjects got a better understanding of the task sequence. The effect of compatibility and the significant difference ($p < .005$) between compatible and incompatible task performance shows that even when the subjects already knew the task sequence, there were difficulties in accepting the task sequence, which resulted in excess time between actions.

5.6.2.3 Evaluation of Task Performance Based on Mouse Traversal Rate Measure

This measure evaluates mouse movement in pixels per second. The major hypothesis when we use this measure is that various task sequence requirements and interface relationships can affect the mouse traversal rate. The analysis of this measure demonstrated the following: The main effect of the compatibility or tool arrangement on the mouse distance traversed per unit of time was not observed. Hence, $HB1_0$ and $HB2_0$ were accepted. The interaction effect was also not observed. Within-group tests showed a significant effect of the learning phase [$F(1, 28) = 9.08$, $p < .01$]. Therefore, $HB3_0$ was rejected at 0.05% level of significance.

The mouse traversal rate of the group performing compatible from the top tasks changed from ($M = 308.7$, $SD = 80$) at the exploratory learning phase to ($M = 366.9$, $SD = 67.2$) at the postlearning phase. For the group performing the compatible from the top tasks, there was an increase in the mouse traversal rate from the exploratory phase ($M = 363.7$, $SD = 71.7$) to the postlearning phase ($M = 381.9$, $SD = 60.9$). For the group performing the task using the incompatible from the top interface, the mouse traversal rate increased ($M = 328.2$, $SD = 73.9$) to the postlearning phase ($M = 346.7$, $SD = 65.8$), while for the group that used the incompatible from the bottom interface, there was also an increase in the mouse traversal rate from the exploratory learning ($M = 328.2$, $SD = 101.6$) to the postlearning phase ($M = 381.4$, $SD = 129.3$). Therefore, this study discovered that subjects required less time in the postlearning phase to traverse the same distance. This measure is particularly sensitive to the effect of skill acquisition.

For the study of individual differences in skill acquisition, the acquisition curves of the individuals are useful. For example, a subject's individual

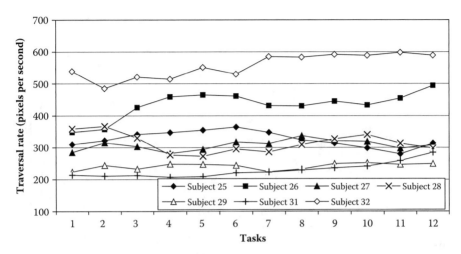

FIGURE 5.12
Subjects' individual curve developed based on mouse traversal rate (incompatible from the bottom interface).

curve developed based on the mouse traversal rate is presented in Figure 5.12. This figure shows that the 26th and 32nd subjects demonstrated individual differences in performance compared with other subjects according to the considered measures. For a correct interpretation of these differences, we should compare the individual curves in relation to other measures. Moreover, a more accurate interpretation of the data can be obtained during comparative analysis of acquisition curves based on diverse measures.

The material presented above is restricted to measures that are based on the mouse movement and the mouse log data. This is not an exhaustive list of measures, and other measures can be used. We consider the measures that use eye movement data below.

5.6.3 Main Set of Experiments in the Study of the Acquisition Process Based on the Eye Movement Data

Here we consider the possibility of using eye movement data to analyze the acquisition process. The objective is to analyze the relationship between eye movement and task graphic interface features.

5.6.3.1 Percentage of Dwell Time in the Tool Area

The formula to be used is $\dfrac{\text{total dwell time in tool area}}{\text{total dwell time}}$ (percents)

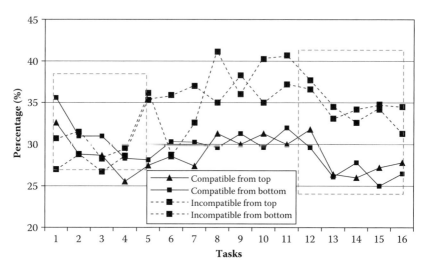

FIGURE 5.13
Percentage of the dwell time in the tool area.

A comparison within the groups for the difference in the exploratory and the postlearning stages and the effect on the eye movement due to learning has been analyzed at the first stage. The percentage of the dwell time (PDT) in the tool area is presented in Figure 5.13.

The ANOVA for the between-groups effect demonstrates that there was no significant interaction effect, and there was a significant effect of compatibility [$F(1, 28) = 12.1, p < .001$]. The tool arrangement did not have any significant effect on the PDT in the tool area. No significant between-subjects effect was observed. However, interaction of the learning stage and the compatibility seemed to have an effect [$F(1, 28) = 5.91, p < .05$].

The ANOVA showed the within-subject effect. The learning stage had significant effect on the PDT in the tool area [$F(1, 28) = 68.8, p < .0001$]. There was also an interaction effect of the compatibility and the learning stage [$F(1, 28) = 33.2, p < .0001$], indicating different dynamics in PDT for groups performing various tasks in terms of their compatibility. No other interaction effects were observed.

Incompatible task groups had significantly higher PDT in the tool area than the compatible ones. The null hypothesis for PDT in the tool area was rejected at 0.05% level of significance. This significance was observed in the postlearning stage.

5.6.3.2 Percentage of Dwell Time in the Object Area

The formula to be used is $\dfrac{\text{total dwell time in object area}}{\text{total dwell time}}$ (percents)

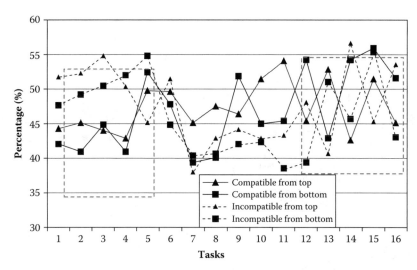

FIGURE 5.14
Percentage of the dwell time in the object area.

The percentage of the dwell time in the object area (PDTO) is presented in Figure 5.14.

The ANOVA for the between-groups effect demonstrated that there was no significant interaction effect. However, there was a significant effect of the tool arrangement [$F(1, 28) = 16.8, p < .001$]. Compatibility did not have any significant effect on the PDTO. Null hypotheses for the tool arrangement (H_0: $\mu_{\text{From top}} = \mu_{\text{From bottom}}$) have been rejected. An interaction effect [$F(1, 28) = 6.84, p < .01$] of the compatibility and the tool arrangement has been also observed. The stage of learning had significant effect on the PDTO [$F(1, 28) = 7.9, p < .0001$]. No interaction effects for the between–within factor were observed. Null hypotheses for the effect of the compatibility (H_0: $\mu_{\text{Compatible}} = \mu_{\text{Incompatible}}$) have been rejected for both the stages of learning. Null hypotheses for the tool arrangement have been accepted for the both stages of learning. The PDT in the goal area has also been analyzed. The within-group null hypotheses for the learning stages has been rejected [$F(1, 28) = 36.8, p < .001$].

5.6.3.3 Aggregated Eye and Mouse Movement Measures

We can determine the ratio of eye visits to mouse visits in the area of interest. As an example, the tool area acquisition curve that uses this measure is presented in Figure 5.15. There was a higher number of eye visits than mouse visits at the exploratory learning stage. The within-subjects ANOVA shows main effect of the learning stage [$F(1, 28) = 36.67, p < .001$]. An interaction effect of the learning stage and the compatibility [$F(1, 28) = 8.26, p < .01$] was also observed. As an example, we also present a means plot of the eyes-to-mouse visit ratio (compatibility stage) in the tool area in Figure 5.16.

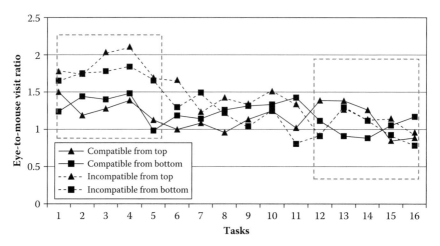

FIGURE 5.15
Ratio of eye-to-mouse visits in the tool area.

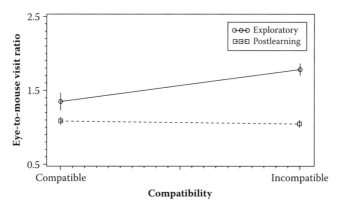

FIGURE 5.16
Means plot of the eye-to-mouse visit ratio (compatibility stage) in the tool area.

Groups that were using the incompatible interface had higher means in the exploratory learning stage than in the postlearning stage. The groups that were using the compatible interface had less of a difference between eyes-to-mouse visit ratio at the exploratory stage than the groups using the incompatible interface.

5.6.3.4 Errors Analysis

Error analysis can be particularly informative in the study of the activity acquisition process. Here we demonstrate how this method can be adapted for this purpose. Error analysis was conducted based on the comparison of

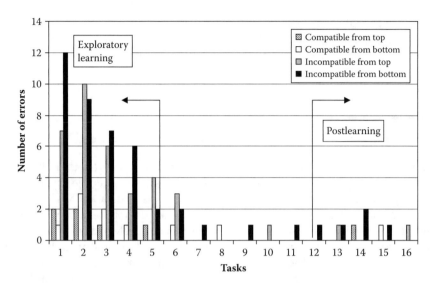

FIGURE 5.17
Errors due to the treatment of the interface for various groups.

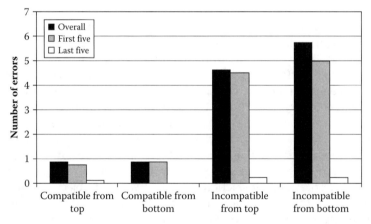

FIGURE 5.18
Average number of errors for different sets of tasks and errors in different groups across tasks.

the number of errors at the various stages of the skill acquisition process. Figure 5.17 depicts how the error rate changes across trials for groups who performed the tasks with different levels of complexity. After six trials the number of errors due to the incompatibility of the interface reduced considerably across the groups. However, there was a significant difference in the number of errors between the groups.

Figure 5.18 shows different types of errors overall, in the first five sets and the last five sets. We can observe a significant effect of the learning stage on more complicated tasks in particular. The ANOVA showed a significant effect of the compatibility [$F(1, 28) = 72.7, p < .001$], whereas there was no effect on the tool arrangement [$F(1, 28) = 2.1, p > .05$].

5.7 Discussion

The purpose of this work was to demonstrate the microgenetic method of study as applied in the evaluation of the task usability, complexity, reliability of performance, and so on. The essence of this principle consists of the study of human activity in the formation and development process. This method is particularly relevant for the study of computer-based tasks where users constantly acquire new tasks and more often than not through the self-learning process in which they utilize explorative strategies derived from the self-regulation principles of activity organization. Thus, learnability is an important feature of a user interface. The more complex the task is, the less its learnability. Hence, in this study we've utilized functional analysis, considering activity to be a self-regulative system.

In this study, we developed four versions of task with various levels of complexity. The more complex the task is, the less its task usability. Analysis of the task learnability is an important experimental approach to the evaluation of the task complexity and therefore the usability of the interface. According to ISO 9241, some usability measures, such as evaluation of the effectiveness of performance, efficiency of performance, and satisfaction, have been suggested. Some of these measures are not precise and are ambiguous. For example, effectiveness can be evaluated based on the calculation of percentage of goals achieved. However, in activity theory, there is an objectively given goal, a subjectively accepted or formulated goal, and so on. Similarly, the time to complete a task as a measure of efficiency is a very general criterion. Satisfaction measures without objective data can be very subjective. In general, these measures do not give us sufficient information about activity structure during the task performance. Therefore, in this work, we described the usage of the microgenetic method and derived from it the measures of usability evaluation of the computer-based tasks.

To use the genetic method, we have developed methods of registration and analysis of the activity acquisition process during performance of computer-based tasks. Development of the learning curves that depict the activity acquisition process and their interdependent analysis is an important method of the genetic study. Learning curves are widely used for analysis of training but not for analysis of task complexity and evaluation of the usability of the interface. Usually, in psychology, integrative learning curves are not sufficiently informative. Moreover, there is no well-defined method for development of acquisition curves for computer-based tasks. In contrast to the study of learning process, in our work, we used acquisition curves for the usability evaluation, which has helped us evaluate the task complexity and understand some changes in structural characteristic of the task.

Activity during performance of computer-based tasks can be considered a complicated structure comprised of logically, hierarchically, and functionally organized elements. According to the functional analysis principles, the same elements of the screen, depending on its purpose in the activity structure, can be related to different functional elements of activity. Therefore, hand and eye movement data in computer-based tasks should be related not only to particular elements of the screen that have some technological characteristics, but also to some elements of activity that have a particular functional purpose in the activity structure.

The important functional elements of activity are the goal, object, and tool. The principle of the extraction of such elements in activity theory is related to different areas when compared to cognitive psychology. The same elements of the screen, depending on its functional purpose in the structure of activity, can be related to different areas at different stages of the task performance. Therefore, the analysis of eye and mouse movement for different areas of the screen at different phases of learning helped us understand the functional structure of activity depending on the stage of the skill acquisition. Our study demonstrated a combination of genetic method of study with the functional analysis of activity.

The other principle that we used in this study was the principle of the unity of cognition and behavior. This has been facilitated by combining eye movement registration with mouse movement registration and the analysis of the interdependency of the obtained data. Yarbus was the first to introduce direct eye movement registration. He demonstrated that behavioral activity is closely connected with cognitive activity, and how cognitive processes depend on the goal, motives, and strategies of activity performance. In the West, this aspect of his study of eye movement was never introduced. In our work, we paid attention to these factors by using some elements of the functional analysis of activity by dividing the screen into three functional areas. An analysis of total eye visits to different areas of interest at different phases of learning, calculation of percentage of distribution of eye visits to these areas (goal, object, and tool areas), calculation of PDT in different areas of interest at the different stages of learning, and so on demonstrated how subjects changed their strategies during the acquisition of various tasks. These data allowed us to evaluate various versions of the interface.

There is no consensus regarding the level of familiarity of users with a particular task during the study of the acquisition process. In our experiment, subjects received significant training for performing separate elements of tasks. Even in such conditions, acquisition process is a sufficiently sensitive method for the task analysis of computer-based tasks. In any practical situation, scientists should specify the level of users' familiarity with particular tasks during the analysis of acquisition process. Such basic concepts as self-learning, the acquisition process, learnability, functional analysis, and so on are critically important when studying computer-based tasks.

References

Anan'ev, B. G., L. M. Vekker, B. F. Lomov, and A. V. Yarmolenko. 1959. *Sensing by Touch in Cognition and Work*. Moscow, Russia: Academy of Pedagogical Science of SSSR.

Bedny, G., W. Karwowski, and M. Bedny. 2001. The principle of unity of cognition and behavior: Implications of activity theory for the study of human work. *Int J Cogn Ergon* 5:401–20.

Bedny, G. Z. and S. Harris. 2005. The systemic-structural activity theory: Application to the study human work. *Mind Cult Act* 12(2):128–47.

Bedny, G. Z., and W. Karwowski. 2007. *A Systemic-Structural Theory of Activity. Applications to Human Performance and Work Design*. Boca Raton, FL: Taylor & Francis.

Bedny, G. Z., and D. Meister. 1997. *The Russian Theory of Activity. Current Application to Design and Learning*. Mahwah, NJ: Lawrence Erlbaum Associates.

Bedny, G. Z., M. Seglin, and D. Meister. 2000. Activity theory, history, research and application. *Theor Issues Ergon Sci* 1(2):168–206.

Bedny, I. S., W. Karwowski, and A. Ya Chebykin. 2006. Systemic-structural analysis of HCI tasks and reliability assessment. *Triennial IEA 2006 Sixteenth World Congress on Ergonomics*, Maastricht, the Netherlands.

Bedny, I. S., and T. Sengupta. 2005. The study of computer based tasks. *Sci Educ* 1–2(7–8):82–4.

Carroll, J. M., ed. 1987. *Interfacing Thought*. Cambridge, MA: MIT Press.

Leont'ev, A. N. 1972. *The Problem of Psychic Development*. Moscow: Moscow University Publishers.

Rubinshtein, C. L. 1973. *The Problems of General Psychology*. Moscow: Pedagogical Publishers.

Sokolov, E. N. 1969. Study of the speech mechanisms of thinking. In *A Handbook of Contemporary Soviet Psychology*, ed. M. Cole and I. Maltzman, 531–73. New York: Basic Books, Inc.

Turvey, M. T. 1996. Dynamic touch. *Am Psychol* 51(11):1134–66.

Vygotsky, L. S. 1960. *Developing Higher Order Psychic Functions*. Moscow: Academia of Pedagogical Science.

Yarbus, A. L. 1967. *Eye Movement and Vision*. New York: Plenum Press (translation from Russian).

Zinchenko, V. P. and A. G. Ruzskaya. 1966. Interaction of sense of touch and vision in pre-school of children. In *Development of Perception in Childhood*, ed. A. V. Zaporozhets, 85–99. Moscow: Educational Publishers.

6

Abandoned Actions Reveal Design Flaws: An Illustration by a Web-Survey Task

I. Bedny and G. Bedny

CONTENTS

6.1 Introduction

In their professional activities, users have to perform multiple tasks, and some of them are performed only once. In most cases, computer-based tasks do not possess rigorous standardized methods of performance as is observed in mass production processes. In human–computer interface (HCI) tasks, a user often does not know in advance the sequence of actions he or she must take, and even experienced users should discover the details of the task performance through an exploratory activity. Hence, continuous self-learning through explorative activity is an important component of users' professional activity. Even when there are standardized requirements for task performance, users still have some degree of freedom in task performance. HCI always involves self-learning, exploration, and individualized strategies of tasks performance (Sengupta and Bedny 2008). The more complex the task is for the user, the more important self-learning and explorative strategies are during task performance.

Explorative activity consists of correct and incorrect actions that provide users with information that helps them understand the system and correct performance strategies. The explorative stage allows the user to examine the situation and the consequences of his or her own actions, when they often utilize "reversible errors." Such errors can be eliminated without negative consequences for the task performance because they perform an informational function. If the task is extremely difficult for the user, his or her goal-directed activity can be transformed into chaotic behavior. Explorative behavior is a self-regulative process; that is, according to the functional analysis of activity, it is the basis for learning, which in turn can be considered a transformation of performance strategies (Bedny and Meister 1997). The more complex the task, the longer it takes the user to find a truly effective strategy. If the user performs similar tasks multiple times, she or he goes through the intermediate strategies until she or he approaches the optimal one. Hence, learning and self-learning in particular are the transformation from less efficient strategy to more efficient ones.

The user explores her or his options; this can be an external or internal mental exploration. The user can observe the result of externalized explorative actions on the screen and evaluate them as positive or negative. Explorative action is a cognitive function of transforming the situation and evaluating the impact of this transformation. Internal explorative actions lead to an increase in the duration of cognitive activity. If the tasks' complexity increases, then the number of explorative actions also increases. Explorative actions can be erroneous and lead to erroneous external changes, and the user has to return to the previous or initial screen. We call cognitive and motor explorative actions that give an undesirable result "abandoned" actions; the goal of HCI design is to reduce them. The less abandoned actions are used, the better the efficiency of the task performance. The users correct their strategies of task performance based on the evaluation of the result of abandoned actions and select the actions they evaluate as positive. HCI tasks include explorative activity that cannot be eliminated totally. In a production environment, some externalized explorative components of an activity can lead to corruption of the database, deletion of important information, and other undesirable results. The goal of the design of HCI tasks is to eliminate or reduce the possibility of such actions.

The emotional-motivational factor is important in the learning and self-learning processes. The user is particularly sensitive to the influence of feedback at the explorative stage of task performance, when she or he promotes the hypothesis, formulates the goal, evaluates the difficulty and significance of the task, examines the consequences of his or her actions, and so on. The feedback influences the user's emotional-motivational state. Activity theory (AT) underlines the complex relationship between the explorative, motivational, executive, and evaluative components of the task performance.

In our study, we used the methods developed in systemic-structural activity theory (SSAT). These methods are organized into stages of analysis such as qualitative, quantitative, and algorithmic (Bedny and Karwowski 2007; Bedny et al. 2010).

6.2 General Characteristics of a Web-Survey Task

Electronic mail (e-mail) is an important communication media in the business environment. People are overwhelmed with the number of e-mails they have to read and reply to. The time someone has to read every e-mail is very limited. The time-restriction factor becomes especially important if the e-mail subjectively has a low priority and distracts from the main duties. There are group e-mails that are distributed to hundreds of employees to take surveys and so on, which are mandatory and are important for business organizations. Here, we encounter a situation in which the same e-mail is important for a sender and is not important for a receiver. The self-regulation concept of motivation (Bedny and Karwowski 2006) considers personal importance or significance to be an emotionally evaluative mechanism of the motivational process. In the model of activity self-regulation (see Chapter 1), the functional block "assessment of the sense of input information" is part of the emotionally evaluative stage of activity. Distributed e-mails that have a low positive or, in some cases, even negative, significance are perceived as a source of waste of time and money. Business e-mail designers should consider the effectiveness of such e-mails because proper design will save companies a lot of money.

We will demonstrate that not only the cognitive but also the emotional-motivational aspects of the user's activity are important for task analysis. Cognitive aspects of activity should be studied in unity with emotional-motivational components. The more complex and lengthy the supplemental task, the greater negative emotional effect it has on the user who loses the connection with the main task, and it becomes more difficult to return to it. "Where was I?" is the first question of the user after completing the supplemental task. Distributed e-mails are the ones that are sent to multiple employees. Sometimes this is done with a regular frequency, once a quarter, once a year, and so on. Every employee has to complete the distributed task that contains a questionnaire. These e-mails require careful reading and answering of a number of questions. The questions can be wordy and numerous. Until the employee completes the questionnaire, the same e-mail keeps coming back. If such communication happens without considering some psychological factors, it can cause confusion, loss of time, and a negative emotional state that affects productivity. We have chosen the e-mail-distributed tasks that are associated with poor emotional and motivational states because these conditions influence cognitive strategies.

In our example, the described task had to be performed by several thousand employees. Even a few minutes reduction in the task performance time had a significant economic effect. An additional purpose of this study was to demonstrate the principles of SSAT that are instrumental in the enhancement of the HCI task's design.

6.3 Description and Qualitative Analysis of the Web-Survey Task

As an example, we have chosen a real Web-survey task. The task has been slightly modified for our research purpose. More than 5000 employees received the e-mail with the content shown in Figures 6.1 and 6.2.

Each employee must read this e-mail and fill out the questionnaire. As can be seen from Figures 6.1 and 6.2, this e-mail does not fit on one screen. As a result, the attached file cannot be observed without scrolling down. In this case, there is an interesting psychological factor associated with the motivation of the employees. This e-mail-distributed task is not a high-priority task. Most employees consider it annoying, with low personal significance, because it takes time from the main duties, and as a result, it is accompanied by low motivation. However, they are required to complete this survey. Moreover, the busier the employee is, the less he or she is motivated to take on this task. Often, performance of such tasks causes irritation and a negative emotional state. From the functional analysis perspective, this task has negative significance for the employee (Bedny and Karwowski 2006; Bedny and Karwowski 2007). This has been observed during the discussion

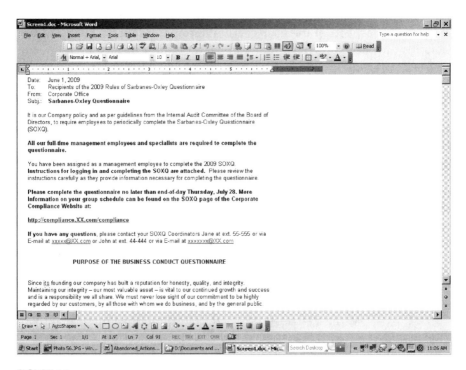

FIGURE 6.1
First page of the e-mail.

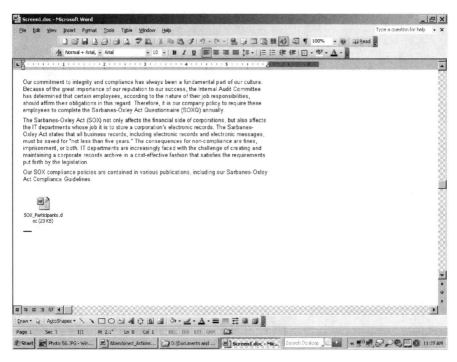

FIGURE 6.2
Second page of the e-mail.

and analysis of the task performance. The general model of self-regulation of activity outlines the following motivational stages: (1) the preconscious motivational stage, (2) the goal-related motivational stage, (3) the task evaluative motivational stage, (4) the executive or process-related motivational stage, and (5) the result-related motivational stage (Bedny and Karwowski 2006). In the task under consideration, the goal-related motivational stage is in conflict with the executive stage because the task has to be completed but is boring and out of the scope of the main duties.

Our observation demonstrates that employees attempt to complete this task as quickly as possible. They select a strategy to move through the content of the e-mail without carefully reading it and to get the questionnaire just as quickly as possible. For the employees, the most significant identification elements of the task are those that provide the direct link to the questionnaire, so they are looking for the links to the questionnaire. These most significant elements are the identification features or the task attributes. Our analysis of the e-mail and the user activity demonstrates that these attributes are not organized very well. The first page of the e-mail has only one link, which immediately attracts the attention of the majority of employees. Thus, they quickly click on the first link they see. When the Web page opens up, they do not see the expected log-in screen. The Web page that opens up is shown in Figure 6.3.

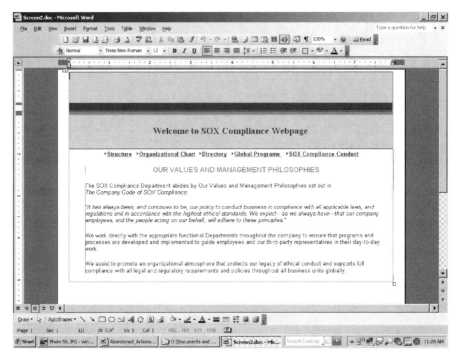

FIGURE 6.3
Screen after clicking at the first link.

The purpose of this page is to give some general information on the topic of the questionnaire. This Web page also includes links to some other pages with additional information. The employees again concentrate their attention only on the available links. Five of the links are shown in Figure 6.3 (see the top line of the screen). These links are the most significant identification features or attributes of the screen. The employees examine the links presented on the screen and click on the one that will most probably lead to the log-in screen. Usually, the fifth link is selected (SOX Compliance Rules) in accordance with the formulated goal. The next screen that opens up is shown in Figure 6.4.

After looking at this screen, employees realize that they are on the wrong pass and start asking each other for help. It is usually just a waste of time, and the employees decide to go back to the e-mail. They scroll down the email and discover an attachment at the bottom of it (Figure 6.2). Most of the employees click on the "Click here" link because it stands out by having two attention attracting features: it is in blue and underlined. There is also a motivational factor that plays a role here: a desire to find the link and to achieve the goal of the activity (a cognitive, imaginative component of a desired future result). Clicking on the link "Click here" eventually brings the employees to the log-in screen they were looking for (Figure 6.6).

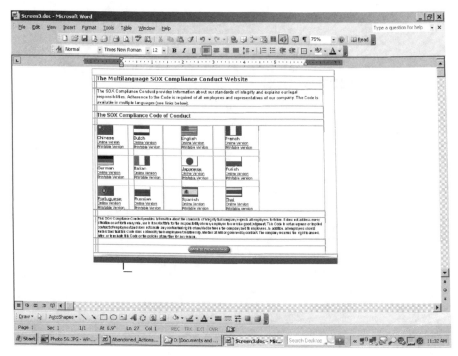

FIGURE 6.4
Screen after clicking at the second link.

To log in, they need to key in the log-in ID and PIN. Now, they are confused again, as they do not have this information. Thus, the employees either go back to the attachment (Figure 6.5) or they start asking each other for help again. Paragraph 2 of the attachment has the log-in information. Employees do not know in advance that this information will be required at the next step. Moreover, even if they did know, they still have to keep this information in working memory until they open the log-in screen and key it in. Keeping information in working memory is an undesirable factor for any task. The unnecessary mnemonic actions could be avoided if this information was presented on the log-in screen itself.

Thus, there are several factors that lead to the inefficient strategies of the task performance. One of them is the low subjective significance of the task, which results in a negative emotional-motivational state. The second factor is the inefficient task design, in which the identification features of the task are not adequate with the strategies of activity. The workers did not use the strategy of carefully reading the screen's content. The developers of the e-mail had two broad goals (objective requirements) in mind: familiarizi themselves with the presented information and completing the questi naire. However, the objectively defined goals (requirements) may not coin

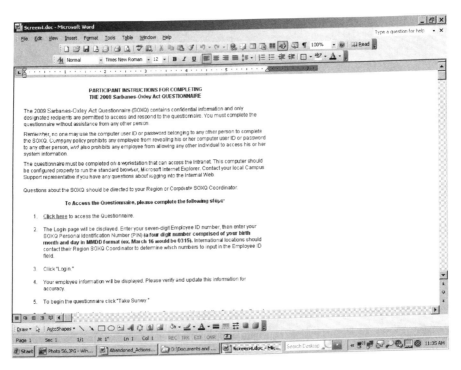

FIGURE 6.5
Attachment.

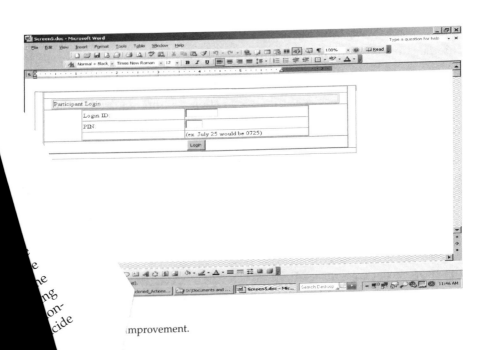

mprovement.

with subjectively formulated or accepted goals (Bedny and Karwowski 2007). Employees rejected the first goal and formulated their own general goal in order to complete the questionnaire as quickly as possible and return to their primary duty. In accordance with this goal, they developed an adequate strategy of task performance.

Only after several failures to find a log-in screen do the employees gradually change their strategy and start paying more and more attention to the content. However, this does not guarantee success, because the structure of the task does not provide an understanding of its identification features necessary to carry out the actions. A qualitative analysis of this Web questionnaire task demonstrates that the authors of the e-mail had completely unsubstantiated the mental picture of how it is going to be used by the employees. Moreover, no feedback information on how this e-mail has been used was given to the sender for the future improvement of similar tasks.

There might be a deceptive impression that this task can be easily evaluated by experimental methods. In the experimental conditions, it is very difficult to simulate emotional-motivational state that arises during this task performance. It is clear that in the experimental conditions, the subjects realize that they are under observation and their attitude to the task totally changes. In the experimental conditions, an e-mail task would be seen as a test of the subject's ability to perform it. As a result, the strategies of the task performance might change completely. Thus, the experimental methods would not be useful. Analytical methods such as algorithmic and quantitative analysis are a much better fit for studying this type of task.

6.4 General Principles of Algorithmic Description

Our qualitative analysis showed that the difficulties in performing the task under consideration are hard to predict at the development stage. Such difficulties are very common and can be avoided by using some analytical procedures. According to SSAT, algorithmic analysis follows the qualitative stage. It includes subdivision of the activity into qualitatively distinct psychological units and determines their logical organization and sequence. A number of methods can be used for qualitative analysis (Bedny and Karwowski 2007). We used objectively logical analysis and some elements of functional analysis, in which activity is considered as a self-regulative system and main units are functional blocks or functional mechanisms such as goal, sense (significance), and motivation etc (for more details see Chapter 9 of this book). For algorithmic analysis of activity, the main units of analysis are cognitive and behavioral actions and members of the algorithm. These units of analysis distinguish a human algorithm from other types of algorithms. The description of these units of analysis is given in Chapter 1. Algorithmic description allows us to describe the preferable strategies of activity during task performance (G. Bedny and Karwowski 2003).

Every activity varies; this is especially true for the performance of computer-based tasks. To analyze human activity, one does not need to describe all the possible strategies of task performance but rather consider the most typical ones and ignore minor variations in task performance. The most representative and important strategy of task performance should be selected. The combined probability of considered strategies should be equal to one. This means that the selected strategies absorb all the other strategies. Such an approach is justified because the designed activity only approaches the real task performance and describes it with a certain level of approximation as does any model in the design process. The algorithmic description of a task in SSAT is a stage in the morphological activity analysis.

In algorithmic analysis, a member of an algorithm is an element of activity, which integrates one or several actions united by a higher-order goal. It consists of interdependent homogeneous actions (only motor, only perceptual, only thinking, decision-making actions, and so on), which are integrated by a particular goal into a holistic system. Subjectively, a member of an algorithm is perceived as a component of activity, which has a logical completeness. The feeling of a logical completeness of actions included in one member of an algorithm is usually associated with a subgoal of the task that integrates several actions. The number of actions that can be included in one member of an algorithm also depends on the capacity of short-term memory. When actions are performed simultaneously or require keeping their order in working memory, each member of an algorithm is limited to one to four homogeneous actions due to limitations on the capacity of working memory. Classification of cognitive and behavioral actions and methods of their extractions can be found in the studies by G. Bedny and Karwowski (2003; 2007), Bedny et al. (2008), and Bedny, Karwowski, and Chebykin (2006). Members of the algorithm are classified according to psychological principles and are designated by special symbols. The following is a short synopsis of the algorithmic description.

Each member of the algorithm is designated by a special symbol; operators are designated by the symbol O and logical conditions by the symbol l. All the operators that are involved in receiving information are categorized as afferent operators (O^α). If an operator describes extracting information from long-term memory, the symbol is O^μ. The symbol $O^{\mu w}$ is associated with keeping information in working memory, and the symbol O^ε is associated with the motor components of activity also known as efferent operators. O^ε cannot include any cognitive actions, O^α can include only perceptual actions, and operator O^μ can include only mnemonic actions. Thus, the superscript determines what kind of actions can be included in a particular member of the algorithm.

For the description of decision-making and thinking actions, the logical conditions and the members of the algorithm that describe thinking actions are used. Logical condition is a special kind of thinking action that requires selection of one out of at least two existing alternatives. Logical conditions can also be used when there is only one possible choice that involves emotional and volitional components, such as a decision: "I have to act" or "I don't have to act." There is a conflict of motives involved in this kind of decision. Such elements of activity can be considered logical conditions although there is only one possible alternative to act. There might be a need to make a decision without a clear understanding of any other alternatives. In such cases, it is an intellectual conflict rather than an emotional conflict. Logical conditions are used in the activity's algorithmic description. Logical conditions are designated in algorithmic description by the letter "l" or "L."

There are also thinking actions that usually precede decision-making actions. Such actions include categorizing objects by certain criteria, making inferences, taking actions that reveal the functional relationship between the elements of the situation, and so on. Members of an algorithm that integrate thinking actions are designated by the symbol O^{th}.

We should distinguish between the two types of algorithms: deterministic and probabilistic. In deterministic algorithms, the logical conditions have only two alternatives that are designated by "l" and have two possible values—0 or 1. The symbol "l" for a logical condition always includes an associated arrow with a number on the top that corresponds to the number of logical conditions. For example, logical condition l_1 is associated with a number on the top of the arrow $\uparrow^1$. A downward arrow with the same number is presented in front of another member of the algorithm to which the arrow makes reference, $\downarrow^1$. Thus, the syntax is based on a semantic denotation of a system of errors and superscript numbers. An upward pointing of the logical state of simple logical conditions, "l" when, "l" = 1, requires skipping the members of the algorithm until the next appearance of the superscript number with a downward arrow (e.g., $\uparrow^1\downarrow^1$). Thus, the operation with the downward arrow with the same superscript number is the next to be executed.

In some cases, logical conditions can be a combination of simpler ones. These simple logical conditions are connected through "and," "or," "if–then," and so on logical rules. Complex logical conditions are designated by a capital "L." They are particularly important in diagnostic tasks in which it is necessary to determine whether a particular phenomenon belongs to a certain category. The algorithmic description is more complicated when the same type of members of the algorithm follow one another: several afferent members of an algorithm such as O_1^α, O_2^α, and O_3^α or several efferent members of algorithm such as O_1^ε, O_2^ε, and O_3^ε. When dividing the above fragment of the task into members of an algorithm of the same type, we should follow the above-described rules of qualitative analysis (Bedny and Karwowski 2007).

In a probabilistic algorithm, logical conditions may have two or more outputs with probability varying from 0 to 1 (Bedny and Karwowski 2003). For example, suppose we have a logical condition with three outputs and distinct probabilities of occurrence. In such a case, a logical condition would be designated as $l_1 \uparrow^{1\,(1-3)}$, which symbolizes three potential values with three versions of output: $\uparrow^{1(1)}$, $\uparrow^{1(2)}$, $\uparrow^{1(3)}$. Each output has its own probability: 0.2, 0.3, and 0.5.

Probabilities of the output should be considered in studying variables such as the probability of performance of various members of algorithms, strategies of performance, calculation of performance time of the algorithm or its components, analysis of errors, and evaluation of task complexity. There might be a need to utilize an always-false logical condition, defined by the symbol ω. This logical condition is introduced only to make it easier to write the algorithm. It does not describe any real operations performed by the subject and always defaults to the next member of the algorithm, as indicated by the arrow included in its specification.

An arrow designates the logic of transition from one member of an algorithm to the other. Thus, an algorithm exhibits all the possible actions and their logical organization and constitutes a description of human performance. It describes the activity of a subject in terms of actions through which the subject attains the goal of activity. An algorithm can be presented in a tabular form or as a formula. We recommend combining these two forms of description, which significantly simplifies an algorithmic description of the task. On the left side of Table 6.1, there is a column where the above-described symbols are placed. It is a symbolic description of the algorithm or its formula. On the right side, there is a verbal description of the members of the algorithm.

Human activity is extremely variable. Therefore, we describe only the most representative strategies of task performance. Knowing the likelihood of each strategy, we can evaluate the efficiency of the activity in general. The algorithmic description of activity is just one of the ways of activity description, which are only models of real activity during task performance. Real activity approaches these models when it is performed repeatedly. Without identifying preferred strategies of activity, we cannot develop adequate models of activity. Qualitative analysis precedes algorithmic analysis.

While analyzing the task performance, simultaneously performed actions and combined actions should be identified. Each of the simultaneously performed actions has its own goal. Actions can be performed simultaneously (the subject moves two arms simultaneously and grasps the objects), in sequence, or independently. A combined action, such as "move arm with an object and turn it in a vertical position at the same time," has one goal. If a subject moves an arm with an object in the given position and then turns it, this would be one action with two sequentially performed motions that have one goal. Cognitive actions can be also combined (in thinking actions, a subject uses external visual information or information from memory). Without combining these cognitive processes, such actions are impossible to perform. Members of an algorithm that include combined cognitive actions are designated as $O^{\alpha\mu}$, $O^{\mu th}$, or l^μ (successive perceptual action combined with mnemonic action, thinking action performed based on information from working memory, and logical condition that includes decision-making based on information extracted from memory). Cognitive actions that have their own goal very seldom can be performed simultaneously (Bedny and Karwowski 2007).

6.5 Algorithmic Description of the Web-Survey Task

Let us algorithmically describe the considered task. This task is multivariant and cannot be treated as deterministic. The objective of this study was not only to analyze the task as a whole but also to analyze each member of the algorithm as a quasisystem. In order to do that, we have to find answers to the following questions: What actions does the algorithm consist of? What operations are parts of the action? What is the duration of each action or operation? What is the level of concentration of attention during the execution of a particular member of an algorithm (according to developed five levels scale in SSAT)? Can the actions or operations be performed simultaneously or sequentially, depending on the level of concentration of attention? What is the probability of each member of the algorithm or its components? And so on. These data are used to develop the time structure of the activity and quantitative measures of the complexity of task performance, to evaluate the reliability of task performance, and so on (Bedny, Karwowski, and Bedny 2010).

If the same screen is used differently, the next line of the table with the new member of the algorithm is used. Every time the same screen is used the same way, a compact description of the algorithm is possible. In such cases the next logical condition brings us to the place in the table where the first interaction with the screen was described, which means that the same member of the algorithm is used repeatedly. Despite the fact that the compact algorithm has fewer lines, the number of activity steps remains the same.

In the algorithmic analysis of the task performance, the probability of each logical condition output should be determined. We can obtain such data through observation or an experiment. In the study of operator's performance, it was shown that the experts can remember or estimate the likelihood of the events (Kirwan 1994). Hence, probability judgment can be used in our example where we assess the likelihood of only two possible outputs within precision of one or two digits after the decimal point. The experts' assessment of the probability of events is accurate enough for such approximation. We have used a modified table of transition from subjective judgment about the frequency of events to the quantitative data suggested by Zarakovsky and Pavlov (1987). In Table 6.1 we present an algorithmic description of the considered task.

In Table 6.1, the third column on the right presents a description of the cognitive and motor actions as per SSAT methods (Bedny and Karwowski 2007). Members of the algorithm and actions performed by the individuals are classified in a standardized manner. Each member of an algorithm usually includes from one to four motor or cognitive actions that are integrated by a high-order goal. The fourth column in Table 6.1 shows the duration of actions. For the algorithmic description of the activity as shown in Table 6.1, we use various units of analysis. In the first and third columns, there are psychological units of analysis, and the technological units of analysis are in the second

TABLE 6.1

Algorithmic Description of E-Mail-Distributed Task Performance

Members of Algorithm (Symbolic Description)	Description of Members of Algorithm	Classification of Actions	Time (in seconds)
O_1^α	Browse initial email.	Successive perceptual actions (separate actions are not considered).	6
$\omega_1 \uparrow_{\omega 1}$	Always false logical condition ($p = .1$).	No actions	0
O_2^ε	Move the mouse to the first link and click to open screen 2 ($p = .8$). (Only this strategy would be considered further.)	Motor action (mouse movement and click)	1.2
O_3^α	Search for questionnaire on the opened screen (Figure 6.3) (sequential examination of five choices, see subalgorithm description Table 2).	Five successive perceptual actions; each action grasps single meaningful output of verbal expression ($*O_g^\alpha, *O_1^\alpha, *O_3^\alpha, *O_5^\alpha, *O_7^\alpha, *O_9^\alpha$)	2.3
$l_1 \uparrow^{(1-5)}$	Choose one (see subalgorithm description Table 2).	Decision-making actions at a sensory-perceptual level (also performed five times; *l1; *l2; *l3; *l4; *l5)	0.3
O_4^ε	Click on one selected choice (SOX Compliance Conduct).	Motor action (mouse movement and click)	1.2
O_5^α	Check to open the screen (Figure 6.4). (There are two most preferable strategies: select language or read the text on top of the screen.)	Successive perceptual actions during reading (separate actions does not considered) or four successive perceptual actions	5 or 4. 0/4.5
$l_2 \uparrow^2$	Select English language screen ($p = .8$) or go back to the e-mail (Figure 6.1; $p = .2$).	Decision-making actions at a sensory-perceptual level	0.3
O_6^ε	Move mouse to required position and click to select language (English).	Motor action (mouse movement and click)	1.2
O_7^α	Browse the text (no login is found; there is only description of procedure).	Successive perceptual actions (separate actions are not considered)	3
$O_8^{\alpha th}$	There is no questionnaire and, hence "I have to go back to e-mail screen."	Deducing action	0.35

$\overset{2}{\downarrow} O_9^\epsilon$	Go back to e-mail in Figure 6.1 ($p = .2$)	Motor action (move mouse and click)	1.2
O_{10}^α	Look for another link in the email	Thinking action based on visual information	0.35
$\downarrow_{\omega 1} O_{11}^\epsilon$	Scroll down to the bottom of the e-mail and go to O_{12}^α.	Motor action (mouse movement click and hold; this member of algorithm partly overlaps with O_{12}^α)	2.5
O_{12}^α	Notice the attachment on the bottom of the screen.	Simultaneous perceptual action (this member of algorithm partly overlaps with O_{11}^α)	0.35
O_{13}^ϵ	Double-click on the attachment.	Motor action (mouse movement and double click)	0.25
O_{14}^α	Browse and detect the link (see Figure 6.5).	Successive perceptual actions (separate actions are not considered).	4
$O_{15}^{\mu th}$	Here is the link to the questionnaire, hence "I need to click here."	Thinking action based on visual information	0.35
O_{16}^ϵ	Click on the link.	Motor action (move mouse and click)	1.2
O_{17}^α	Examine the log-in Figure 6.6.	Three simultaneous perceptual actions ($0.3\ s + 0.25\ s + 0.3\ s$)	0.85
$O_{18}^{\mu th}$	What login ID and PIN should be?	Explorative thinking actions based on information extracted from long-term memory	0.7
$O_{19}^{\mu th}$	I need instruction for the ID and PIN, hence I go back to attachment of Figure 6.5.	Logical thinking actions	0.5
O_{20}^ϵ	Go back to e-mail attachment Figure 6.5.	Motor action (move mouse and click)	1.2
O_{21}^{qu}	Read the instruction how to log in and keep the information in the working memory.	Successive perceptual actions combined with mnemonic action (separate actions are not considered).	11
O_{22}^{qu}	Click on the link to the questionnaire and keep the information in the working memory.	Motor action (mouse movement and double click) combined with mnemonic action	1.2

(Continued)

TABLE 6.1 (*Continued*)

Algorithmic Description of E-Mail-Distributed Task Performance

Members of Algorithm (Symbolic Description)	Description of Members of Algorithm	Classification of Actions	Time (in seconds)
O_{23}^{μ}	Recall instruction for login ID.	Mnemonic action	0.3
$O_{24}^{e\mu}$	Recall and type the employee number and hit Tab.	Combined action (executive action performed based on information extracted from memory; typing and recalling performed at the same time)	3.4
O_{25}^{μ}	Recall instruction for PIN.	Mnemonic action	0.3
$O_{26}^{e\mu}$	Recall and type employee birthdates and hit Login.	Combined action (executive action performed based on information extracted from memory; typing and recalling performed at the same time)	3
	Read and answer the questions from questionnaire.		

column. This combined utilization of units of analysis allows us to provide the most accurate description of the activity and its clear interpretation.

When performing computer-based tasks, users often need to read or print text. There are two strategies for reading. The first strategy involves careful reading of the entire text that is similar to reading a book (detail reading). The second involves browsing the text when certain parts of text are scanned (the user looks at a piece of text, captures the main idea of the fragment, and moves to the next piece of text). Browsing allows a user to get quickly acquainted with the main idea of a particular fragment of the text. After browsing the whole text, the user can return to some parts of the text for more details or move to another screen. Browsing the text usually requires simultaneous perceptual actions. Before describing algorithmically the process of reading, we should understand the relationship between these two strategies.

If necessary, segmentation of the reading process into separate verbal actions can be conducted. Each verbal action correlates with an elemental phrase, each of which represents a separate meaningful unit of information. Separate verbal motor actions determine meaningful typing units such as typing a word or several interdependent words that convey one meaning. This is in line with Vygotsky's (1978) idea about meaning to be one of the basic units of activity analysis. Similarly, cognitive psychology shows a possibility of segmentation of verbal activity in verbal protocol analysis (Bainbridge and Sanderson 1991). According to SSAT (Bedny and Karwowski 2007), reading and typing normally require the third level of concentration of attention and have the third-level complexity of activity. Sometimes, processing the most subjectively significant units of text can be transferred into higher level of complexity. If the text is relatively homogeneous, there is no need to extract separate actions. The time for processing the text can be measured according to the above-described strategy.

In this study, we measured the duration of reading or typing the text and evaluated their complexity based on the level of concentration of attention. The time performance for some simple cognitive and behavioral actions was taken from previous studies. For example, the duration of decision-making action at the sensory-perceptual level "if–then" was estimated at 0.350 milliseconds (Zarakovsky and Pavlov 1987; Lomov 1982). For simple motor actions such as mouse click and mouse movements (average), 0.1 seconds and 1.1 seconds were assigned, respectively (see Card et al. 1983; Kieras 1994). Other data was obtained from chronometrical study. In more complicated cases, eye movement registration can be used (Bedny et al. 2008).

Let us consider some members of the algorithm (Table 6.1). The first step of task performance suggests that employees read all information on screen 1, which has two pages and attachment on the bottom of the second page. Our experimental data demonstrated that reading the whole first page takes 31 seconds on average (60 seconds for two pages). However, because of the low motivation to perform this task, employees are looking for the shortest way for task performance. Instead of reading the e-mail, they select a browsing strategy to quickly find the link for the questionnaire. It takes an average of 6 seconds to find out the first link available. After the performance of O_1^α, there is always a false logical condition ω_1 that simply designates that there is a possibility of either O_2^ε or O_{11}^ε. This is the way to demonstrate the transition from one member of the algorithm to another that does not involve any action. The probability of O_2^ε per the experts' analysis is 0.9 (the first basic strategy) and the probability of O_{11}^ε is 0.1 (the second basic strategy).

Now, we will describe the most probable strategy. After browsing the e-mail, employees quickly move their mouse to the first link they spot on the page and click. According to the experts' analysis, the probability of this strategy is 0.9. This step of task performance is described by two members of the algorithm (O_1^α and O_2^ε). This choice of action greatly affects the strategy of task performance as follows: O_3^α and l_1 are complex members of the algorithm, and their detailed description can be found in Table 6.2.

TABLE 6.2

Subalgorithm of E-Mail-Distributed Task Performance

Members of the Algorithm (Symbolic Description)	Description of Members of the Algorithm
$*O_1^\alpha$	Read the first header.
$\overset{1}{*l_1\uparrow}$	Can the first header lead to the questions? If "YES" click on it. If "NO" look at the second header.
O_2^ε	Click on the first header (this step is omitted).
$\overset{1}{\downarrow}*O_3^\alpha$	Look at the second header and read it.
$\overset{2}{*l_2\uparrow}$	Can this header lead to the questions? If "YES" click on it. If "NO" look at the second header.
O_4^ε	Click on the second header (this step is omitted).
$\overset{2}{\downarrow}*O_5^\alpha$	Look at the third header and perform according to the formula (see formula in the next row of the table)[a]. $$\left(*O_5^\alpha * l_3 \overset{3}{\uparrow} * O_6^\varepsilon \overset{3}{\downarrow} * O_7^\alpha * l_4 \overset{4}{\uparrow} * O_8^\varepsilon \overset{4}{\downarrow} * O_9^\alpha * l_5 \overset{5}{\uparrow} * O_{10}^\varepsilon \overset{5}{\downarrow} \omega \uparrow \right)^a$$
$*O_{10}^\varepsilon$	Click on last selected choice "SOX Compliance Conduct" (the title of the last choice makes it possible to assume that one can find a link to the questionnaire).

[a] This is the second part of the subalgorithm described as a formula. Afferent operator O_5^α in Table 6.2 in the left column (vertical formula) and $*O_5^\alpha$ in parenthesis (horizontal formula) is the same member of the algorithm.

The * symbol is used to distinguish between the members of the algorithm in Tables 6.1 and Table 6.2. Operators $*O_9^\alpha$ and $*O_{10}^\varepsilon$ in Table 6.2 are the same as operators O_3^α and O_4^ε in Table 6.1.

Table 6.2 includes $*O_{10}^{\varepsilon}$, which is the same as O_4^{ε} in Table 6.1. The stars next to the operators (in Table 6.2) are used to distinguish the symbols in Tables 6.1 and 6.2. This is an example of the decomposition of activity and how the algorithmic description of task performance helps in its understanding. In Table 6.2 we use two methods of symbolic description of the algorithm. The left-hand side of the table depicts the algorithm as a vertical formula. In parentheses we continue the description of the second part of this algorithm as a horizontal formula. This formula demonstrates the sequence of the execution, going from left to right.

Before we describe activity algorithmically, we must discuss possible strategies of activity performance. They can be discovered at the preliminary stage of design based on logical or functional analysis of activity. In the last case, activity is considered to be a self-regulative system. When we attempt to describe the above subalgorithm of activity, it is sufficient to use logical analysis. Table 6.2 describes the most preferable activity strategy when users open the screen depicted in Figure 6.4. Employees tried to find the questionnaire or at least the link to it on the opened screen. Thus, for employees who formulate such a goal, the most significant and the most attractive are the five underlined headers (the identification features of this Web page, Figure 6.4). Employees assume that the five underlined headers can bring them to the next Web page, and one of these links would open up the desired questionnaire. Therefore, we can infer that in our example employees scan five headers. The common habit is to browse information from left to right. Out of five headers, only the last can be considered to have some relation to the goal of the subtask (to find the link to the questionnaire). Therefore, this sequence of actions should be described in our subalgorithm. Thus, after discovering the headers, they start reading them from left to right. They guess that only the last choice implies the desired result and decide to click on it. This subalgorithm is described in Table 6.2.

For a description of further steps of task performance, we need to refer again to Table 6.1. The analysis of the performed subalgorithm shows that employees use six successive perceptual actions. Each perceptual action grasps a single meaningful output or verbal expression. The following members of the algorithm describe these perceptual functions: $(*O_g^{\alpha}, *O_1^{\alpha}, *O_3^{\alpha}, *O_5^{\alpha}, *O_7^{\alpha}, *O_9^{\alpha})$. The first member of the algorithm $(*O_g^{\alpha})$ is for receiving the information "there are headers." The other symbols are for the possible perceptual steps associated with the five possible perceptual analyses of the headers. The decision-making action at the sensory-perceptual level is selected from the following five possible actions: $(*l_1, *l_2, *l_3, *l_4, *l_5)$. Logical conditions $*l_1 - *l_4$ have a value of 1. Only the last logical condition $*l_5$ has a value of 0. Hence, only the operator $*O_{10}$ is executed. Only the last efferent operator $*O_{10}$ has been selected because its title, in contrast to four others, gives hope to find the link to the questionnaire and the employees click on the last header (simple motor action).

The logical condition with an arrow means that some steps might be omitted. The first member of the second part of the algorithm is $*O_5^\alpha$. It symbolizes the perceptual action "look at the third header." If the value of the logical condition $*l_3$ is 0, then O_6^ε is performed. If $*l_3$ is 1, then O_6^ε is bypassed and O_7^α is performed, and so on. The first four logical conditions have a value of 1, and only the last one has a value of 0. Employees click on the last header. There are always false logical conditions such as ω after each operator, that is, after making a selection, the switch is made back to Table 6.1. Thus, employees perform six perceptual actions, one decision-making action, and one motor action, "click." The last member of the algorithm is always a false logical condition ω that signals the switch back to the main algorithm. The description of the algorithm by horizontal formula is suitable for simple tasks; otherwise it becomes difficult to read.

Figure 6.4 depicts the Web page that has "SOX Compliance Rules" in multiple languages because the company has branch offices in multiple countries. When employees open this screen, they once again hope to find the link to the questionnaire. The information presented on the screen does not meet their expectations, and they explore the screen to find out its purpose. There are several preferable strategies for using this Web site. The first preferable strategy is to find the corresponding language and click on it, without reading the information on the top of the screen; the second is to read the information at the top of the screen and decide to go back to the e-mail (Figure 6.1); the third strategy is to carefully read the information on the screen. This last strategy contradicts the dominating motivational state and the goal associated with it; the second is most likely to be rejected because employees already made a number of steps, and it is now subjectively quite risky to go back without checking the next screen. They select the first one: "I made multiple steps before getting to this screen; it seems quite risky to go back without checking the next screen with the appropriate language. Maybe I have to return to this screen later on, so I better check what's on the next screen." The employees open up the screen, quickly look at it, find no link to the questionnaire, and therefore go back to the e-mail. The probabilities of each of the two main strategies as per the experts' estimate are presented in Table 6.1. The given probabilistic characteristics are based on a qualitative analysis of possible strategies of the task performance. Analysis of the algorithm shows that there are two possible ways to go back to the email (Table 6.1, member of algorithm O_9^ε). When calculating the time of task performance, we need to consider these two possible strategies.

When employees get back to the e-mail (Figure 6.1), their goal is different because now they want to explore it in detail to find the link to the questionnaire; they eventually discover the attachment at the bottom of the e-mail (Figure 6.2) and open it (Figure 6.5). They browse the attachment, detect the link, and decide to click on it. The next screen opens up (Figure 6.6).

Employees examine this screen and find out that they do not have the necessary information to log in. This is an unexpected situation that activates users' explorative activity, and they attempt to find related information in their long-term memory. When the employees realize that they do not possess such information, they have to return to the attachment (Table 6.1; $O_{18}^{\mu th}$ and $O_{19}^{\mu th}$). This analysis demonstrates that employees perform a number of unnecessary actions that do not lead them to the desired goal, that is, "to log in," and such actions are the examples of abandoned actions. When employees return to the attachment, they carefully read the log-in instructions and keep the information in working memory until the completion of the log-in (Table 6.1).

Once the algorithmic description is done, each member of the algorithm should be considered beginning with qualitative analysis. The possibility of errors and ways of their elimination should be analyzed. The performance time for each member of the algorithm, the level of concentration of attention (according to the five-point scale) during its performance, the fraction of time spent on the perception of information, the fraction of time spent on decision making, and other factors should be determined. The actions that can be performed simultaneously or sequentially, work complexity, reliability of performance, and so on can be identified.

An analysis of Tables 6.1 and 6.2 demonstrates that the designer of this e-mail did not envision the real execution strategies of this task and was never informed about the issues with this e-mail. The designer of this task assumed that the employees would open the e-mail and take the following steps: (1) read the text carefully; (2) open the attachment; (3) read the attachment carefully; (4) memorize the information about the ID and Pin; (5) click on the link; and (6) if any additional information is required, use the link to the Web page for SOX compliance information.

In fact, a completely different strategy has been observed (see algorithmic description in Tables 6.1 and 6.2). It is not a rational one, but the users are looking for the shortest way to open the questionnaire. They click on the first link they see and find themselves in a maze of Web pages they were not looking for. Such strategy contains a lot of unnecessary steps that we categorize as abandoned actions. Some of these actions require memorization and keeping information in the working memory, they deviate attention from main elements on the screen, and produce irritation and a negative emotional-motivational state.

We have described the first main strategy, which has some variations as has been shown in Table 6.1. There is also the second main strategy of this task performance, whose probability is 0.1. For practical purposes, the second strategy can be neglected, but its consideration might be useful for understanding the applied method of study. Thus, we will consider it briefly. The false logical condition ω_1 always demonstrates a possibility of going directly to O_{11}^{ε} to perform the second part of the algorithm. If we want to determine the time performance of the considered task, we should consider two basic strategies. The first basic strategy has a probability of 0.9, and the second one has a probability of 0.1. Thus, task performance time should be determined using the following formula:

$$T = \sum p_i t_i \tag{6.1}$$

where p_i is the probability of the ith member of algorithm and t_i is the performance time of ith member of the algorithm.

Then, the task performance time equals the following:

$$T = p_1 \times T_{ST1} + p_2 \times T_{ST2} \tag{6.2}$$

where T_{ST1} and T_{ST2} are the performance times of the first and second basic strategies, respectively, p_1 and p_2 are the probabilities of performing these strategies.

In our case, the task performance time equals

$$T = .9 \times T_{ST1} + .1 \times T_{ST2} \tag{6.3}$$

This example demonstrates that, if necessary, we can determine all the required quantitative characteristics of the task under consideration based not only on the analysis of separate strategies but also on the analysis of all possible strategies of the task performance. For simplification, we limit further discussion to the quantitative measures of abandoned actions for the first basic strategy.

At the last stage of task performance (Table 6.1, $O_{23}^{\mu th} - O_{26}^{\varepsilon \mu}$), the employee reads the instructions in the attachment (Figure 6.5), keeps them in the working memory, goes to the login screen (Figure 6.6), recalls her or his employee number, birth date, and the rule of transforming the birth date into the required format, and then keys the information in. This is a sequential mnemonic action that includes several mental operations.

6.6 Analysis of Abandoned Actions

Here we consider the abandoned actions for the first main strategy. The most common abandoned actions for a particular task should be presented in algorithmic analysis. Depending on the purpose of the study, the performance time of the abandoned actions can be considered or ignored. An increase in the number of unnecessary explorative actions not only complicates and prolongs the performance of the computer-based tasks but also has a negative effect from a technical viewpoint. The more switchings from one screen to the other are performed, the more time delays are associated with such switchings.

We will consider the quantitative evaluation of the efficiency of the computer-based task performance using the measures we have developed for the assessment of abandoned actions. The efficiency measures are derived from the evaluation of the task performance time and the duration of various types of abandoned actions. The following symbols will be used: A is the general time for all abandoned actions, A^α is the time required for afferent abandoned actions, A^ε is the time required for efferent abandoned actions, A^l is the time required for abandoned logical conditions, and A^μ is the time required for abandoned actions associated with keeping the information in the working memory.

The first step in the assessment of task performance efficiency is the evaluation of the task performance time (Equation 6.1). The time taken for all described abandoned actions can be determined as follows:

$$A = A^\alpha + A^\varepsilon + A^l + A^\mu + A^{th}; \quad A^\alpha = \sum p_i^\alpha \times t_i^\alpha; \quad A^\varepsilon = \sum p_b^\varepsilon \times t_b^\varepsilon; \quad A^l = \sum p_r^l \times t_r^l$$

$$A^\mu = \sum p_j^\mu \times t_j^\mu; \quad \hat{A}^{th} = \sum p_k^{th} \times t_k^{th}$$

where p_i^α, p^ε, p^l, p_j^μ, and p^{th} are the probabilities of ith abandoned actions of the corresponding types and t_i^α, t^ε, t^l, t_j^μ, and t^{th} are the performance times of ith abandoned actions of the corresponding types. This time can be obtained based on existing studies or experimentally based on chronometrical analysis.

In the next step of the evaluation of the task performance efficiency, the following measures of efficiency are used:

$$\hat{A} = \frac{A}{T}$$

$$\hat{A}^\alpha = \frac{A^\alpha}{T}$$

$$\hat{A}^\varepsilon = \frac{A^\varepsilon}{T}$$

$$\hat{A}^l = \frac{A^l}{T}$$

$$\hat{A}^\mu = \frac{A^\mu}{T}$$

$$\hat{A}^{th} = \frac{A^{th}}{T}$$

These are the measures of various types of abandoned actions. The less the value of these measures, the more efficient is the performance. In any given study, all these measures or just the most suitable ones should be used. When mnemonic, thinking, or decision-making actions are performed with motor components of activity simultaneously, each type of activity is accounted for separately in the calculations. Let us determine the efficiency measures for the above-considered task.

The first step is to define the task performance time T and the performance times for the various types of abandoned actions. The performance time for each member of the algorithm is presented in the fourth column of Table 6.1. Using Table 6.2, we have evaluated the performance time of the subalgorithm in which the users performed an average of six successive perceptual actions in sequence, one decision-making action at sensory-perceptual level, and one simplest motor action (click and release). Table 6.1 gives the performance times of O_3^α and l_1.

This example shows that if chronometrical measurements are used, the preferred strategies of activity performance should be identified first (Bedny and Karwowski 2007). During chronometrical analysis, it is useful to collect subjective assessments of the performance pace by subjects involved in the experimental study. For measurements of duration of automatic mental operations, the methods developed in cognitive psychology can be used (Sternberg 1969, 1975). In more complex tasks, eye movement registration may be necessary (Bedny et al. 2008).

When assessing the performance time of the first strategy, the outcomes of various options for logical conditions should be taken into account because they define the logic of the transfer to the individual members of the algorithm. Table 6.2 demonstrates the method of using members of the algorithm O_3^α and l_1. The logical condition l_2 has two outputs. The probability of the first output is 0.8, and the second is 0.2. These two lead to the following two possible strategies of performance (see l_2):

Strategy 1 that has the probability of 0.8 can be described as

$$STR(1) = p_1 \times \left(tl_2 + tO_6^\varepsilon + tO_7^\alpha + tO_8^{\alpha th} + tO_9^\varepsilon \right)$$

where tl_2 is performance time of logical conditions. Strategy 2 that has the probability of 0.2 can be described as

$$STR(2) = p_2 \times \left(tl_2 + tO_9^\varepsilon \right)$$

The performance time for the considered fragment of the algorithm is

$$MT = STR(1) + STR(2)$$

where MT is the mean time of these two strategies:

$$\text{MT} = p_1 \times \left(tl_2 + tO_6^\varepsilon + tO_7^\alpha + tO_8^{\alpha th} + tO_9^\varepsilon \right) + p_2 \times \left(tl_2 + tO_9^\varepsilon \right)$$

In the above equation, the performance times for the corresponding members of the algorithm are given in parentheses.

By substituting symbols with the values, we get the following result:

$$\text{MT} = 0.8 \times \left(0.3 + 1.2 + 3 + 0.35 + 1.2 \right) + 0.2 \times \left(0.3 + 1.2 \right) = 5.14 \text{ seconds}$$

The probability of each member of the algorithm except $l_2 - O_8^{\alpha th}$ is 1. Summarizing the performance time of each member of the algorithm including MT (performance time of the fragment of algorithm from l_2 to O_9^ε) gives the task performance time (first main strategy):

$$T = 6 + 1.2 + 2.3 + 0.3 + 1.2 + 4.5 + 5.14 + 0.35 + 2.5 + 0.35 + 0.25$$
$$+ 4 + 0.35 + 1.2 + 0.85 + 0.7 + 0.5 + 1.2 + 11 + 1.2 + 7 = 49.84 \text{ seconds}$$

In this calculation, we ignored the second main strategy that has a low probability ($p = .1$).

At the next step, the performance time of all abandoned actions that are included in the following members of the algorithm is calculated:

$$A = O_2^\varepsilon;\ O_3^\alpha;\ l_1;\ O_4^\varepsilon;\ O_5^\alpha;\ l_2;\ O_6^\varepsilon;\ O_7^\alpha;\ O_8^{\alpha th};\ O_9^\varepsilon;$$
$$O_{10}^\alpha;\ O_{15}^{\alpha th};\ O_{16}^\varepsilon;\ O_{17}^\alpha;\ O_{18}^{\mu th};\ O_{19}^{\mu th};\ O_{20}^\varepsilon$$

The combined time for the abandoned actions should be determined considering that some abandoned actions or mental operations are performed simultaneously. All members of the algorithm that have a combination of several qualitatively different superscripts such as $O^{\mu th}$, l^μ, $O^{\varepsilon th}$, or $O^{\varepsilon \mu}$ are examples of combined actions, where $O^{\mu th}$ is the combination of mnemonic and thinking actions or operations, l^μ is the decision-making action performed based on information extracted from memory and/or requires keeping information in the working memory, $O^{\varepsilon th}$ is the combination of executive or motor actions with thinking actions or operations, and $O^{\varepsilon \mu}$ is the combination of executive or motor actions with mnemonic actions or operations.

Thus, we will count the time when actions overlap separately for each type of these actions. The member of the algorithm $O_{19}^{\mu th}$ is an example of combining mnemonic and thinking actions. The total time for abandoned actions can be determined by summarizing their performance times. O_6^ε, O_7^α, and $O_8^{\alpha th}$

have a probability of 0.8. Hence, the time for all abandoned actions can be determined as follows:

$$A = 1.2 + 2.3 + 0.3 + 1.2 + 4.5 + 0.3 + 0.8 \times (1.2 + 3 + 0.35) + 1.2$$
$$+ 0.35 + 0.35 + 1.2 + 0.85 + 0.7 + 0.5 + 1.2 = 19.79 \text{ seconds}$$

Therefore, the fraction of time spent on abandoned actions in the whole task performance time is

$$\hat{A} = \frac{A}{T} = \frac{19.79}{49.84} = 0.4$$

Thus, the abandoned actions take approximately 40% of this task performance time.

We can also determine the fraction of perceptual actions, thinking, decision-making, and mnemonic abandoned actions in the total task performance time by calculating the times for various types of abandoned actions: A^α, A^ε, A^l, A^μ, A^{th}.

Let us determine the performance time of the motor components of activity. All motor members of the algorithm have a probability of 1, excluding O_6^ε, which has a probability of 0.8. The following efferent members of the algorithm are abandoned actions:

$$O_2^\varepsilon, \ O_4^\varepsilon, \ O_6^\varepsilon, \ O_9^\varepsilon, \ O_{16}^\varepsilon, \ O_{20}^\varepsilon$$

Therefore, the performance time for the abandoned efferent actions (motor) is

$$A^\varepsilon = 1.2 + 1.2 + 0.8 \times 1.2 + 1.2 + 1.2 + 1.2 = 10.95 \text{ seconds}$$

The following are the afferent abandoned actions (including reading): O_3^α, O_5^α, O_7^α, O_{10}^α, O_{17}^α.

The performance time of the afferent abandoned actions is determined as follows:

$$A^\alpha = 2.3 + 4.5 + 0.8 \times (3 + 0.35) + 0.85 = 10.33 \text{ seconds}$$

Sometimes, the reading time should be considered separately from the other types of afferent actions.

Two logical conditions l_1 and l_2 are also abandoned actions.

Hence, the performance time for decision-making abandoned actions is

$$A^l = 0.3 + 0.3 = 0.6 \text{ seconds}$$

All the members of the algorithm that include thinking actions are abandoned actions: $(O_8^{th}, O_{15}^{th}, O_{18}^{\mu th}, O_{19}^{\mu th})$.

The performance time for these actions is

$$A^{th} = 0.8 \times 0.35 + 0.35 + 0.7 + 0.5 = 1.83 \text{ seconds}$$

The members of the algorithm that include mnemonic abandoned actions are $O_{18}^{\mu th}, O_{19}^{\mu th}$, and their performance time is

$$A^{\mu} = 0.7 + 0.5 = 1.2 \text{ seconds}$$

The following are the coefficients for various types of abandoned actions:

$$\hat{A}^{\alpha} = A^{\alpha}/T; \; \hat{A}^{\varepsilon} = A^{\varepsilon}/T; \; \hat{A}^{l} = A^{l}/T; \; \hat{A}^{\mu} = A^{\mu}/T; \; \hat{A}^{th} = A^{th}/T$$

$$\hat{A}^{\alpha} = 0.207; \; \hat{A}^{\varepsilon} = 0.22; \; \hat{A}^{l} = 0.012; \; \hat{A}^{\mu} = 0.024; \; \hat{A}^{th} = 0.037$$

The sum of these five fractions is greater than the fraction of the combined abandoned actions because some of them are performed simultaneously. We don't consider the second strategy because it has a low probability. We calculate the above-described measures for only one basic strategy that has the probability of 0.9. However, if it is required, we can calculate these measures, taking into consideration two basic strategies—one with a probability of 0.9 and the other with a probability of 0.1. If these two main strategies are considered, the formula for the task performance time would be

$$T = 0.9 \times T_{ST1} + 0.1 \times T_{ST2}$$

Using the same principle, we can calculate measures that have more than two strategies. The analysis of the considered coefficients shows that some of them have insignificant value. The coefficients that describe the fraction of the abandoned actions associated with logical conditions and thinking can be calculated together as an integral factor because the logical conditions that describe the decision-making process are one of the stages of the thinking process. The members of an algorithm that includes mnemonic, decision-making, and thinking actions are critical points of any computer-based task, and therefore, specialists should pay particular attention to these members of the algorithm. The analysis of the members of the algorithm such as $O_{18}^{\mu th}$ and $O_{23}^{\varepsilon \mu}$ demonstrates that there is unnecessary load on the working memory. Some of these members include abandoned actions that should be eliminated. Moreover, even if the members of the algorithm do not include the abandoned actions, the strategies of their performance should be changed to reduce the temporal load on the working memory as much as possible.

In our example, to reduce the load on the working memory, information about the ID and password should be transferred to the appropriate screen as shown in Figure 6.7.

When on the log-in screen, employees should be able to type the required ID and password based on the visually presented instructions. This is particularly desirable because it is difficult to retain this information in the working memory during possible distractions, such as the need to answer phone calls. Keeping information in the working memory increases the likelihood that it can be forgotten, and the user has to return to the original screen that contains the information about the ID and password (performance of $O_{18}^{\mu th} - O_{23}^{eu}$ is repeated), and the number of abandoned actions increases. The members of the algorithm that includes thinking actions and/or logical conditions should also be reduced as much as possible. It is best for the user to perform the required actions based on simple perceptual information.

A quantitative analysis of the abandoned actions shows that they constitute a significant portion of the task, which means that a significant period of time is spent unproductively. Unproductive activity includes not only the motor but also the cognitive elements. An analysis of the abandoned actions indicates that the objectively given goal "to take the Web-survey questionnaire in accordance with the given e-mail" was subjectively reformulated into a goal "find

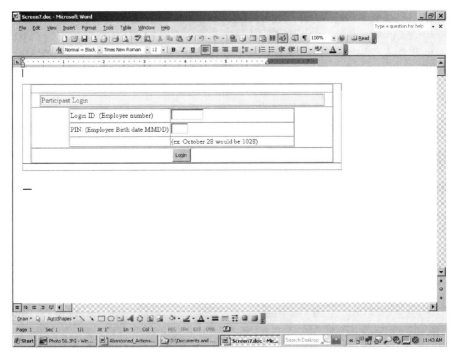

FIGURE 6.7
Log-in screen for the improved version.

where the log-in screen is." When employees have a very low level of motivation to perform the task, such reformulation of the goal is understandable.

By trial and error, the employees try to find a way to reach the goal. However, their actions are not blind trials and errors as described by Skinner (1969). Employees act based on the promoted hypotheses that are evaluated and adjusted during task performance. Cognitive and behavioral actions are interconnected and perform cognitive, executive, and evaluative functions. Abandoned explorative actions test the promoted hypothesis and perform these functions. The result of transformations on the screen is the source of information for further actions. This is an example of explorative activity during interaction with the computer. This type of activity is important in the computer-based tasks. If task performance is inefficient, then the computer-based task has not been efficiently designed. The coefficients for various types of abandoned actions are useful in the analysis of the abandoned actions and explorative activity in general.

Let us consider an algorithmic description of the optimized version of this task performance (Figure 6.8). Qualitative and algorithmic analyses of the task at hand help us develop the improved version of the task under consideration, which involves the use of a more efficient strategy. The most

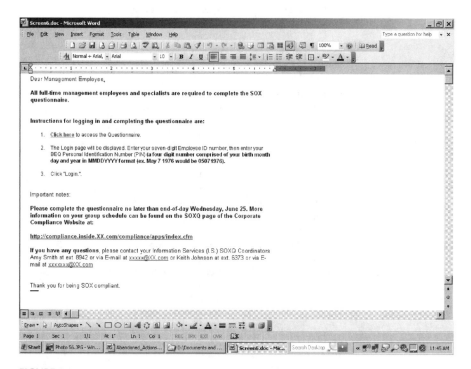

FIGURE 6.8
E-mail for the improved version of the Web-survey task.

important information is now presented directly on the Web page containing the questionnaire if the employees need to get additional information.

Table 6.3 shows that the algorithm of the task performance has changed. The number of the members of the algorithm is significantly reduced from 29 to only 5. Most of them are easy to perform, and their duration is reduced. Logical conditions are eliminated, that is, the task does not require the decision-making process. This implies that the initially considered version of the task was related to the algorithmic- or rule-based tasks, and the improved version belongs to the skill-based tasks.

TABLE 6.3

Algorithmic Description of E-Mail-Distributed Task Performance (Optimized Version)

Members of the Algorithm (Symbolic Description)	Description of Members of the Algorithm	Classification of the Actions	Time (in milliseconds)
O_1^α	Read email.	Successive perceptual actions (separate actions are not considered)	6
O_2^ε	Click on the link to the questionnaire.	Motor action (mouse movement and double-click)	1.2
O_3^α	Read login ID instruction on the screen (Figure 6.7).	Perceptual action	0.35
$O_4^{\varepsilon\mu}$	Recall and type employee number and hit Tab.	Combined action (executive action performed based on the information extracted from the memory; typing and recalling performed at the same time)	3
O_5^α	Read PIN instruction on the screen (Figure 6.7).	Perceptual action	0.35
$O_6^{\varepsilon\mu}$	Recall and type the employee date of birth and hit Login.	Combined action (executive action performed based on the information extracted from the memory; typing and recalling performed at the same time)	3
O_7^α	Read and answer the questions from the questionnaire		

Table 6.3 presents the algorithmic description of the optimized version of the task performance.

It is interesting to compare the performance strategy for the optimized algorithm when the employee performs $O_3^\alpha - O_6^{\varepsilon\mu}$ with $O_{23}^{\mu th} - O_{26}^{\varepsilon\mu}$ in the real algorithm. These members of the algorithm describe the same stage of the task performance. In the optimized algorithm, employees read the instructions for ID, recall the employee number, key it in, then read instruction for PIN, recall the birth date, transform the birth date into the required format, and key it in. In the optimized version of the task performance, this stage includes two afferent operators: O_3^α and O_5^α. Only two members of the algorithm $O_4^{\varepsilon\mu}$ and $O_6^{\varepsilon\mu}$ are performed based on the information from the working memory, whereas in the real algorithm all four members of the algorithm $O_{23}^{\mu th} - O_{26}^{\varepsilon\mu}$ are performed based on information extracted from memory. If this type of action is repeated many times, it can lead to premature fatigue because the regulation of actions based on the information extracted from memory is more complicated than the regulation of similar actions based on the visual information.

The performance time of the optimized algorithm is $T = 13.9$ seconds. The times for mnemonic and executive (behavioral) actions ($T\mu$ and $T\varepsilon$) are the same and equal 6 seconds, and they are performed simultaneously. The time for afferent operators is $T\alpha = 6.7$ seconds. The abandoned actions are eliminated, and other positive structural changes in activity performance can be observed.

We have considered the explorative activity efficiency. In SSAT, there are also methods of quantitative evaluation of complexity and reliability of computer-based tasks performance. In the considered Web-survey task, the workload on the working memory and thinking mechanism could be assessed. Procedures of complexity evaluation are described in the study by Bedny and Karwowski (2007) in which 20 measures of complexity are suggested. The reliability evaluation method is described in work of Bedny, Karwowski, and Bedny (2010).

6.7 Conclusion

Users are faced with a lot of computer-based tasks that do not have strictly specified strategies of performance and often do not know in advance how the task should be performed. Thus, they explore to find a way to perform the task. The more uncertain the task, the more complicated is the explorative activity. In conditions when uncertainty about possible strategies is significantly increased, explorative activity can approach a chaotic mode. The concept of abandoned actions and a method for their analysis are important for the development of methods for studying explorative activity in performance of computer-based tasks.

There are various methods for reducing the number of abandoned actions. You can change the way you describe the task, change the location of its elements on the screen, develop special instructions, and so on. In the task considered here, access to the questionnaire should be facilitated. First, the employees should be able to open the questionnaire, and then if necessary, turn to the instructions or explanations. The most important information should be placed at the top of the screen. It is also necessary to eliminate, if possible, situations when users have to keep intermittent information in their working memory. Decision making and thinking actions also have to be reduced as much as possible.

Usually, total elimination of the explorative action is not achievable, but it can be reduced. It is necessary to reduce the actions that include the processes of thinking, decision-making, and memory workload. It is desirable that actions are performed based on perceptual information, and this is recommended for production tasks. In contrast, for entertainment tasks, we often need to introduce explorative actions, which include decision-making, thinking, and so on. This leads to the conclusion that there is a need for quantitative analysis of explorative activity.

It is not only cognitive and behavioral components that are important in computer-based task performance. Emotional-motivational factors can also totally change the strategies of task performance. Low-level motivation or even negative motivation in performance of the described task makes it difficult to study this task experimentally. If subjects are aware that they are involved in an experiment, their motivation could be different, and therefore subjects would implement a different strategy for task performance. This suggests that in such cases, analytical methods of study are necessary.

The emotional and motivational components of activity are critical factors in the goal-formation process. Unity of cognitive and emotional aspects of activity is the base for the goal-formation and goal-acceptance processes. The factor of significance is the mechanism that links the cognitive and emotional-motivational components of activity. There are no unmotivated goals. The goal and motives create a vector that gives the activity its direction and meaning.

In addition to the final goal of the task, there are intermediate goals, which are formed at different stages of the task performance. In relation to the overall goal of the task, such goals have a subordinate role. The formation of subgoals depends on the specificity of the task, the methods of its presentation, and the user's individual features. The formation of intermediate goals is a critical component in the formation of activity strategies.

According to SSAT, there are different stages of motivational processes. In this study, a conflict between process-related and goal-related stages of motivation has been observed. On one hand, the employees want to complete the task as quickly as possible because this is a mandatory task. This provides the positive goal-related stage of motivation. On the other hand, this task is perceived as insignificant and distractive in relation to the main duties. As a result, process-related motivation is very low, and this factor influences the cognitive process strategies. In most cases, employees look for the link to the questionnaire they can click on. Because the e-mail is not effectively designed and is stipulated by negative emotional and motivational factors, the cognitive strategies lead to a sharp increase in unwanted explorative actions.

Every computer-based task should be considered in terms of how important it is to the user. It is necessary to determine not only the significance of the task as a whole but also the significance of its individual elements to determine the possible cognitive strategies to be selected by the users. The most significant elements of the task should be identified and optimized because these elements have a decisive importance for the preferred strategies. Therefore, we cannot agree with the opinion of some experts in the field of HCI that the emotional-motivational factor is not important in the production environment. The emotional-motivational components of an activity are closely connected to the cognitive components of the activity. Humans are not simple logical devices and always have a predilection to the events, situation, or information. In the activity approach, the design process always includes analysis of the emotional-motivational component of an activity in both production and entertainment environments. When designing HCI tasks, one should strive to reduce abandoned actions in order to increase productivity and mitigate negative emotional impact.

References

Bainbridge, L., and P. Sanderson. 1991. Verbal protocol. In *Evaluation of Human Performance. A Practical Ergonomics Methodology*, eds. J. R. Wilson, and E. N. Corlett, 2nd ed, 169–201. Boca Raton, FL: Taylor & Francis.

Bedny, G. Z., and W. Karwowski. 2003. A systemic-structural activity approach to the design of human-computer interaction tasks. *Int J Hum Comput Interact* 2:235–60.

Bedny, G. Z., and W. Karwowski. 2006. The self-regulation concept of motivation at work. *Theor Issues Ergon Sci* 7(4):413–36.

Bedny, G. Z., and W. Karwowski. 2007. *A Systemic-Structural Theory of Activity. Applications to Human Performance and Work Design*. Boca Raton, FL: Taylor & Francis.

Bedny, G. Z., W. Karwowski, and T. Sengupta. 2008. Application of systemic-structural theory of activity in the development of predictive models of user performance. *Int J Hum Comput Interact* 24(3):239–74.

Bedny, G., and D. Meister. 1997. *The Russian Theory of Activity: Current Application to Design and Learning*. Mahwah, NJ: Lawrence Erlbaum Associates.

Bedny, I. S., W. Karwowski, and A. Ya Chebykin. 2006. *Systemic-Structural Analysis of HCI Tasks and Reliability Assessment, Triennial IEA 2006 Sixteenth World Congress on Ergonomics*, e-book. Maastricht, the Netherlands.

Bedny, I. S., W. Karwowski, and G. Z. Bedny. 2010. A method of human reliability assessment based on systemic-structural activity theory. *Int J Hum Comput Interact* 26(4):377–402.

Card, S., T. D. E. Moran, and P. G. Polson. 1983. *The Psychology of Human-Computer Interaction*. Hillsdale, NJ: Erlbaum.

Kieras, D. E. 1994. Towards a practical GOMS model methodology for user interface design. In *Handbook of Human-Computer Interaction*, ed. M. Helander, 135–56. Amsterdam: North Holland.

Kirwan, B. 1994. *A Guide to Practical Human Reliability Assessment*. London: Taylor & Francis.

Lomov, B. F., ed. 1982. *Handbook of Engineering Psychology*. Moscow: Manufacturing Publishers.

Sengupta, T., and I. S. Bedny. 2008. Study of computer based tasks during skill acquisition process, *Second International Conference on Applied Ergonomics jointly with Eleventh International Conference on Human aspects of Advanced Manufacturing* (July 14–17, 2008).

Skinner, B. 1969. *Contingencies of Reinforcement: A Theoretical Analysis*. New York: Appleton Century Crofts.

Sternberg, S. 1969. Memory-scanning, mental processes revealed by reaction-time experiments. *Am Sci* 57:421–57.

Sternberg, S. 1975. Memory scanning: New findings and current controversies. *Q J Exp Psychol* 27:1–32.

Vygotsky, L. S. 1978. Mind in Society. *The Development of Higher Psychological Processes*. Cambridge, MA: Harvard University Press.

Zarakovsky, G. M., and V. V. Pavlov. 1987. *Laws of Functioning Man-Machine Systems*. Moscow: Soviet Radio.

Section III

Evaluation of Computer Users' Psychophysiological Functional State

7

Optimization of Human–Computer Interaction by Adjusting Psychophysiological State of the Operator

A. M. Karpoukhina

CONTENTS

7.1 Introduction

Optimization of the interaction between humans and computers is important for increasing the human operator's effectiveness when a computer is the major means of work. Until recently, all efforts were focused on the study of the ergonomic link between the user and the computer. However, professionals overlook another important problem of human–computer interaction (HCI), which includes the analysis and regulation of the psychophysiological functional state of the user or simply the psychophysiological state (PPS) during interaction with the computer (Bedny and Meister 1997; Genkin and Medvedev 1973; Leonova 1984). A functional state is formed based on the interaction between the physiological, psychological, and behavioral subsystems of an organism. This state is dynamic and changes over time. During the study of the functional state of an organism, scientists pay attention to the registration of changes in the state of the organism rather than to the regulation of the functional state in the work process. This is because no method has been developed for the regulation of the human functional state in the work process, which is particularly important in the study of computer-based work.

In ergonomic and engineering or psychological design, the following characteristics of humans are considered: anatomic, neurodynamic and psychological properties, attention parameters (scope, switching capacity), and characteristics of visual and auditory analyzers. At the same time, any effort to improve other links of the system with humans have been "pushed to the back burner." However, it remains a well-established fact that in the process of human activity, the PPS is subject to significant changes in both the level of individual characteristics and overall labor effectiveness (Bedny and Seglin 1999; Genkin and Medvedev 1973). These changes can produce not only a positive effect during warming up but also an unfavorable one when developing negative praxis states such as stress, fatigue, or monotony.

Such a "technical" bias was caused not by the underestimation of the significance of the human link but rather by the difficulties in implementing a controlled PPS adjustment in the process of real interfacing between humans and computers. The major requirement of adjusting the PPS effect is that besides being adjusted and normalized, it does not constitute a diversion, that is, it does not interfere with the operator's activity. In essence, this requirement means that the influence thus produced is below the sensitivity threshold.

The objective of our research was to develop a method for adjusting the PPS to increase the efficiency and potency of the interface between humans and computers, which is done in the following three stages:

1. Searching for and selecting the adjusting PPS effect and developing a method that would meet health and safety requirements and not interfere with the operator's activity
2. Verifying the method's efficiency by experiments
3. Working out practical recommendations for the practical use of this method in the future

7.2 Method of Adjusting the Operator's Psychophysiological State

When developing the PPS adjustment method, the most important issue was to determine the "target" of the impact. From the systematic approach viewpoint (Anokhin 1969; Karpoukhina 1985; Kokun 2004), the human PPS is considered a dynamic functional system with a multilayer hierarchical structure in which all the elements interoperate to achieve the desired future result, ensuring motivated human operations. Such an interface is based on self-regulation, which can be performed at both physiological and psychological levels (Anokhin 1969; Bedny and Karwowski 2007).

The human PPS hierarchy includes social–psychological, psychic, physiological, biochemical, biophysical, and bioenergetic levels. Of these, the bioenergetic level is the fundamental level that underlies the functioning of all superior PPS system hierarchy levels. The lack of bioenergy affects the implementation of the internal informational processes and, consequently, the functioning of all superior hierarchical structure levels.

Today, to a certain degree, the metaphoric notions of "psychic energy" (Freud 1975) and activation concept, with its representations of one-dimensional homogenous series of functional statuses varying only in the nonspecific activation level, are replaced with the concept of "specific structure of nonspecific activation" (Aladjanova et al. 1979) and the belief that the specifics of psychic energy are determined not only by common physics of interface between the objects but also by particular information patterns through which the exterior object and the body's internal state are represented in the nervous system.

Modern neurophysiology interprets the informational aspects of physiological and psychological processes not only in terms of conducting nerve impulses. This is explained not only by the existing limitations of the electrophysiological approach but also apparently by the objective extent to which the human mind can go in using the bioelectrical impulse code of psychic processes (Bekhtereva and Budzen 1974). The experimental data and, based on such data, the theoretical concepts regarding the role of the brain's neuro-electric states and biological fields (Livanov and Ananiev 1955; Schuldt 1976) are also critically important. Such a double-process mechanism (Pribram 1975) for characterizing the activity of the nervous system and the dialectic principle of matter based on the wave-corpuscle duality lead to a belief that the bioenergy of informational processes should not be considered only from the perspective of the electrical capacity required to distribute nerve impulses and therefore to ensure the body's regulatory processes. Rather, the energy consumption with regard to the regulation processes in the human body should also be considered in this case. Another important aspect is the role of energy in the functioning of nerve tissue, that is, the dynamics of the brain's electrical state, up to the level of neuronal synapse and slow gradual potentials and subsynaptic processes, represents the substratum of data coding and conversion.

This realization lends a fresh view of the mechanism behind the psychic processes and states—thinking, perception, memory, attention, emotions— by binding this mechanism with conversions of the brain's electrical states, that is, with conversions of its field and not only with the generation of nerve impulses.

This leads us to perceive an extremely important role of the PPS bioenergetic level, whose nonspecific activation entails—based on autoregulation— specific positive changes in other levels and in the PPS system as a whole. The

PPS bioenergetic level of all the nervous system bioenergy was selected as a "target" of the regulating action. The immediate impact had to be directed at biologically active points (BAPs) or acupuncture points that are featured by the elevated concentration of nerve elements. The next issue, in order of importance, was to consider an influencing factor, that is, to select the effect that could enrich the human bioenergy. The choice fell on low-intensity laser radiation with the wavelength of the red (0.6328 micrometers) or infrared (0.8–0.94 micrometers) range (soft laser). Such radiation is generated by He–Ne or semiconductor lasers. Their radiation capacity does not exceed and—even more—by far falls short of the threshold established in the Laser Operation Safety Rules. In general, the laser beam is characterized by a nonionizing, normalizing, and nondoping effect on the target.

The normalizing effect was achieved by bringing the parameters of the targeted functions to the normal levels under the impact of soft laser regardless of the initial hypofunction or hyperfunction. It proves that the laser does not produce any irritant or stimulant effect but rather generates a surge of energy that feeds the autoregulation processes.

The nondoping effect was corroborated through a series of research in which the effectiveness of the operators in defining (selecting) the signals was compared between the experimental (previously subjected to laser radiation) and control (placebo) groups. These groups operated in three modes: (1) stress-free, (2) under stress, and (3) under time pressure. In the first mode, no significant variance was manifested itself for the groups, but in the stress mode and especially in the time-pressure mode, the work parameters of the operators exposed to the laser were undoubtedly higher. Apparently, in the stress-free mode, the initial energy level was sufficient, and no energy "surplus" was required, whereas in the stressful modes, the operators in the experimental group received an energy boost that allowed them to significantly exceed the results achieved by their counterparts in the control group who did not undergo any laser treatment (Karpoukhina and Jeng 2003). Remarkably, in none of the multiple experiments was the so-called negative effect stage observed.

The impact of a low-intensity laser on BAP is called laserpuncture, which is a largely successful technique to treat a wide range of pathological disturbances. The use of laserpuncture to adjust the PPS of a healthy human was justified by an analysis of published data (see "Bibliographical Index" in Karpoukhina 1985) and a 30-year track record of experimental research and practical applications performed by numerous psychophysiologists at the Institute of Psychology, Ukrainian Academy of Pedagogic Sciences that involved tens of thousands of subjects (including schoolchildren and students, teachers, sportsmen, cosmonauts, pilots and parachutists, metro train and truck drivers, divers and scuba divers, radar operators, and precision engineering operators).

Let us briefly review the major stages of multilayer laser influence on the human body to better understand how the impact of lasers on BAP is translated into an improved PPS and overall human body efficiency. We will discuss red and ultrared lasers.

In essence, a laser beam is coherent polarized light. It is a flux of energy quanta—photons—that is accepted by live cells, particularly neurons; it is absorbed mostly by subcellular organelles such as mitochondria, also known as the cell's "power stations," resulting in the stereoconformation of mitochondrial membrane molecules, thus determining the direction of energy metabolism: either to conserve or to spend energy (Skulachev 1969; Skulachev and Kozlov 1977). These authors have also proven that if a portion of a nerve fiber or axon is exposed to an He–Ne laser, the membrane potential immediately increases throughout the entire neuron cell membrane so that the neuron is ready to generate and transmit nerve impulses and, consequently, implement the information and regulatory processes.

Rusakov (1988) in his research on the impact of lasers on neuromuscular and neuroneuronal synapses proved that they contributes to a much faster restoration of acetylcholine bubbles in presynaptic plaque (that at the previous transfer of nervous impulse was poured into a synaptic cleft) and that such synapses increase and transfer acetylcholine bubbles into the active area of the presynaptic membrane. This is evidence for the decrease in the time of synapsis refractoriness and the possible increase in the frequency of nerve impulse synaptic transmission, that is, improved nervous regulation.

On the physiological level, the following statistically significant improvements were found to occur in the cardiovascular system parameters under the influence of laserpuncture: normalization of pulse and arterial blood pressure, restoration of the ECG T-wave decreased under the influence of fatigue, and a lower cardiovascular system (CVS) tension index estimated by the ECG Baevsky method (Karpoukhina and Kaliberdin 1981; Karpoukhina et al. 1989). It was also proven that laserpuncture leads to the amelioration of the muscular system and motility. Of particular interest is the opposite direction of changes that the muscle tone assumes as the human body tenses or relaxes: after laserpuncture, the muscle tone measured at the command "Strain the muscle" significantly increases, whereas the tone of relaxed muscle abates, which leads to a larger lumen in blood microvessels and favors the reduction processes inside the muscle. This is further evidence of the laser acting not as an irritant but rather as a mechanism of autoregulation (Karpoukhina and Ignatenko 1985).

On the PPS neurodynamic level, the influence of laserpuncture manifests itself in a shorter time of simple and complex (with choice) sensorimotor reaction and a higher liability and efficiency of the brain (Chaichenko et al. 1981). Under the influence of lasers, the brain activation rate (index) calculated by electroencephalogram increases.

On the level of psychic processes, the lasers' impact on a human body results in higher parameters of mnemic and intellectual processes such as attention span and precision; indices of immediate, short-term, and long-term memory; and cogitation parameters (Karpoukhina et al. 1982).

Such a multilayer mechanism of laserpuncture adjustment brought about the optimization of the PPS system as a whole, which correspondingly led to a significant growth of efficiency and reliability of training and occupational activity; increase in overall proficiency, quality, precision, and speed of work performance; and higher learning capabilities.

Finally, laserpuncture meets the major requirements applicable to the human PPS adjustment before or in the process of subjects' activity:

- Not only due to the lack of pain or damaging effects as it might happen if subjected to acupuncture or electropuncture, but also due to the lack of any sensations, laserpuncture is not an interfering factor and does not distract subjects from the tasks they may pursue.
- The laser can be administered without an assistant and can be administered single-handedly.
- Laserpuncture can be automated.
- It is possible to install special hardware to run the PPS laserpuncture adjustment at the operator's workplace.

7.3 Experimental Research on the Effect Produced by the Psychophysiological State Laserpuncture Adjustment on Better Interface between Humans and Computers

To measure the effect produced by PPS laserpuncture adjustment on the efficiency of the HCI, an experiment that meets the requirements listed below was conducted to ensure the required precision of the experiment and authenticity of the results obtained:

- *Representative choice of subjects*: This allows us to apply the statistical analysis of data acquired by using student's *t* test and measure their authenticity by applying significant difference index *p* between the experimental and control groups. For this purpose, 224 subjects were selected to participate in this experiment.
- *Reliability of any findings*: We ensure that this is achieved based on the results and by conducting repetitive experiments with different subjects within two subsequent years (109 subjects and 115 subjects) and by comparing their dynamics.
- *Elimination of all the other factors*: This eliminates the influence of all the other factors except the one under research, that is, laserpuncture. It was achieved by selecting only male subjects within the same age group of 18–19 years, graduate students applying for postgraduate

studies in the Naval Academy who at the time of entry tests and occupational selection tests were staying in the summer camp; and thus all of them were in the same residential, nutrition, occupational, and environmental conditions.

- *High motivation of subjects to achieve the objective of the experiment*: This was obtained by informing the entrants that their computer test grades would be accounted for as part of their acceptance score.

All of the subjects were divided into three groups. Two experimental groups underwent laserpuncture: one group in uninterrupted radiation mode and the other one in a discrete-continuous frequency-modulated (110 Hz) mode because in some literature the discrete-continuous frequency-modulated mode is referred to as having better efficiency. For the subjects in the control group, simulated laserpuncture treatment was given. The subjects were not aware of which group they were in as long as the whole environment of laser-puncture procedures was similar across all the groups (affixing flexible light conductors on BAPs, measuring the duration of "exposure" by a stopwatch, illuminating the operating Ge–Ne laser). The only difference of the control group from the other two was that the beam was intercepted at the entry into light conductors by a "trap" that was kept invisible to the subjects.

Laserpuncture was carried out by a Ge–Ne laser beam with a wave-length of 0.6328 mkm and a density of energies on the border of the skin in the BAP area equal to 0.5–2 $\mu W/mm^2$. By using the flexible dissector light guides, the beam was broken into separate flows and brought to the skin in the BAP area. The skin was exposed immediately before working on the computer and steadily over the entire process. In the latter case, the light conductor was fastened by an elastic "bracelet" with a yellow tape in the form of a hollow tube positioned over the BAP area where the light conductor was inserted before getting in touch with the skin BAP. The BAP areas were selected at the subjects' shanks to prevent any interference with their working at the PC. The so-called generic BAPs and those that were of importance to the PPS adjustment were exposed. The duration of exposure was 30 seconds, 1 and 2 minutes before work (indicated in parentheses in the following list of BAPs).

The BAP points used were indicated by Arabic numerals and the merid-ians were designated by Roman numerals in accordance with the recog-nized international classification (Tabeeva 1980). The English classification was put in parentheses. Before work, the applied BAPs were symmetrical 4 II (4 LI), located in the depression between the thumb and the index finger at the dorsal parts of hands (2 minutes); symmetrical auricular BAPs AT7 located in the area of the antitragus (1 minute); BAP 20 XIII (20 GV) located at the top of the head (1 minute); and 10 extrameridional points 86 PC at the fingertips (30-second parallel simultaneous impact). In the process of the operators' work, laserpuncture was administered by an experimen-tal multichannel laser stimulator MLS 4/24 to symmetrical BAPs 36 III

(36 St) located below the knee at the outside of the participants' shanks and symmetrical BAPs 6 IV (6 SP) located above the inner ankle bone at the inner surface of the participants' feet.

The program of the experiment was based on a case of HCI with the operator having an objective to determine a set of "target" coordinates. The simulator of the operator's objective was downloaded into the PC. The software used was developed by the Experimental Designer Bureau at Taganrog Radio-Technical Institute. Its objective was the following: The sign matrix was shown in the display (in our experiment 6 × 5), one of which, picked at random, started to "blink" for 1–2 seconds, and then vanished, only to reappear after awhile. The complete signal transformation period (T) was 4 seconds. The subject had to determine the coordinates of the vanishing sign and enter them into the computer, first the line number followed by the column number. The number of signals was 120. The subjects had to operate in an uninterrupted mode. Each twentieth result was processed automatically by the software, which allowed tracing of the dynamics of the results broken down into six stages. The HCI performance measurements were used to judge the efficiency of the HCI. At the same time, though the objective was agreed by the subjects because they had no experience working with the software, the stage-by-stage results can be interpreted as the dynamics of grasping the new types of software.

The subject's activity was evaluated by the following 10 measurements:

1. Number of correct answers ($X1$ by lines and $X2$ by columns)
2. Number of false answers ($X3$ by lines and $X4$ by columns)
3. Number of omissions ($X5$)
4. Time of correct line determination ($T1$)
5. Time of correct column determination ($T2$)
6. Efficiency $E = (X1 + X2)/(X1 + X2 + X3 + X4 + X5)$
7. Final score FS $= E [1 - (T1 + T2)/2T3]$ where $T3 = 4$ seconds
8. Entropy in its informational aspect as a measure of reliability
9. Performance (conventional value)
10. Load under established parameters of the operator's task

The analysis of the results of subjects' activity showed a gradual improvement of measurements across all the groups, which implied the training process. However, in the experimental groups (with laserpuncture), there was greater improvement when compared with the control group. The vacillation between the stage-by-stage measurements and, in particular, individual measurements can be explained by the systematic nature of structure of the performance measurement, reconfiguration of this structure, and change in the weight of this measurement in terms of global performance. This explains a more pronounced improvement of integral measurements such as final score and efficiency. The group average data obtained during the determination of target coordinates for both the experimental and control groups are shown in Table 7.1.

It is evident that in the first stage, the measurements of the experimental group exceed those exhibited by the control group, which can be explained by the effect produced by the laserpuncture before the test operation. The immediate effect of the laserpuncture was also achieved in our early experiments conducted to test the extent of the laser's effect on the coefficient of cortical arousal. This coefficient increased right after the exposure.

Due to the lack of homogeneity between the groups (according to the Parzen estimation of evaluation of frequency distribution for estimating the efficiency of activity result and the procedure of determining elementary statistics), which could have been caused by the wide range of individual properties across the subjects, the measurements of subsequent stages were normalized against those of the first stage that were taken for 100%.

The dynamics of the group's average final score measurements (the second-year experiment series) are represented in Figure 7.1, which shows higher results in the experimental groups when compared to the control group. The

TABLE 7.1

Group Average Measurements of Operators Having a Goal to Determine a Set of "Target" Coordinates

Index	Group	Human–Computer Interaction					
		1	2	3	4	5	6
Amount of right answers	EG	8.81	9.7	11.17	10.97	11.51	11.54
	C	7.43	9.46	9.17	10.76	10.65	11.43
Effectiveness	EG	0.54	0.64	0.68	0.66	0.7	0.69
	C	0.53	0.61	0.61	0.66	1.57	0.67
Amount of gaps	EG	3.03	2.21	1.67	1.87	1.57	1.44
	C	3.5	2.17	1.91	1.89	1.33	1.52
Final score (total assessment)	EG	0.4	0.46	0.49	0.48	0.52	0.5
	C	0.36	0.44	0.44	0.47	0.47	0.49
Amount of wrong answers	EG	8.49	9.29	7.21	7.03	6.68	6.75
	C	8.93	8.37	8.5	7.35	8.41	7.13
Entropy	EG	0.79	0.87	0.84	0.83	0.82	0.82
	C	0.8	0.85	0.83	0.84	0.83	0.81
Time of correct determination of line	EG	1.97	1.87	1.8	1.76	1.77	1.73
	C	1.99	1.81	1.88	1.86	1.74	1.76
Productivity	EG	0.28	0.18	0.25	0.27	0.3	0.31
	C	0.19	0.15	0.22	0.27	0.26	0.27

EG = experimental group with laser regulation (63 subjects); C = control group (46 subjects).

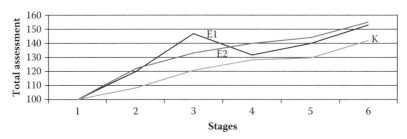

FIGURE 7.1
The dynamics of group average final scores.

statistic reliability (p) of intergroup deviations was 0.05 for groups E1 and K
and 0.02 for groups E2 and K.

Characteristically, the results of both years—in terms of concrete data and
their ratio in the experimental and control groups and in terms of measurement curves—are virtually identical, which strengthens the reliability of
major findings stating that the exposure to laserpuncture not only enhances
the performance but also accelerates the training process. In the third stage,
the performance exhibited by the operators in the experimental group
became statistically significant as opposed to the results shown by their
counterparts in the control group.

The question we wanted to answer was whether the better results achieved
in the laserpuncture groups were indeed caused by raising the intensity of
the operators, that is, a more pronounced psychophysiological "value" of
their activity; in other words, whether the laser produced a doping effect. To
answer this question, over the entire process of the experiments, the subjects
were tested for cardiac rate and pulse by an earflap photoelectric sensor. In
addition, immediately before and after work, an electrocardiographic rhyth-
mogram was taken to register the variations between durations of 100 con-
secutive cardiac cycles, and a histogram of R-R intervals was created. Based
on the parameters of the histograms, the stress index was calculated using
the Baevsky (1980) formula. The shift of these PPS measurements under the
impact of the working load reflects the psychophysiological "value" of the
operators' activity.

The group average dynamics of operators' cardiac beats over the time of
working with computers is represented in Figure 7.2. These data are normalized in percentage against the background that was taken as 100%. In the
laserpuncture groups, the cardiac beat rates, and consequently the intensity,
are lower than those in the control groups.

Figure 7.3 shows the shift in the group average intensity index under the
influence of operating the computer. The extent of measurement shift in the
control group significantly exceeds the one exhibited in the experimental
groups. The statistical confidence of divergence between groups E1 and K
at level p was 0.01 and between E2 and K at level p was 0.05. These statistics

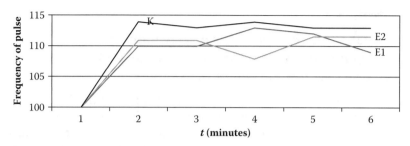

FIGURE 7.2
The group average dynamics of the operators' cardiac beats.

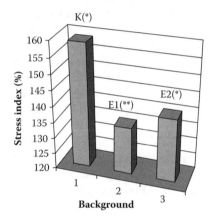

FIGURE 7.3
The shift in the group average intensity index. * means $p \leq 0.05$; ** means $p \leq 0.01$.

point to the fact that laserpuncture produces a regulating effect on the PPS and drives down the psychophysiological value. Apparently, this effect can be explained by the increase in the bioenergy of the PPS system by laserpuncture and thus ensuring the best mode of self-regulation in the system, which, finally, leads to enhancing the operator's overall performance.

By comparing Figures 7.1 through 7.3, we can observe a trend of frequency-modulated laser impact prevailing over that produced in uninterrupted mode.

In all cases in which the above modes and parameters of laserpuncture were used, neither any negative effects nor any psychological complaints were revealed.

We found that under the influence of PPS laserpuncture adjustment, we can increase the performance of the PC operator and simultaneously decrease the psychophysiological "value" of the operation, that is, increase the effect of HCI.

7.4 Approbation and Prospects of Developing Laserpuncture-Based Human–Computer Interface Performance Enhancement Methods

Today, we already have a positive experience of applying laserpuncture PPS adjustment to improve HCI performance. For instance, the method was tested in real conditions on radar operators who performed surveillance over spacecrafts. The operators' activity was characterized by a high degree of responsibility, which was equated to activity under stress conditions in which any fault in the operators' actions could entail the loss of complex technical systems and human lives. The operators worked 24-hour shifts. The laserpuncture equipment was located in the immediate vicinity of the operators' workplace, and the adjustment procedure was carried out every 4 hours. Structurally, the laserpuncture unit incorporated an He–Ne laser acting as a multichannel laser stimulator, which ensured a simultaneous impact on all chosen BAPs, that is, short delays in procedure implementation (1–2 minutes). The beam was conducted to the BAPs through flexible light conductors or dissectors that were fastened on a headset (five light conductors at each side), which allows a three-dimensional hinged adjustment relative to the operator's BAP. To make sure that the treatment was given simultaneously in 10 fingertip BAPs, special supports, left- and right-handed, were manufactured with sockets to accommodate the fingers, with light conductors inserted at the bottom.

Approbation of the method in such real conditions proved its capability for improving the HCI efficiency and reliability. Laserpuncture treatment resulted in better operational parameters and objective and subjective improvements in the operator's state and performance.

In the other case, the method was successfully used to test the artificial-intelligence equipment designed at the Kyiv Institute of Cybernetics, Ukrainian Academy of Sciences. This computer-based equipment was designed to assist the pilots of large sea and ocean ships in negotiating straits and channels under extreme weather conditions. In this case, the design and technical solutions behind the laserpuncture effect were different. The radiation sources were tiny semiconductor lasers (up to 0.125 cubic centimeters; uninterrupted action) inserted into the finger sockets. Five sockets were positioned to the left and to the right of the monitor screen and were conveniently inclined to accommodate the operator's fingertips. They allowed the operator to get a bioenergetic boost while reading data or making decisions that did not require any manual actions. The analysis of the operators' work showed that those treated with laserpuncture performed better and that rookie operators got the basics two to three times faster before achieving the required skill level.

We conclude that the positive effect of the laserpuncture adjustment on the speed and effectiveness of acquiring computer skills, safety, speed, and better possibility of implementing technical solutions to achieve the desired effect (without interfering with operators' main activity) offers vast opportunities for using this method to address issues related to the efficiency and reliability of HCI.

Taking into account today's headlong progress and the expansion of remote computer training methods, we can safely predict the use of PC-inbuilt laserpuncture PPS-adjustment tools in the near future to enhance HCI not only when developing the operators' computer workstations or technical training techniques but also when designing new types of computers for the broader public.

7.5 Conclusion

We proposed a method that increases the effectiveness of HCI by means of optimization of psychological and physiological conditions (PPC) of the user or operator. Assuming that PPC is a functional system, we proved that increasing the bioenergetic level of the system provides the regulation of the PPS of the user during computerized task performance. For this purpose, the laser acupuncture method (LA) of PPC regulation before and during interaction with a computer was developed. Based on theoretical and experiment studies, we proved that the suggested method does not produce any negative impact on humans. This is explained by the fact that we utilized a low-intensity laser that has a nondoping and nonionizing effect on the human body. Also, this method does not interfere in the operator's task performance or produce any undesirable feelings or sensations. The developed procedures last for a short period and provide a possibility for automatic influence on the operator. The mechanisms of LA regulatory influences on various psychological and physiological processes were explained.

The experiment with 224 participants proved the effectiveness of the method. As an example, we selected a task that requires the determination of target coordinates on the screen. Changes in the heart rate and cardiovascular system tension index, which reflect the psychological and physiological "price" of task performance, were measured. The comparison of experimental and control groups' data clearly demonstrated the positive effect of LA on operators' PPS and the improved efficiency of their performance. The suggested method demonstrates the possibility of increasing the user's efficiency and reliability in terms of the performance of the computerized task.

References

Aladjanova, N. A., T. V. Slotintseva, and E. D. Khomskaya. 1979. Relationship between voluntary attention and evoked potentials of brain. In *Neuropsychological Mechanisms of Attention*, ed. E. D. Khomskaya, 168–73. Moscow: Science.

Anokhin, P. K. 1969. Cybernetic and the integrative activity of the brain. In *A Handbook of Contemporary Soviet Psychology*, eds. M. Cole and I. Maltzaman, 830–57. New York: Basic Books, Inc.

Baevsky, R. M. 1980. *Forecasting Organism's States Between Normal and Pathological Level*. Moscow: Medicine.

Bedny, G. Z., and W. Karwowski. 2007. *A Systemic-Structural Theory of Activity. Applications to Human Performance and Work Design*. Boca Raton, FL: Taylor & Francis.

Bedny, G., and D. Meister. 1997. *The Russian Theory of Activity: Current Application to Design and Learning*. Mahwah, NJ: Lawrence Erlbaum Associates.

Bedny, G., and M. Seglin. 1999. Individual style of activity and adaptation to standard performance requirements. *Hum Perform* 12(1):59–78.

Bekhtereva, N. P., and P. V. Budzen. 1974. Neurophysiological organization of human psychic activity. In *Neurophysiological Mechanisms of Human Psychic Activity*, ed. N. Bekhtereva, 42–60. Leningrad: Nauka.

Chaichenko, I. A., I. A. Mitronova, and G. S. Gorkavenko. 1981. Impact of laserpuncture on functional state of central nervous system. In *Theory and Practice of Reflexotherapy*, ed. R. Durinian, 121–2. Moscow: Saratov.

Freud, S. 1975. *Beyond the Pleasure Principle*. New York: Norton.

Genkin, A. A., and V. I. Medvedev. 1973. *Forecasts of Psychophysiological States. Issues of Methodology and Algorithmization*. Moscow: Nauka.

Karpoukhina, A. M. 1985. *Control and Adjustment of Human States as Factor of Enhancing Labor Efficiency*. Kiev: Znanie.

Karpoukhina, A. M., and L. N. Ignatenko. 1985. Laserpuncture method of muscle tone adjustment. In *Physiological Fatigue and Rehabilitation Problems*, ed. P. Kostiuk, 182–5. Cherkassy.

Karpoukhina, A. M., I. A. Mitronova, and M-L. A. Chepa. 1989. Efficiency of laser-puncture in human functional adjustment when modeling blood circulation disturbances. In *Treatment-and-Prophylactics Work for Medical Institutions in Coal Industry*, ed. Y. Podshivalov and A. Medelianovsky, 86–91. Moscow: Central Research Institute of Coal Industry.

Karpoukhina, A. M., and O.-J. Jeng. 2003. Systems approach to psycho-physiological evaluation and regulation of the human state during performance. In *Proceedings of the Fifteenth Triennial Congress of the International Ergonomics Association and the Seventh Joint Conference of Ergonomics Society of Korea/Japan Ergonomics Society*, vol 6, 451–4. Seoul, Korea: International Ergonomics Association.

Karpoukhina, A. M., and G. V. Kaliberdin. 1981. Specifics of cardiovascular system reaction to impacts on BAP. In *Theory and Practice of Reflexotherapy. Medical, Biological and Physicotechnical Aspects*, ed. R. Durinian, 25–9. Moscow: Saratov.

Karpoukhina, A. M., N. K. Isayevskiy, and M-L. A. Chepa. 1982. Laserpuncture as factor of optimizing neurodynamic, perceptual and cognitive mnestic processes. In *Methods of Reflector Diagnostics, Therapy and Rehabilitation to Improve Health in Coal Industry*, ed. V. Pushkin, 19–38. Moscow: Central Research Institute of Coal Industry.

Kokun, O. M. 2004. *Optimization of Human Adaptation Capabilities: Psycho-Physiological Aspect of Supporting Activity*. Kiev: Millennium.

Leonova, A. B. 1984. *Psycho-Diagnostic of Functional States of Human*. Moscow: Moscow University Publishers.

Livanov, M. N., and V. M. Ananiev. 1955. Electrophysiological study of activity spacious distribution in cerebral cortex of rabbits. *USSR Physiol J* 41(4):461–9.

Pribram, K. 1975. *Brain Languages*. Moscow: Progress.

Schuldt, H. 1976. Condensations of field force in biologic systems. An interpretation of acupuncture lines. *Am J Acupunct* 4(4):344–8.

Skulachev, V. P. 1969. *Accumulation of Energy inside Cells*. Moscow: Nauka

Skulachev, V. P., and I. A. Kozlov. 1977. *Proton Adenosine Triphosphates. Molecular Biological Power Generators*. Moscow: Nauka.

Tabeeva, D. M. 1980. *Manual of Acupuncture*. Moscow: Medical Publishers.

8

Psychophysiological Analysis of Students' Functional States during Computer Training

D. A. Yakovetc and I. Bedny

CONTENTS

8.1 Introduction

The functional state (FS) of a human being is an integral component of the physiological and psychological processes of adjusting to ever-changing work conditions. The main purpose for studying FS is to determine how these changes impact productivity during training or work shifts (Bedny and Meister 1997). In activity theory, the following types of FS are distinguished: fatigue, monotony, stress, anxiety, tension, and functional comfort and discomfort.

Fatigue is a natural reaction of the human body, manifested as a decreased ability to function accompanied by subjective feelings of tiredness. There is mental fatigue, associated with the deterioration of cognitive functions (changes in perception, memory, attention, and thinking ability), and physical fatigue, associated with decreased energy resources and ability to perform physical work.

Emotional stress is a general reaction in terms of the cognitive functions and behavior of the subject, accompanied by a deterioration of mental abilities caused by a sudden emergence of very significant information and a lack of ways and time to resolve the situation. Monotony is the psychological state of the subject, characterized by a general decline in the level of activation, loss of ability to sustain attention and conscious control over the execution of actions, deterioration of short-term memory, reduction in the level of motivation to achieve the goal of activity, and an increasing desire to stop working. It is characterized by a low subjective value of content, the nature of the work, and underestimation of the significance of the subject's own efforts to continue the work, which is usually the result of a lack of variety, sameness, and a high repetition of relatively simple tasks.

Anxiety is a temporal state of the subject, characterized by unpleasant emotional coloring, when a person is inadvertently focused on allegedly inappropriate events due to a lack of understanding and inability to anticipate and manage future events.

Psychic tension is the temporal state of a subject, which can be divided into operational and emotional tension:

- Operational tension is determined by a combination of task difficulty and a lack of available time for task performance. The work intensity depends on the pace of performance. Studies demonstrate that work intensity is reported by workers who experience exposure to a high pace or time restrictions in completing their jobs.

- Emotional tension is caused by a combination of task difficulty and personal significance of work activity. It is reported by workers who experience a high level of responsibility in performing their required duties.

These two types of tension are closely interrelated, and under certain conditions, one type of tension might cause the other. For example, a lack of available time to perform an important task causes more tension than a lack of time for an insignificant task. Tension should be treated as the leading FS that accompanies any purposeful activity, because it defines the relationship between activity conditions and a subject's functional abilities (Chainova 2003).

The psychophysiological "price" of activity can be used as a quantitative measure of the tension level. The nonproductive (nonoptimum) form of tension is caused by labor conditions that are not adequate to the functional ability of a subject. It is characterized by increasing physical and psychological expenditures, inadequate activation of the functional physiological system, and high psychophysiological "price" of the activity. The nonproductive form of tension is manifested by a state of functional discomfort. It is experimentally proven that nonproductive forms of tension can appear in both easy and difficult work conditions. In the first case, functional resources are not used at the maximum level, and in the second case, functional resources do not ensure the required level of accomplishment. Experiments have shown that the first form of functional discomfort is psychophysiologically costlier than the second one.

The productive form of tension increases the level of functional comfort, which is understood as the optimum FS of a person involved in the activity. Here, the FS results in a positive attitude toward activity, provides adequate mobilization (activation) of subject's psychophysiological processes, slows down exhaustion, improves efficiency, and prolongs the ability to work without any detriment to health.

Functional comfort has psychological and psychophysiological components. Work satisfaction and an adequate attitude toward the goal of the activity, its duration, content, and conditions are the psychological components of FS. The leading component is the attitude toward the goal of the activity. The psychophysiological components of FS manifest themselves as the productive tension and mobilization (activation) of the functional physiological systems. Productive tension leads to the minimal psychophysiological price of the activity.

Experiments have shown that the FS in various types of activity is characterized by a moderate electroencephalogram rhythm and frequency, high efficiency of sensorimotor operations, rational distribution of bioelectric activity between leading and supporting muscles, high consistency of visual and motor systems, and stabilization of the heart rate.

In real work settings, when it is impossible to get an instrumental record of psychophysiological parameters to estimate the FS, functional and psychophysiological tests of several kinds are carried out at fixed time intervals during the entire working day. The obtained generalized indicator of the FS reflects the nature of fatigue development. However, an indirect measure of the relationship between the productivity and the functional comfort can be obtained. The less the psychophysiological characteristics change during work, the later fatigue is felt, and the better is the productivity.

The FS can also be used as a criterion for the optimization of any activity and for a comparative ergonomic evaluation of different versions of production processes (Zarakovsky and Kazakova 2004).

The functional comfort for any activity can be represented by four groups of indexes:

1. The first group of indexes describes the relationship between a subject and the substantial or material environment, its aesthetic properties, the functional purpose of using the biomechanics and anthropometric procedures, interview, and questionnaires.

2. The second group consists of indexes that indicate the activity's psychophysiological "price." This set of indexes is obtained by

assessing the FS using a variety of psychological and psychophysiological methods such as measuring the heart rate, blood pressure,
critical light flicker fusion frequency sensitivity, hand tremors, and
Lyusher's test.

3. The third group is the measure of performance efficiency.

4. The fourth group includes indexes that are based on data from questionnaires that characterize the subjects' attitude toward activity,
work processes, pace and intensity of work, its content, the results of
the activity, and the level of workload.

An integrated index of functional comfort is the result of integration of the
four above-described indexes.

Maximizing the efficiency of work activity with a minimal psychophysiological price is the main principle of FS analysis. However, when applying
this principle to the educational process, some additional factors should be
considered. For educational activity, the concept of efficiency differs from the
similar concept in ergonomics.

Chainova (1998) evaluated the FS of preschool children while they were
playing computer games. He discovered that the children's FS was better
when the games were not very simple. If the game was of optimal complexity, the FS was the best. By overcoming obstacles, children not only acquire
knowledge and skills but also get positive emotions and feelings of satisfaction, which are important components of the FS. The same data were
obtained in vocational training (Bedny and Karwowski 2007). Introducing
time standards in vocational training not only increased the productivity of
students, but also reduced fatigue. Time standard optimizes the vocational
school students' workloads and performs motivational functions. Similarly,
it is important to create an optimum workload while teaching technical disciplines utilizing computers.

8.2 Description of the Experiment

Indicators of the FS such as indexes of the central nervous system, cardiovascular, respiratory, endocrine, and other systems are the most important
indicators. FS is characterized by certain shifts in the main physiological
processes such as perception, attention, memory, thoughts, and emotions.
The study and diagnosis of the FS requires the use of objective instrumental
methods and subjective methods that utilize the subjects' opinion about their
own FS.

Most of the objective instrumental methods need a lot of time and complicated equipment. Moreover, these methods are very difficult to use in field

studies. They restrict the movements of the subjects and introduce other inconveniences that interfere with the regular production or educational process. The instrumental methods are more suitable for the laboratory studies, whereas in field studies, only the subjective method of FS analysis is often utilized.

Students' FS during computerized training have been analyzed using the method discussed by Zainutdinova and Yakovetc (2003) and Zarakovsky and Kazakova (2004). We've evaluated ergonomic features of two educational programs used in training of students in general technical disciplines (GTDs). The subjects were 54 full-time sophomore students who had the following majors: automated information processing systems and management (AS) and automation of technological process and enterprises (AP). The average age of the participants was 19 years. Their gender and major composition is shown in Table 8.1. The evaluation of students' FS was conducted during their electrical engineering training utilizing the following programs:

1. Three-phase circuits (TPC)
2. Theory of electrical circuits (TEC)

The TEC program is a very detailed type of modeling educational program. The drawback of this program is the lack of feedback and active involvement of students in educational activity. The students act as observers and are not graded at the completion of this educational program. The TPC program is a different type of computerized educational program that provides continuity and completeness of the didactic cycle of the learning process. It consists of theoretical material, exercises, and interactive feedback and allows control of the level of knowledge acquisition. This program allows us to utilize the method of theoretical imaginative modeling, which improves understanding of a complicated subject matter, an issue in engineering training.

Since physical and mental fatigue, as well as ways of describing them, greatly depend on the subject's personal disposition, we obtained the

TABLE 8.1

Major and Gender of the Subjects

Major Specialty	Total (Students)	Male		Female	
		Number (Students)	Percentage from the Total (%)	Number (Students)	Percentage from the Total (%)
AP	22	18	82	4	18
AS	32	24	75	8	25

AP = automation of technological process and enterprises; AS = automated information processing systems and management.

subjects' personal profiles. The subjects' motivation should also be considered. The study was conducted in two phases. The first phase involved the evaluation of individual features of the subjects. The second phase involved the evaluation of their FS while working with the above-described educational software.

8.3 Evaluation of Students' Individual Differences

The evaluation of students' individual differences includes the following stages:

1. Professional preferences
2. Level of logical thinking
3. Basic properties of the nervous system:

   - The nervous system balance is the ratio of excitation and inhibition in cortical cells.
   - The mobility of the nervous system is the ability to react quickly to changes in the environment.
   - The strength of the nervous system is its robustness and endurance of the neural cells of the cortex (reflecting the ability of neural cells to sustain strong excitation or mid-level excitation during long periods without transforming into inhibition).

The subjects' individual characteristics were assessed once during the class at a specially allotted time, which required a follow-up of the below standards:

- It should be reliable, valid, and suitable for group testing to ensure uniform testing conditions when assessing the results using the software.
- Measuring procedures, evaluation, and interpretation of the obtained data should be standardized.
- Assessment of individual characteristics should be conducted within 70 minutes (the length of lectures).

8.3.1 Definition of Students' Professional Preferences

To identify the professional preferences of the subjects, it is imperative to assess the students' level of interest in studying GTD. The presence or absence of positive motivation when using the software is an important factor influencing the effectiveness of the training, the perception of the computer training software (CTS) properties, and the subjects' FS. The questionnaire developed by Klimov was selected (Gorbatov 2000) to determine the professional preferences of the students.

8.3.2 Degree of Development of Logical Thinking

When studying technical disciplines, such as in electrotechnical courses, students need to form and keep in their memory a sufficient number of concepts. The system of scientific concepts has complex, logically organized abstract components with a high level of hierarchy. Hence, the degree of the logical thinking development is an important factor affecting CTS acquisition. It has been studied using Raven's test. This test complies with the requirements of reliability and validity and is suitable for group testing. An abbreviated version of the test was chosen because of the time limit for testing. Each student received a Raven's test form.

8.3.3 Identifying the Individual Features of the Subjects' Nervous Systems

The features of the nervous system are stable individual characteristics that affect all the psychological features of a person. The combination of neural system features determines the type of nervous system as the basis of individual physiological characteristics. These characteristics influence individual strategies of performance (Bedny and Seglin 1999a) and formation of students' FS (fatigue, stress, functional comfort, productive or unproductive tension, etc.).

Laboratory diagnostic techniques for studying the nervous system's basic properties require special conditions and equipment. They are time consuming and require a lot of effort. So, in this study, to determine the balance and mobility of the nervous system, a questionnaire developed by Strelyau (Stolyarenko 2004) was utilized. Students were given the questionnaire, and they had to fill it in. A tapping test was used to evaluate the strength of the nervous system (Stolyarenko 2004). This test allows the evaluation of the strength of the nervous system with some approximation, and the test determines its properties on the basis of psychomotor performance. This method is based on an evaluation of the dynamics of the maximum pace of hand movements. Using the obtained data, experimental curves can be drawn and inferences about the strength of the nervous system can be made. In the experiment, the students were instructed to use the right hand first and then the left hand. To take a tapping test, the students were given a piece of paper (8 × 11) divided into two rows and six columns. Students had to put as many points as possible in each blank box, working at maximum speed. The transition to the next square was performed according to the command that was given every 5 seconds. Experimental curves were developed based on the number of points in each box and then a conclusion was made about the type of nervous system using these curves.

For statistical analysis, an integrated statistical package called "Statistics" was utilized. The results of this study were evaluated in accordance with the nominal scale (qualitative data were translated into numeric data, Tables 8.2 through 8.3) in terms of the students' professional preferences and the strength of their nervous system.

The results of the study of the degree of logical thinking development, balance, and mobility of the nervous system were numeric. For statistical analysis of obtained data, the mean (average), standard deviation, and confidence interval for the mean were calculated. The data for the first phase of the pilot study is presented in Tables 8.4 through 8.8. Coefficient K is the ratio of the number of points related to the strength of excitation to the number of points related to the strength of inhibition. In this study, $0.5 < K < 1.853$.

TABLE 8.2

Transfer of Nominative Scale into Numeric for the Evaluation of Professional Preferences of the Students

Object of Work	Assigned Number
Technology	1
Sign information	2
Sign Information + technology	3
Sign information + man	4
Man	5
Art	6
Man + technology	7

TABLE 8.3

Transfer of Nominative Scale into Numeric for the Classification of Nervous System Type (Strength of the Nervous System according to the Tapping Test)

Strength of the Nervous System	Assigned Number
Weak	1
Average–weak	2
Average	3
Strong	4

TABLE 8.4

Evaluation of the Students' Professional Dispositions

Object of Work	AP (%)	AS (%)
Technology	55	28
Sign information	32	40
Sign information + technology	0	16
Sign information + man	0	0
Man	0	5
Art	9	7
Man + technology	5	5

AP = automation of technological process and enterprises; AS = automated information processing systems and management.

TABLE 8.5

Evaluation of the Degree of Logical Thinking Development

Score	AP (%)	AS (%)
5	27	5
6	18	5
7	32	30
8	9	35
9	14	26

AP = automation of technological process and enterprises; AS = automated information processing systems and management.

TABLE 8.6

Strength of the Nervous System According to the Tapping Test

Type	AP (%)	AS (%)
Weak	55	37
Average–weak	40	42
Average	5	12
Strong	0	9

AP = automation of technological process and enterprises; AS = automated information processing systems and management.

TABLE 8.7

Mobility of the Nervous System Based on Strelyau's Questionnaire

Level of Development	AP (%)	AS (%)
High level (score over 42)	86	97
Low level (score under 42)	14	3

AP = automation of technological process and enterprises; AS = automated information processing systems and management.

TABLE 8.8

Nervous System Balance Based on Strelyau's Questionnaire

Balance of Nervous System	AP (%)	AS (%)
Nervous system imbalance toward inhibition ($K < 1$)	45	45
Balance nervous system ($K = 1$)	36	39
Nervous system imbalance toward excitation ($K > 1$)	18	16

AP = automation of technological process and enterprises; AS = automated information processing systems and management.

The first phase of a pilot study, with the statistical probability $p = .95$, showed the following:

- Among the students of both specialties (AS—83% and AP—86%), a majority expressed professional disposition to work with technology; that is, using machines, engineering materials, and so on, or to work with sign information such as numbers, letters, codes, and other symbolic data. About 54% of the AP specialty students have shown a propensity for occupations where the object of labor is engineering only, and 39% of the AS specialty students showed a tendency to professions in which the object of labor is sign information only. Graduates of the AP specialty worked mainly with hardware, the technical part of industrial control, and communication systems. Graduates of the AS specialty were computer programmers working with the sign information. It was determined that a contingent of students participating in the experiment selected professions, which in most cases corresponded to their professional preferences. Therefore, we can assume a positive motivation of students in their classrooms during our experimental studies.
- Most students participating in the experiment had a high or medium level of logical thinking: 83% of AP students and 95% of AS students possessed a medium level, and 40% of AS students and 23%

of AP students possessed a high level of logical thinking. Therefore, students participating in the experiment had a high level of logical thinking to understand and learn considered disciplines.

- Students of all considered specialties had mostly weak or average weak nervous systems, with sufficient degrees of mobility. There were quite a number of students with unbalanced nervous systems. The advantage of a weak nervous systems in comparison to a strong one is the ability to respond to stimuli of a low intensity. A weak nervous system is more finely organized and more sensitive. A mobile nervous system reacts quickly to changes in the environment.

Consequently, the influence of certain properties of the CTS on the FS of the students while working with these educational programs is sufficiently clear.

8.4 Methods of Analyzing Students' Functional State during the Experiment

Evaluation of the students' FS participating in computerized training in electrical engineering was conducted during their classes. The FS evaluation did not interfere with the educational process. The time spent on the FS evaluation did not exceed 10 minutes (5 minutes at the beginning of the class and 5 minutes at the end). Physiological and psychological methods were used for evaluation of students' FS.

The physiological method of FS evaluation included measurement of arterial pressure and pulse frequency, which were measured selectively at the beginning and at the end of the class. The students were selected based on the test results of their nervous system. Thirty-four students with weak and unstable nervous systems were selected.

The psychophysiological methods included the evaluation of the features of the students' nervous systems (Bedny and Seglin 1999a, 1999b; Klimov 1969; Nebilitsin 1976). Individual features of personality in general and temperament in particular depend on these features of the nervous system. Resistance to fatigue and stress, specificity of tension during work, and so on also depend on the above-mentioned features of the nervous system. There are instrumental and noninstrumental methods that can be used to evaluate the features of the nervous system. Instrumental methods are more precise, but they require special equipment that makes them very difficult to apply in the field studies. Therefore, to simplify the test procedures, noninstrumental methods were used instead. Features of the nervous system such as strength, mobility, and balance were evaluated.

The strength of the nervous system reflects features such as robustness and endurance of the neural cells and their structure. The ability to function in overload conditions and stressful situations depends on these characteristics of the nervous system. This ability, of course, is moderated by motivational factors. Sometimes, an individual with a weak nervous system can perform better in a stressful situation than a person with a strong nervous system. Similarly, the more skilled person would perform better in stressful conditions. However, a person with a weak nervous system will soon be undermined by the effect of exhaustion and fatigue. At the same time, the weakness of the nervous system cannot be treated as a negative factor. Weakness correlates with sensitivity. Hence, in some situations, a person with a weak nervous system can adapt to a situation better than a person with a strong nervous system. Therefore, there is always a need to consider subject–situation relationship during the study of FS. In the described experiment, a tapping test is conducted to evaluate the strength of the nervous system of the students. This is an express method, which can be used to evaluate the strength of the nervous system with some approximation. This method is based on a special analysis of the frequency of hand movement per 5 seconds. The dynamic of the tapping frequency was evaluated based on the obtained data. Using this dynamic, researchers can infer conclusions about the strength or weakness of the nervous system.

Some features of the nervous system, such as mobility and balance, were measured by using Strelau's questionnaire (Stolyarenko 2004). Mobility reflects the "speed of reconditioning" when the conditioning stimulus is altered. The opposite of "mobility" is "inertness." Mobility is important for adaptation in a dynamic environment. At the same time, mobility is not important in a slow-changing environment. The balance of the nervous system can be evaluated based on the comparison of the strength of excitation and the strength of inhibition of neuroprocesses. Evaluation of these features is also performed based on the calculation of scores according to Strelau's method.

Psychological methods include psychometric procedures (behavioral) and methods of subjective estimation. Psychometric methods are used to estimate changes in the cognitive processes (perception, attention, memory, thoughts, and so on). These changes may be triggered by certain factors. The deterioration of the FS during activity is accompanied by the deterioration of cognitive processes. Therefore, in this experiment, to evaluate the FS, psychological tests that determine the efficiency of cognitive processes were administered. The following test procedures were used: evaluation of the capacity of short-term visual memory (a test for storing numbers) and the ability of distribution and switching of attention ("A Numerical Square"). The tests were carried out at the beginning and at the end of the class using a computerized program. To eliminate the effect of training and getting accustomed to the test material, several batteries of tests were used.

8.4.1 Test to Measure the Ability to Memorize Numbers

All students were presented with a table that had 12 numbers for 20 seconds. The students were asked to memorize as many numbers as they could. Then, the students were asked to write down all the numbers they memorized in any order on the answer sheet. In the beginning and at the end of the class, students were given different versions of these number tables. Different tables were used in different educational programs. Therefore, during estimation of the capacity of the short-term memory, four different tables were utilized.

8.4.2 Test "Number Square"

Each student was given a square with 25 numbers ranging from 1 to 40 and a linear sequence of 40 numbers. From this, the students were asked to cross out the numbers that were missing in the number square. They were given 1.5 minutes for this task. At the beginning of the class, one group of students was presented with the "odd" number square and the other group with the "even" number square. At the end of the class, the tables were switched. Different tables were used in different educational programs. Therefore, during evaluation of their capacity, distribution, and switching of attention, four different number squares were used.

For a subjective evaluation of the student's own FS at the end of the class, the students were offered a questionnaire. This questionnaire was developed especially for this study based on the test of differentiated self-esteem (DSE). The offered questionnaire was an abbreviated version of the DSE test (Stolyarenko 2004). Only a few subjective scales were selected. They included a self-evaluation of the student's health condition, the degree of concentration of attention associated with wakefulness of the subject and the ability to concentrate her or his mental efforts, and the mood associated with emotional feelings. These variables had the clearest and most precise formulation. The subjects were asked to evaluate the following six psychological variables:

1. Self-estimation of the subject's own health (a, feels well; b, bad; c, difficult to answer)
2. Tension (a, tense; b, relaxed; c, difficult to answer)
3. Vivacity (a, vigorous; b, languid; c, difficult to answer)
4. Mood (a, good; b, bad; c, difficult to answer)
5. Satisfaction with the work (a, satisfied; b, not satisfied; c, difficult to answer)
6. Degree of concentration or attention (a, attentive; b, absent-minded; c, difficult to answer)

TABLE 8.9

Transition of Nominative Scale into Numeric Scale

Version of Response	Assigned Rank
A	1
B	2
C	3

The statistical analysis of the results was performed using the Microsoft Excel program and an integrated statistical package called "Statistic 5.5." The results of the students' subjective estimations of their FS were expressed in units of nominative scale and were transformed into a numeric message (Table 8.9).

The mathematical mean (average), confidence interval for the average, coefficient of excess, and asymmetry were calculated for each physiological variable and psychometrical and subjective indexes of the FS. It was discovered that the distribution of the obtained experimental data did not fall into normal distribution. The same sample and the same number of students were used in the statistical analysis for the evaluation of FS in the beginning and at the end of the class. Therefore, during evaluation of changes in a student's FS, we utilized nonparametric methods for interdependent samples (Fisher's criterion $< p$). In the case of comparison of the FS for TPC and TEC educational software, the general number of students was 54. The differences in the changes of the arterial pressure are not statistically significant (the level of significance is $p > .05$), so the result of the pressure measurement was excluded from the analysis of FS.

8.5 Results

The results of the evaluation of the students' FS while working with TEC and TPC educational software using physiological methods are provided in Table 8.10 and Figure 8.1. The evaluation of the students' FS while working with TEC and TPC educational software using psychometrical methods are shown in Table 8.11 and Figure 8.2.

The analysis of the presented data demonstrates that during work with TPC, the majority of students' pulse rates stayed the same (does not change) in 44%, against 26% for TEC. The less the students' psychophysiological characteristics deteriorated during computer training in comparison to the same characteristics before the class, the better the design of the educational software, and the better it corresponded to the students' FS.

The students' pulse rates increased while working with TEC in comparison to working with TPC (42% for TEC against 15% for TPC). This can be explained by the fact that TEC has a more aggressive visual environment (with an increased number of superfluously small bright objects whose position changes more often, and these factors create visual tension during interaction with the screen).

The result of evaluation of the students' FS while working with TEC and TPC using psychometrical methods is presented in Tables 8.11 and 8.12 and Figure 8.2. All the data are presented in both absolute value (frequency)

TABLE 8.10

Students' Functional States Evaluation Applying Physiological Methods

Change of Pulse	TEC (Students)	TPC (Students)	TEC (%)	TPC (%)
Increased	14	5	41	15
No change	9	15	26	44
Decreased	11	14	32	41

TEC = theory of electrical circuits; TPC = three-phase circuits.

TABLE 8.11

Changes in Short-Term Memory Capacity

The Changes in the Capacity of Short-Term Memory	TEC (Student)	TPC (Student)	TEC (%)	TPC (%)
Increased	20	7	37	13
No change	19	16	35	30
Decreased	15	31	28	57

TEC = theory of electrical circuits; TPC = three-phase circuits.

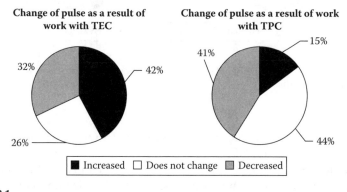

FIGURE 8.1
Change in the pulse rate as a result of working with the theory of electrical circuits and three-phase circuits.

TABLE 8.12

Changes in the Capacity of Distribution and Switching of Attention

The Changes in the Capacity of Distribution and Switching of Attention	TEC (Student)	TPC (Student)	TEC (%)	TPC (%)
Increased	15	13	28	24
No change	23	16	43	30
Decreased	16	25	30	46

TEC = theory of electrical circuits; TPC = three-phase circuits.

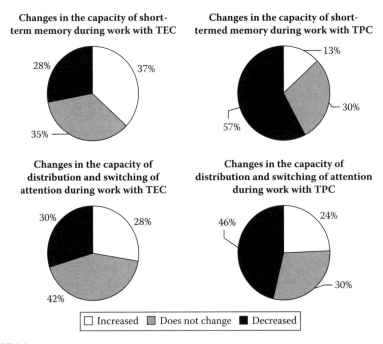

FIGURE 8.2
Evaluation of the students' functional states during work with the theory of electrical circuits and three-phase circuits.

and percentage. The evaluation of the students' FS while working with TEC and TPC using methods of subjective estimate is shown in Tables 8.13 and 8.14 and Figures 8.3 and 8.4. The collected data demonstrate that the capacity of visual short-term memory decreased more significantly while working with TPC (almost 50% of students) than with TEC, in which a decrease in the short-term memory capacity was observed in only 28% of the students.

TABLE 8.13

Subjective Evaluation of the Functional State while Working with the Theory of Electrical Circuits

	Self-Estimation of Health		Tension		Vivacity		Mood		Satisfaction with the Work		Degree of Concentration of Attention
Good	24	Tense	16	Vigorous	17	Good	29	Satisfied	22	Attentive	14
Bad	12	Relaxed	30	Languid	13	Bad	12	Not satisfied	21	Absent-minded	17
Difficult to answer	18	Difficult to answer	8	Difficult to answer	24	Difficult to answer	13	Difficult to answer	11	Difficult to answer	23

TABLE 8.14

Subjective Evaluation of the Functional State while Working with Three-Phase Circuits

	Self-Estimation of Health		Tension		Vivacity		Mood		Satisfaction with the Work		Degree of Concentration of Attention
Good	33	Tense	20	Vigorous	27	Good	38	Satisfied	36	Attentive	28
Bad	7	Relaxed	27	Languid	12	Bad	7	Not satisfied	14	Absent-minded	11
Difficult to answer	14	Difficult to answer	7	Difficult to answer	15	Difficult to answer	9	Difficult to answer	4	Difficult to answer	15

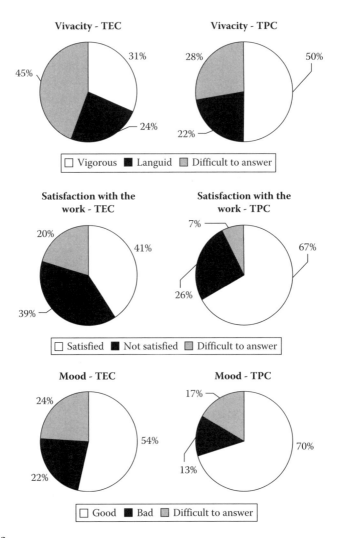

FIGURE 8.3
Subjective evaluation of the students' own functional states while utilizing the theory of electrical circuits and three-phase circuits.

The decrease in the students' abilities to distribute and switch attention was more significant in students working with TPC in comparison to the students working with TEC (43% for those who work with TPC in comparison to 30% in case of TEC). The obtained results demonstrate a greater fatigue and tension while working with TPC than with TEC. This can be explained by the more intensive learning activity of the students who used the TPC program (students calculated electric circuits, draw charts, and vector diagrams and were graded for each task). The TEC modeling program

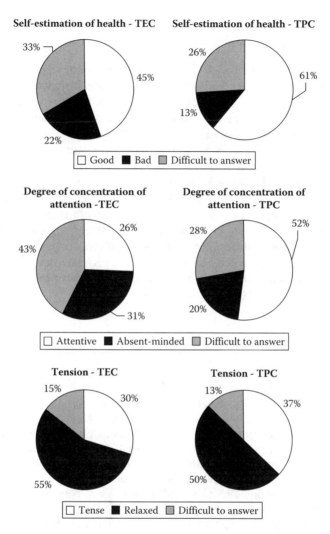

FIGURE 8.4
Subjective evaluation of the students' own functional states while utilizing the theory of electrical circuits and three-phase circuits.

required a simple observation of ready-made charts, and the students were not graded for each task. The analysis of the subjective evaluation of the FS while working with TEC and TPC showed the advantages of TPC over TEC based on differences in "vivacity" (over 19%); "work satisfaction" (over 26%); "mood" (over 16%); and "health estimation" (over 16%). Working with TPC was accompanied by a greater degree of concentration of attention (over 26% compared with that TEC). Working with TPC caused more tension than working with TEC (TPC: 37% and TEC: 30%). This might be explained by the greater intensity of the learning activity.

8.6 Discussion

The differences between the FS of students working with TEC and TPC education software was evaluated by using various methods (physiological, psychometric, and subjective). These differences between groups were statistically significant ($p < .05$). The results of the experimental study demonstrated that working with TPC is accompanied by great tension and intensity of learning activity. These facts, in combination with work satisfaction and positive evaluation of the students' own state, bring us to a conclusion about adequate mobilization of psychophysiological processes and the optimal level of their activation so that tension during work does not exceed a permissible level.

At the same time, working with the TEC program is not as satisfactory. It is accompanied by a negative evaluation of students' own FS, low work satisfaction, low concentration of students' attention, and increasing tension. Hence, the TPC program is better than the TEC program according to the FS criteria. The evaluation of the computerized training program efficiency can be based on the analysis of the students' FS.

References

Bedny, G. Z., and W. Karwowski. 2007. *A Systemic-Structural Theory of Activity. Application to Human Performance and Work Design*. London: Taylor & Francis.

Bedny, G. Z., and D. Meister. 1997. *The Russian Theory of Activity. Current Application to Design and Learning*. Mahwah, NJ: Lawrence Erlbaum Associates.

Bedny, G., and M. Seglin. 1999a. Individual style of activity and adaptation to standard performance requirements. *Hum Perform* 12(1):59–78.

Bedny, G., and M. Seglin. 1999b. Individual features of personality in the former Soviet Union. *Hum Perform* 33(1):546–63.

Chainova, L. 1998. Computer games in preschool training. *An Ind Art* 1:19–21.

Chainova, L. 2003. *The Concept of Functional Comfort: Theoretical Basics of Designing of Subject-Spatial Environment. Subject-Spatial Environment in Sphere of Management,* 79–85. Moscow: ARIIA.

Gorbatov, D. C. 2000. *Practicum on Psychological Testing*. Samara: Bakhram Publishers.

Stolyarenko, S. 2004. *Psychology Practicum*. Rostov: Feniks Publisher.

Klimov, E. A. 1969. *Individual Style of Activity*. Kazan: Kazan University Press.

Nebilitsin, V. D. 1976. *Psychophysiology of Individual Differences*. Moscow: Science Publisher.

Zainutdinova, L. H., and D. A. Yakovetc. 2003. Experimental evaluation of ergonomic parameters of the computer training programs on electro-technical disciplines. *Int J Inf Theor Appl* 10(1):72–9.

Zarakovsky, G. M., and Y. K. Kazakova. 2004. Systemic principles of research of functional state of the gas-man operator while working in desert. *Theoretical Issues in Ergonomic Science*, Vol.5, No.4, 338–57.

Section IV

Work Activity in Aviation

9

Characteristics of Pilots' Activity in Emergency Situations Resulting from Technical Failure

V. Ponomarenko and G. Bedny

CONTENTS

9.1 Introduction

Several theories of activity have been developed in the Soviet Union, the most authoritative being those offered by Rubinshtein (1959) and Leont'ev (1978). These theories involve a high level of generalization and are considered the two main versions of the general activity theory. However, attempts made in the past to apply these theories directly to the study of work activity have met with a number of difficulties because a theoretical and methodical basis for making them applicable has not been developed. Successful use of the activity concept has been achieved in (1) the applied theory of activity (Ponomarenko 2006; Zavalova et al. 1986; Konopkin 1980; Zarakovsky and Pavlov 1987, etc.) and (2) the systemic-structural theory of activity (Bedny and Meister 1997; Bedny and Karwowski 2007). We have used the latter theories in our study of the activity of pilots in emergency situations. We study pilots' activity in emergencies from functional analysis perspectives where activity is considered to be a goal-directed self-regulative system. The theory of self-regulation developed within systemic-structural activity theory is fundamentally different from the theoretical concepts of self-regulation discussed outside the activity approach. The activity approach to the problem of self-regulation has its roots in the work of such physiologists and psychophysiologists as Anokhin (1969) and Bernshtein (1966). The self-regulation approach developed outside of activity theory, on the other hand, originated in the works of Wiener (1948), who is the creator of cybernetic theory.

Models of self-regulation beyond activity theory do not differ much from each other. They include several control-free regulatory mechanisms such as goal-standard, input, comparator, and output (Carver and Scheier 2005; Vancouver 2005). These models are based on the homeostatic principles, the major purpose of which is elimination of deviation from the specified and ready-made standard and reaction to disturbances. Self-regulation is presented as a cycle because of the feedback from prior performance (Zimmerman 2005). Moreover, according to Carver, Scheier, and Vancouver, a self-regulative system can function only after receiving feedback from an externally performed behavior or the appearance of deviations of the controlled variable as a result of interference of disturbances. Such regulation is based on analysis of already-made errors in prior manipulations with external variables. However, a subject cannot just respond to the unacceptable in his view of deviations of variables in the external environment. He or she can also regulate his or her behavior (external and internal) and can prevent unwanted deviations before execution of material actions. Moreover, such content-free regulatory mechanisms do not have sufficient theoretical justification in psychology. The approach is borrowed from the technical disciplines.

From a control theory perspective, emergency situations can be viewed as disturbances. Disturbances are factors or events that affect both the emotional state of humans and the efficiency of their performances. In activity theory, the term "emergency situation" is used instead of the term "disturbance." We give here a general definition of the term "emergency situation" before considering the activity of pilots in such situations. An emergency situation is complex and has an unexpected influence on a person, causing two levels of responses: (1) adaptive and protective reactions (such as orientation-exploratory reaction) and (2) complex intellectual actions that are associated with the analysis of a situation while developing a behavior strategy, in association with attention distribution between (1) the condition of an object that is under control and (2) the formation of a new unpredictable scheme of actions.

Emergency situations differ from negative physical factors, which also can be considered to be particular types of disturbances (temperature, accelerations, vibrations, pressure decrease, and so forth). Any physical factor is precisely characterized by its duration, force, and point of influence. These factors define the response of an organism. An emergency situation mainly influences the human psyche. The same emergency situation may have a different appearance in different persons' minds. For example, from the functional analysis of activity perspective, it is necessary to distinguish between the objectively existing reserve time and the operator's subjective evaluation of this time (Bedny and Meister 1997). As a result, the same duration of time may or may not lead to emotional tension during the flight; this depends on the pilot's mental dynamic model of the situation. The pilot's subjective evaluation of the reserve time does not directly determine the resistance to a perceived situation. The pilot acts against his or her psychological preparedness in an emergency situation. This readiness to go through an emergency situation defines the limits of unexpected and undesired influences.

Mental preparedness consists of three components: (1) psychophysiological stability, which results from an organism's condition; (2) psychic stability, which results from professional training; and (3) the functional level of the general physical properties of the person. Mental preparedness requires special training for executing the actions needed in an unexpected situation, the ability to think in terms of operative action and to urgently actualize prior knowledge for decision making, a high level of motivation combined with a predisposition toward a positive solution while resolving a problem, and a sense of duty.

9.2 Functional Analysis of Activity

In systemic-structural activity theory, functional analysis considers an activity a goal-directed self-regulative system. The main unit of analysis for this method of study is the functional mechanism or functional block. Functional analysis describes the strategies used in the performance of an activity in relation to the different functional mechanisms or functional blocks. The more the functional mechanisms are introduced to the description and analysis of activity strategies, the more detailed the functional analysis. The same psychological processes are included in a particular functional mechanism with a specific organization and carry out specific functions. Moreover, the same function can be achieved by utilizing a different content of cognitive processes. Hence, the stages of activity, which have particular functions in the regulation of activity, are called functional mechanisms. When functional mechanisms are presented as the components of the model of self-regulation of activity, with their feed-forward and feedback interconnections, they are defined as functional blocks.

The functional block can be defined as a coordinated system of subfunctions, which have a specific role in the regulation of activity. All functional blocks are integrated into a self-regulative system. Functional blocks are defined in an unvarying manner, but their content may vary, and it usually does. Every functional block can include the same cognitive processes, such as perception, memory, imagination, and thinking. However, their integration can be carried out in different ways depending on the specificity of the task on hand. The content of the functional block can change, but the purpose of each functional block in the self-regulation model is constant. The notions of goal, conceptual and dynamic models, motivation, assessment of difficulty and significance of the task, and the subjective standards of successful results are considered from the perspective of their functional purpose in the structure of the self-regulation of activity. Each functional block determines the range of issues that are connected with this block and should be considered when specialists study the role of this block in the regulation of activity. The nature of these issues is also associated with the specificity of the activity during performance of a particular task.

A specialist must choose the most important functional blocks for consideration in any particular case. He or she must also pay attention to the relationship between the functional blocks and their influence on each other. Hence, the meaning of the functional block in any specific activity can be understood only in the light of its relationship with other functional blocks. The models of self-regulation of activity demonstrate that cognition is not a linear sequence of the information-processing steps, but rather a self-regulative system. The general model of self-regulation includes 20 functional blocks; in this study, we consider a self-regulatory model of orienting activity by utilizing just 13 functional blocks; this model represents a slightly modified version of the model of self-regulation described in the work of Bedny and Karwowski (2007). The model (Figure 9.1) describes the self-regulation process that precedes the performance of executed actions. Application of these models can be understood if a researcher considers each functional block as a window that can be opened to observe the activity during a task performance.

Functional analysis is a systemic-qualitative analysis of activity, which is critically important because it helps in describing the strategies of task performance. The self-regulation model is also an information-processing model. However, information processing is considered not a sequence of linear steps but a self-regulatory process.

The model presented in Figure 9.1 includes both conscious and unconscious levels of self-regulation, which interact with each other. The dashed box "informational model" is not a functional block; it is the information that is presented to the performer in a normal or emergency situation. Let us consider the unconscious level of self-regulation (channel 2). The incoming, and usually unexpected, information activates the orienting reflex (block 4). The orienting reflex is conveyed by responses such as turning the eyes or head toward the stimulus, altering the sensitivity of different sense organs, changing the blood pressure and heart rate, and changing the breathing pace. At the same time, there appears to be some electrophysiological change in the activity of the brain. The orienting reflex seems to play an important role in the functioning mechanisms of involuntary attention. Specific types of neurons that detect novelty have been discovered in the brain cortex (Sokolov 1963). In healthy people, the electrical activity in the frontal part of the brain significantly increases when they concentrate on certain objects.

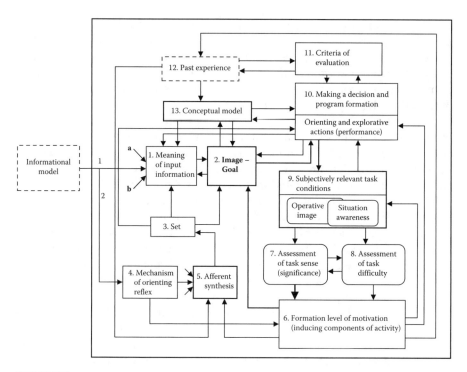

FIGURE 9.1
Self-regulative model of orienting activity.

The orienting reflex provides automated tuning to external influences and influences on the general activation and motivation of a subject (see the connection between blocks 4 and 6). The orienting reflex (block 4) also influences afferent synthesis (see the horizontal arrow between blocks 4 and 5), the latter being a major piece (relevant) of information that initiates a response or orienting reflex. For example, in aviation, this may be instrumental information (the major stimulus), which determines the orienting reflex of the pilot. However, major stimulus never exists in isolation. The environmental background is also a source of additional (situational) information. These situational stimuli have some influence on the response to the major stimulus. Hence, the afferent synthesis mechanism (see block 5) can also receive irrelevant stimuli (diagonal arrows). For example, in aviation, the latter is in the form of noninstrumental information or irrelevant environmental stimuli (noise, vibration, etc.). Thus, instrumental (horizontal arrow) and noninstrumental (diagonal arrows) information can influence afferent synthesis in combination.

Block 5 (afferent synthesis) is also affected by the blocks of motivation (block 6) and previous experience (block 12). Changes in the motivational block can, in turn, influence afferent synthesis. Therefore, the block "afferent syntheses" performs an integrative function by integrating temporal needs and motivation, mechanisms of memory representing relevant past experience, and the effects of both irrelevant (noninstrumental) and most significant (instrumental) information. Hence, the person does not react to some isolated stimulus as described in the classical conditioning theory but reacts to a combination of influences, such as the major stimulus, irrelevant stimulation, temporal motivational state and needs, and previous experience. Afferent synthesis permits the organism to determine and select, with a certain degree of accuracy, the major stimulus from among the infinitely varying influences of the external environment in accordance with the temporal needs of the organism, previous experience, and specificity of the situation. Specific neurons that are involved in the performance of such integrative functions have been discovered in the brain by Anokhin (1969).

Influences from afferent synthesis promote the formation of block 3 (a set). There are different types of sets. A person forms a general, stable system of sets as a result of her or his life experience, which can become an important characteristic of the personality (Uznadze 1961). There is also the goal-directed set, which is formed by the input instructions and specific situations. This set is defined as an internal state of the human organism, which is close to the concept of a goal but is not sufficiently conscious or completely unconscious. Such a set gives a constant and goal-directed character to the unconscious component of activity. This set is the readiness or predisposition of a subject to process incoming information, provided in a specific situation, which she or he is not well aware of. Generally, the set is an unconscious regulator of activity, which allows the activity to retain its goal-directed tendency under constantly changing situations without the subject being aware of the conscious goal.

A goal-directed set manifests itself as a dynamic tendency to complete interruptive goal-directed activity. In our model of self-regulation, these types of sets (block 3) influence the meaningful interpretation of information (block 1) and an image-goal (block 2) by interacting with the conscious channel of information processing. If activity regulation is predominantly unconscious, then a "set" has a direct impact on block 10 (making a decision about the situation or strategy of explorative actions). If the "set" is inadequate, further interpretation of the situation can also be incorrect. As a result, explorative actions associated with functional block 10 cannot have a purposeful or goal-directed character but become chaotic. For example, because of undesirable explorative activity in this type of circumstance, a pilot can decrease or totally lose reserve time in emergency situations. Very often, explorative actions are performed as unconscious mental or motor operations.

If block 3 (set) affects block 1 (meaning) due to the lack of sufficient activation of block 2 (image-goal), the meaning in block 1 is primarily nonverbalized. The concept of nonverbalized meaning has been studied by Tikhomirov (1984) in more detail. This type of nonverbalized meaning is sometimes called the "situation concept of thinking" (Pushkin 1978). In such cases, according to the functional model (Figure 9.1), meaning is formed under the influence of the "set" and through unconscious explorative operations from block 10. Such development of meaning helps to create nonverbalized hypotheses in emergency situations.

During unconscious information processing (channel 2), the goal (block 2) is not yet activated. Furthermore, block 1, which actualizes nonverbalized information, becomes responsible for the preliminary interpretation of information. At this point, signals from various devices are not yet integrated into a holistic system. Their integration and interpretation as a holistic mental model of the situation becomes possible only after activation of the block's image-goal (block 2) and subjectively relevant task conditions (block 9).

We utilize the term "mental model," in which not only the verbally logical component is important, but the imaginative components are also critically important during information processing; these components are subsequently integrated into a holistic mental picture of the situation. Functional blocks 2, 9, and 13 are outlined in bold to indicate that they include imaginative components and can be considered mental models.

The conceptual model (block 13) is relatively stable and changes slowly in the time model. It reflects the various scenarios of possible situations that are relevant to particular duties. The conceptual model, in contrast to previous experience, is more specific to the type of activity a pilot has to perform according to his or her duties. For example, we can talk about the conceptual model of a flight from Moscow to New York, which is developed after previous experience during training.

In contrast to the relatively stable conceptual model (block 13), the dynamic mental model is adequate for particular situations. Functional block 9 (subjectively relevant task conditions) is responsible for the creation of such a model. It provides a reflection of not only the current situation, but also anticipates the near future and infers from the past. Block 9 includes two subblocks, namely, "operative image" and "situation awareness" (SA). Therefore, not only the logical or conceptual components of activity, but also the imaginative components provide a dynamic reflection of reality. Imaginative reflection of the situation can be largely unconscious and is easily forgotten due to the difficulty in its verbalization. Imaginative and conceptual subblocks partially overlap. The operator is conscious of the information being processed by the overlapping part of the imaginative subsystem. The relationship between these two subsystems changes constantly. Functional block 9, which is responsible for the creation of the dynamic mental model, is of particular importance for a pilot during conscious reflection of the flight situation.

In general, during unconscious information processing (channel 2), a goal (block 2) is not yet activated. Block 1 (meaning), which in this case is unconscious at first (nonverbalized) is formed primarily under the influence of the set (goal-directed set, block 3), conceptual model 13, and the decision-making and program-formation block 10. The functioning of the last block is aided not by conscious actions but by unconscious operations. Conscious goal-directed actions have a subordinate role in this situation. Block 1, at this stage of information processing, is responsible for the preliminary interpretation of the information from various devices, which are not yet integrated into a holistic system. Integration and interpretation of this information as a holistic model of the situation becomes possible only after the formation of the goal of activity (block 2).

The conscious level of information processing (channel 1) provides a more detailed interpretation of the task, its reformulation, formation of an elaborate dynamic model of the situation, and a forecast of future events. At this stage, one can observe the intensification of the interaction between blocks 1, 2, and 13. Conscious and unconscious processing of information involves not only cognitive but also emotional-motivational mechanisms. Therefore, the interaction of blocks 6 and 2 becomes especially important at this stage. This interaction facilitates the formation of the vector "motives → goal."

The functional block "image–goal" is constructed as a model for the desired future result of the current activity or actions. A "goal" in activity theory is a conscious image or logical representation of a desired future result of the current activity. The future result emerges as a goal only when it is jointed with motives. The vector "motives → goal" provides a direction to the self-regulation process (bold line). Objectively presented requirements should then be interpreted and accepted by the subject and transformed into the individual goal. Different individuals may have an entirely different understanding of the goal, although objectively identical requirements or instructions are given. The goal performs an integrative function in self-regulation. It integrates all other functional blocks into a holistic self-regulative system. This representation of the goal is entirely different from its interpretation in cognitive psychology. For example, Preece et al. (1998) consider a goal to be a state of the system that the subject wishes to achieve. This goal is externally given in a ready form to the subject. In activity theory, the goal is always associated with some stage of activity (interpretation, acceptance, formation, etc.).

An unconscious set can be transferred into a conscious goal and vice versa. For instance, when a subject drives home and discusses some problem with his or her passenger, the goal "to drive home" is transformed into an unconscious set. The driver shifts his or her attention and formulates different goals associated with conversation. At the same time, the goal "drive home" does not disappear. It is simply transformed into an insufficiently conscious set. At any particular time, when taking a certain exit is required, this set is transferred back into the conscious goal. This proves an ability of the subject to switch from one task to another or from an unconscious to the conscious level of self-regulation and vice versa.

When a major channel is conscious, information goes directly from channel 1 through block 1 to block 2. Goal 2 influences block 10 and activates the conscious explorative actions and conscious decisions. Under the influence of blocks 10 and 6, functional block 9 (subjectively relevant task conditions) is activated. As a result, a dynamic model of the situation is developed. In contrast to cognitive psychology, where a dynamic mental model is a result of purely cognitive functions, in activity theory, a dynamic mental model of the situation is developed during the interaction between the cognitive mechanisms of blocks 10 and 9 and the motivational mechanisms of block 6. The motivational mechanisms of activity are included in block 6 and are considered the inducing components of activity. This block influences the conscious and unconscious aspects of information processing. Without motivation, there can be no "afferent synthesis" or "set." This block is involved in the creation of the vector "motives → goal" and formation of the dynamic model of the situation (block 9).

Block 6 is tightly connected with block 7 (assessment of the task sense or significance) and block 8 (assessment of task difficulty). These three blocks are tightly interconnected and influence each other. Block 7 (sense) is responsible for evaluation of the significance of the situation (emotionally evaluative mechanism). These two blocks are so tightly interconnected that their relationship is designated by a bold arrow. The other block, block 6 (motivation) refers to the inducing components of activity. Block "sense" (block 7) has a direct influence the logical and meaningful interpretation of the situation. The more significant the situation is for the subject, the higher the level of motivation. Block 6 is involved in the goal-formation process and in the switching of attention from one feature of the situation to another in the dynamic model. In other words, the factor of significance is involved in extracting the adequate features of the dynamic model of the situation (see the interaction between blocks 7, 6, and 9).

Functional block 11 is involved in the formation of subjective criteria of success and the evaluation of the activity result. Note that the objective requirements for the results of activity and the subjective criteria of success are not the same. Moreover, the goal of activity and the criteria of evaluation are often not the same (Bedny and Karwowski 2006).

9.3 Emotional-Motivational Aspects of Self-Regulation

In this section, we discuss the emotional-motivational aspects of task analysis from a functional-analysis perspective, in which activity is considered a self-regulative system with its main units of analysis being functional blocks. When studying motivation, functional mechanisms such as "goal" (block 2), "assessment of task difficulty" (block 8), "assessment of task sense (significance)" (block 7), "formation of level of motivation" (block 6), and "criteria of evaluation" (block 11) are particularly important. Motivation is considered a process of dynamic interactions between these blocks. Let us briefly consider the functional block "difficulty" (block 8) and its interaction with the other above-mentioned blocks.

Functional analysis distinguishes between the objective complexity of the task and the subjective evaluation of task difficulty. A subject can evaluate the same task as being more or less difficult depending on the complexity of the task, past experience, differences between individuals, and even the person's temporal state. The more complex the task, the more probable it is that the task will be evaluated as difficult. Cognitive task demands during task performance depend on the task complexity. A subject does not experience the task complexity but its difficulty.

A method for quantitative evaluation of the task complexity can be found in the work of Bedny and Karwowski (2006, 2007). We are considering these characteristics in our discussion from a functional analysis point of view, where it is important to find out how a subject evaluates the task difficulty. An individual might under- or overestimate the objective complexity of the task. For example, a subject can overestimate the task difficulty and may reject it in spite of the fact that objectively he or she would be able to perform it. Moreover, overestimation of a task difficulty produces emotional stress and, even if the task is accepted, the quality of performance can suffer. On the other hand, if a subject underestimates the task difficulty, he or she may fail to perform the task. Psychological concepts, such as the subject's evaluation of her or his own abilities in comparison to the task requirements, self-efficacy (Bandura 1997), and self-esteem, are useful for the analysis of the functional block "difficulty," which is considered the cognitive mechanism of self-regulation due to its influence on motivation.

The functional block "difficulty" is a task-specific entity. The subject either can correctly estimate or can over- or underestimate the difficulty of a task. However, the reason for the over- or underestimation of a task can depend on stable personal features or rest solely on a task-specific situation. The evaluation of a task difficulty can also be a function of the previous experience of the subject. This cognitive mechanism (functional block 8) is critical to the motivational process because the block "difficulty" interacts with a number of other blocks in the self-regulation process.

Here, we consider the interaction of the functional block "difficulty" with the block "sense" (here and later, "sense" is understood as a person's evaluation of the subjective significance of a task or situation). Hence, the block "sense" is tightly connected with the motivational block (the connection between blocks 7 and 6 is designated by the bold line). These two blocks include the emotional, evaluative, and inducing mechanisms of activity. The individual's sense creates the predilection of the human consciousness (Leont'ev 1978). The functional block "sense" (block 7) predetermines the significance of the task, its elements, situation, and a subjective value of both the elements of the task in obtaining the desired result and of the task as a whole, which provides a sense of achievement. The blocks "difficulty" and "sense" have a complex relationship; the interaction between them influences the process of motivation and can explain motivation in a very different light in comparison with the existing theories. Let us consider the construct "self-efficacy" developed by Bandura (1997). According to Bandura, the stronger the belief in self-efficacy, the greater the chance a person will pursue the desired result. He suggests that people with high personal efficacy set more difficult goals and show greater persistence in pursuing them. People with low efficacy set lower goals and will abandon an externally given goal in the face of adversity. It implies that all motivational manipulations are effected through self-efficacy. From the self-regulation or functional activity analysis point of view, if a person evaluates a goal as a very difficult one due to her or his low self-efficacy, the resulting negative influence on motivation (inducing the component of motivation) increases the probability that the goal will be avoided or abandoned. On the other hand, if a particular goal of a task is significant or has a high level of positive subjective value, those with even low self-efficacy can nevertheless be motivated to strive for the goal.

Let us analyze another example. The basic postulate of goal-setting theory is that difficult goals, if accepted, lead to greater job performance than easier goals do (Locke and Latham 1990). We would say that goal-setting theory merely substitutes the rather complex issues of motivation with the simple statement "if the goal has been accepted." According to the general model of self-regulation (Bedny and Karwowski 2007) and the model of self-regulation of orienting activity, described in this book, there is a complex relationship between difficulty and motivation. Increasing the difficulty of the task does not always lead to an increase in the level of motivation, as stated in goal-setting theory. Locke and his colleague used a static explanation: an increase in the level of motivation always follows an increase in the task difficulty. However, the level of motivation depends on a complex relationship between the functional blocks "assessment of task difficulty" and "assessment of the sense of task" (significance).

If the task is evaluated as having a high difficulty but its significance is very low (attainment of the task's goal is not subjectively important), the subject is not motivated to perform the task. In such situations, a subject does not have any reason to spend her or his efforts on the task. On the other hand, if the task is evaluated as very difficult and, at the same time, it is very significant for the subject, he or she is motivated to perform the task even with the risk of failure. There could be various other scenarios. For example, if the difficulty and significance of the task are low, it means that the work is very boring, and the subject has a low level of motivation.

Other functional mechanisms also influence motivation. Motivation, among other things, depends on a subjective criterion of successful result (see the functional block "criteria of success"). This subjective standard can deviate from objective requirements; thus, the satisfaction derived from attainment of a goal depends on this criterion. If the subject achieves a required goal but her or his level of aspiration exceeds the goal, the subject will not be satisfied by the obtained result. The concept of subjective standards of successful result is also deemed important in the social-learning theory. However, in this theory, there is no clear understanding of the difference between the goal and the subjective standard of success. A subjective standard of success can deviate from a goal, particularly at the final stage of task performance. A subjectively accepted goal can be used as a subjective standard of success. However, a goal—by itself—might not contain enough information to evaluate the result of task performance. This standard has a dynamic relationship with the goal and previous experience (Bedny and Meister 1997) and can be modified during goal acceptance and task performance. For instance, a person might be tired during the second part of the shift but does not want to decrease productivity. As a result, the qualitative criteria of success can be lowered and the requirements of quantitative criteria can be increased. Therefore, here, the objectively established goal deviates from the subjective criteria of success.

Functional analysis of activity describes human cognition and motivation during work more comprehensively than cognitive task analysis and existing motivational theories. Functional models of activity allow a more detailed analysis of performance strategies, integrating motivation and cognition. The explanatory and predictive features of these models can be explained by the fact that, in the self-regulation model, activity, goal, motive, meaning, sense, and so on are considered parts of a system of interconnected mechanisms that have specific functions. This allows a description of activity as a goal-directed self-regulative system. For example, the self-regulative models derived from control theory (Carver and Scheier 2005; Vancouver 2005) assume the existence of a fixed goal as a standard, whereas the functional model emphasizes the process of goal acceptance or formulation and its specific functions in the formation of the strategies for receiving information and activity performance. Our model emphasizes the existence of differences between the goal and the subjective standard of success.

Furthermore, the model of self-regulation explains why an operator can neglect safety requirements when he or she has high aspirations of reaching the goal, based on her or his subjective criteria or standard of success. A subjective standard of success can also influence the precision of an operator's performance. Another example is that the accuracy with which a pilot can read aviation instruments often depends more on the significance of the instrument than on its visual features. It was discovered that depending on the goal created by a pilot, the same display can perform different functions. For example, depending on the activity goal, the same apparatus can be used for evaluation of the flight parameters in one situation and for evaluation of the functioning of other apparatuses in other situations. Further, the concept of SA, which has been developed in cognitive psychology (Endsley 2000), can be understood as a functional mechanism of activity regulation. Finally, we want to stress that ergonomics and industrial-organizational psychology do not study separate cognitive processes; these applied fields study human work as a whole, wherein cognitive processes are integrated into the system. Functional analysis of activity helps us to study cognition as a system where all cognitive processes are considered in their entirety.

As already discussed, application of these models can be understood if we consider each functional block to be a window that can be opened to analyze the same activity during task performances. For example, a researcher can open a window called "goal" and, at that stage, pay his or her attention to aspects of activity such as perception, interpretation, formation, and acceptance of a goal, in addition to the relationship between verbal and imaginative components.

On the other hand, when paying attention to functional block 13, the researcher would study a stable mental model of the flight. This stable mental model can change slowly over time and may include different scenarios of possible duties. This model is created by a pilot before the flight and includes the basic scenarios of a possible flight mission. It also includes the knowledge about possible tasks that may need to be performed, the final goal of the mission, understanding of the possible constraints and difficulties, specificity of team performance, assumptions made about changes in flight mission, environmental changes, and so on. Procedural knowledge is a more important component of the conceptual model than declarative. Hence, the conceptual model is more specific in comparison with previous experience.

At the next step, the researcher can open another window, for example, the window called "subjectively relevant task conditions" (functional block 9). At this stage, one would study those aspects of activity that are responsible for the creation of a dynamic mental model of a given situation. This block includes not only conceptual but also imaginative mechanisms. They are regarded as two subblocks that overlap. One of these blocks, according to the existing cognitive psychology tradition, is called the SA block, and the other is called the operative image block. The SA the block includes verbalized processes and involves consciousness. The operative image subblock largely provides an unconscious dynamic reflection of reality. The conscious and unconscious components of dynamic reflection can transform into each other to some degree. Feed-forward and feedback connections between these blocks demonstrate how these blocks coordinate their functioning.

Based on the suggested model, we can also consider how blocks such as "goal" and "subjectively relevant task conditions" influence each other. Similarly, we can consider other functional blocks that are most important for a particular task. The relationships between blocks should be also considered during task analysis. These brief discussions demonstrate that self-regulation cannot be reduced to the concepts of goal, self-efficacy, motivation, and other separate psychological mechanisms and processes, as presented in a handbook of self-regulation (Boekaerts et al. 2005). We also don't agree with such authors as Vancouver (2005) who explain self-regulation based on some mechanistic models that derive from analysis of technical systems. Vancouver defines self-regulation as maintaining a variable at some value despite disturbance to the variable. This homeostatic definition of self-regulation is acceptable for explaining the functioning of technical systems or physiological processes. His understanding of self-regulation of human activity or behavior is incorrect. The homeostatic self-regulative process cannot prevent human error. Such a process does not provide the ability to regulate activity without external disturbances, ignores the conscious goal-directed level of self-regulation, and so on. We also disagree with those who try to explain the process of self-regulation without presenting any specific models of self-regulation (Boekaerts et al. 2005).

9.4 Structure of the Pilots' Dynamic Mental Model

The functional model of activity helps us to describe the pilots' dynamic mental model of flights more precisely. Zavalova and Ponomarenko (1980) developed the structure of an operative image of the flight. In our work, we slightly modify this model. According to systemic-structural activity theory, it is a dynamic mental model of a flight in which there is a combination of imaginative and verbally logical components. A pilot, using both instrumental and noninstrumental signals, creates a dynamic mental model of a flight. The creation of an adequate mental model becomes more complicated when instrumental and noninstrumental signals contradict one another. The reliable control of an aircraft based on the data from the equipment alone has its limitations. An adequate mental model provides the correct interpretation of the input information in ambiguous situations. Before takeoff, a pilot creates a relatively stable conceptual model of the flight (see function block 13, Figure 9.1). The dynamic mental model is situation specific. The functional blocks "image-goal" (block 2), "conceptual model" (block 13), and "subjectively relevant task conditions" (block 9) are the major mechanisms responsible for creation of these dynamic mental models of a flight. The imaginative components are critically important in this mental model. The basic components of a pilot's dynamic mental model are presented in Figure 9.2.

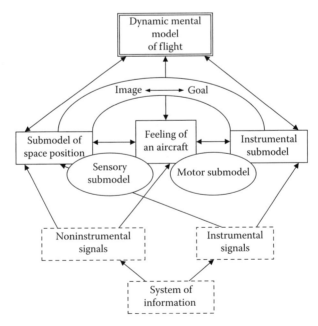

FIGURE 9.2
Structure of the pilot's dynamic mental model of the flight.

The concept of image-goal is specifically relevant to a pilot's tasks. In a pilot's activity, the requirements of tasks are embodied in the goal, and these requirements can be presented not only verbally, but also in an imaginative form. An image-goal is the desired future result of a pilot's own activity. The image-goal determines the specifics of the pilot's selection of information. It is a relatively stable component until the task is completed. Not only memory, but also operative thinking, is important in the process of goal formation or acceptance. If a pilot switches from one task to another, it means that he or she changes the image-goal.

Occasionally, it is difficult to retain the image-goal in the pilot's mind for a sufficiently long period of time. In these circumstances, it is necessary to develop specific external means that help sustain the image-goal during task performance. This is particularly relevant when the pilot has to interrupt his or her current task, switch to another one, and then return to complete the interrupted task. To facilitate performance of the interrupted task, the goal of this task can be presented to the pilot as a picture or verbal description or as a voice-induced command. The submodel of the space position regulates the special orientation of the pilot and helps him or her to be aware of the aircraft position in relation to the earth. A contradiction (1) between perception and thinking and (2) between external, visual, and internal receptor signals can produce illusions. In such circumstances, the pilot tries to repress the false sensations by concentrating his or her attention and forcing activation of the logical components of thought.

The instrumental submodel is created by a pilot based on the perception of information from instrument displays. This model reflects the discrepancy between the required and real flight regime based on perceived differences between the set parameters of the flight and the real position of the indexes. The instrumental model is very pragmatic, and it regulates motor actions. It facilitates the automaticity of activity, which, although providing speed and accuracy to the performance, may lower the reliability of a task performance in an unexpected situation.

Another component of the dynamic model is the "feel" of the aircraft. Its formation is based on the influx of noninstrumental signals. Although these signals may cause errors, they are important for the feel of the aircraft. For example, the lack of muscle feeling during piloting with the automatic stabilizer of the aircraft's position increases the duration of visual fixation on the displays. This can be explained by the fact that the use of a stabilizer disrupts the level of resistance of the controls, affecting the pilot's feel of the aircraft. The controlling actions of the pilot change the position of the aircraft, which leads to changes in the instrumental and noninstrumental signals. Such signals give information not only about the position of an aircraft, but also about the result of the pilot's motor actions. Based on the consequences of the pilot's own actions, she or he can change his or her strategy of task performance and correct the dynamic mental model of the flight. Thus, we can see here that a pilot works as a self-regulative system.

The feel of an aircraft and instrumental model are basic sources for formation of motor model. A model of a space position, in combination with the feel of the aircraft, is the basic source of the sensory submodel. Finally, there are instrumental and noninstrumental sources of information for the creation of different models of the flight (Figure 9.2). The pilot does not simply perceive the output of the instruments but interprets them in accordance with the existing models and the dynamic mental models in general. All the points discussed previously allow the creation of a general dynamic mental model of the pilot-aircraft interaction. As can be seen from Figure 9.2, a pilot's dynamic mental model has a complicated structure. This is a very useful tool for the study of pilots' activity in emergency situations. Let us consider some basic characteristics of a pilot's work in an emergency situation.

9.5 Basic Characteristics of Information Presented in an Emergency Situation

The characteristics of emergency information are important for pilots to choose the correct actions. At the first stage, we should distinguish the physical characteristic of signals in an emergency situation, which can be high, average, and low. For example, a high attraction effect is attributed to physically strong noninstrumental signals (angular rotation of a plane with acceleration exceeding $10°/s^2$, jolting, vibration of a plane, and shrill sounds) and instrumental audio signals (siren, ring, buzzer, and voice). An average attraction effect is attributed to alarm lights, displays located in the central area of sight, and some noninstrumental signals (asymmetrical traction, angular accelerations ranging from 5 to $10°/s^2$, and increase or reduction of the force on an operating handle). Such an average attraction effect may remain unnoticed by a pilot if the pilot is focused on plane control and is not expecting a failure. This is particularly relevant for signals with a low attraction effect. The value of the attraction effect determines how quickly the process of information reception and interpretation starts. The initial stage of this process is the orienting reaction ("something happens"). From a physiological point of view, it is an orienting reflex. If the attraction effect is sufficient, then it causes the involuntary switching of attention to unexpected signals or tasks.

The regularity of the orienting reflex as a functional mechanism of involuntary attention should be taken into consideration during the design of emergency signals. Sometimes, displays may attract our involuntary attention processes. However, in a few cases, involuntary attention can distract an operator, which can lead to errors. A repeated stimulus weakens and loses its influence over the involuntary attention process, causing the operator to ignore it. If, after attracting our attention, information is evaluated as significant (subblock "sense"), involuntary attention can be transformed into voluntary attention. The major purpose of the orienting reflex or reaction in an emergency situation is the detection of unexpected signals.

At the next stage, the pilot recognizes the situation and makes a decision. During this stage, a pilot compares the detected signal with a dynamic model of the situation (block 9, Figure 9.1). Based on these data, the pilot formulates the hypothesis of "what has happened." Signal detection depends on the attraction effect, whereas the second step of information processing depends on the content of the signal and on its certainty and ability to be interpreted correctly.

We have to distinguish the terms *certain, contradictory,* and *uncertain* information in emergency situations. Certain types of information let us recognize the situation accurately, which is done simultaneously with signal detection. The certainty of the information is technically provided with a representation of the signal on a display or as a voice message in earphones. Contradictory information is usually provided by signals associated with the event indirectly. An example of such a signal is a mismatch in the displays due to a failure in one of them. This contradictory signal will not show the cause of mismatches directly, and it hinders signal interpretation and decision making. Most of the noninstrumental signals are characterized by contradictory information; to detect them, a pilot has to search for definite information actively and use his or her experience and knowledge about similar signals. Uncertain information cannot be unequivocally interpreted. Uncertainty is a characteristic of noninstrumental information.

The pilot's information processing depends not only on the signal characteristics but also on the pilot's mental preparedness and foremost on the content of his or her mental dynamical model. Emergency situations can be divided into five classes as follows:

1. *Conflict situations*: The pilot chooses to form opposite but subjectively equally significant decisions. The choice is made without a clear prediction about the consequences of each decision.

2. *Situations with unexpected results*: Here, the pilot makes a purposeful action but meets unexpected results. In most cases, such a situation is the result of an uncertain, though physically intense, stimulus (noninstrumental signal). The situation may be aggravated by a pilot's insufficient training for emergencies.

3. *Situations with time and information deficits*: The pilot has to make a decision correctly and promptly in spite of a lack of information. This situation is the most complex. The pilot's ability to act safely depends only on his or her heuristic, creative decisions.

4. *Ambiguous situations*: The pilot misinterprets controversial signals and guides his or her actions based on the misinterpretation. This situation is more prolonged, and errors are revealed more gradually than in class 2.

5. *Definite (certain) situations*: The pilot knows what to do, and the results match his or her predictions.

The objective complexity of these five classes is different. However, the objective evaluation of the complexity in these five classes and the subjective evaluation of their difficulty are not the same. The latter is influenced by the pilot's performance. For example, a situation with uncertain signals may be easiest for a trained pilot (because it falls into the fifth class). These circumstances often make the interrelation of the information model quality and the pilot's reliability and safety performance latent. There is a probabilistic relationship between complexity and difficulty. The more complex a situation, the higher is the probability that it will be more difficult for a pilot (relationship between complexity and difficulty associated with the functional block "assessment of task difficulty," which is presented in the general model of self-regulation activity).

This analysis of an emergency situation demonstrates that a pilot's interpretation of information depends not only on the cognitive components, but also on the emotional-motivational components of activity. A person creates complicated self-regulative strategies of activity performance, which determine the specifics of activity interpretation (Bedny and Karwowski 2006). The extent of emergencies influence a person, and the reliability and safety of a person's actions are defined by the emergencies that arise both by the content of a person's internal psychic pattern and the dynamic model of the situation (Ponomarenko et al. 2006).

9.6 Pilots' Actions during Display of Uncertain Information with a High Attraction Effect

A typical emergency situation with high attraction effect and ambiguity of information is autopilot failure, which results in wheel deviation and aircraft rotation around the x- or z-axes. Failure is not displayed, but the pilot feels the physical influence of angular acceleration, which is perceived as jerking, pulling of the handles out from the hands, bumpy flight, and pulling and pushing pressures on the handle.

The pilots' actions are evaluated at two levels of physical influence. In the first condition, the plane rotates around the longitudinal x-axis with an acceleration of 40–95°/s^2; in the second condition, the plane rotates around the z-axis with an acceleration of 10–30°/s^2. To prevent this rotation, the pilot should turn off the autopilot (in this case, the wheels will be set in the neutral position) or use the handle to overcome pressure (in this case, the neutral position of the handle will shift from the centre).

In accordance with the instructions, after the emergence of the filling that the aircraft is gaining or losing altitude, and positive or negative pitching movements, the pilot should lead the plane toward horizontal flight and turn off the autopilot at the same time. Pilots can operate the plane even with an activated autopilot, but in this case, he or she has to exert additional effort because of both handle resistance and hampering of feedback from the plane.

We have assumed that it would be difficult for the pilot to identify the failure because noninstrumental signals do not represent clear information and do not point to failure directly. This will result in mistakes. Below, we describe an experiment conducted with pilots in emergency situations. One of the main difficulties in the experiment is to simulate the emergency situation so that the pilot does not suspect the artificial cause of the faults and does not expect them. The unexpectedness of the faults is one of the main conditions of the experiment because it retains the ambiguity of information received by the pilot. For the experiment, a plane was equipped with a fault-emulation facility, and the plane's control panel was placed in the instructor's cabin. The flight was carried out on an aircraft equipped with a remote control panel located in the cabin of the instructor. There was also recording equipment: an oscilloscope to record flight parameters and the motor reactions of pilot, intensifier of biopotentials to record heart rate and breath rate, a video camera for recording sight movements, and a tape recorder for registering the radio exchange between the pilot and instructor.

Fifty-seven skilled pilots participated in the experiment (12 flight instructors, 20 test pilots, and 25 regular pilots). Test pilots and regular pilots did not know about the forthcoming faults. A flight instructor accompanied them on every flight. Four to seven faults were applied in every flight, but a detailed analysis of the pilots' actions was carried out for several initial faults only. The actions of the pilots and the 12 informed flight instructors during repeated failure were analyzed to compare them with the actions of the pilots who were unaware of the faults (timing performances and some characteristics of signal perception were analyzed). Flight instructors did not intervene in the pilots' actions and did not comment on the changes in the flight mode, at least not until the pilots reported the faults (instructor intervention was allowed only if there was danger for the flight).

After the pilot reported the fault, the instructor interviewed the pilot in the air with a questionnaire designed by the experimenters. The purpose of the interview was to find out the signs the pilot used to identify the fault, the difficulties the pilot met during the control of the plane and fault recognition, and the other faults this fault could be confused with. The experimenter interviewed the pilot right after the flight to find out whether the pilot had had any previous experience in plane control with refusal of a position stabilizer, what signals (physical influence appearing at faults) the pilot considered most typical, what suggestions about the cause of situation arose during decision making, how the pilot checked the accuracy of his or her suppositions, and what other malfunctions resembled the malfunction in this fault. To get additional information about the pilot's tension, a biochemical analysis of urine and blood was conducted.

One of the most typical and effective actions of pilots was hand pressure on the handle in the direction opposite the rolling-off of the plane. Comparison of the reaction time during anticipated (repeated) and unexpected faults showed the unconditional nature of a pilot's motor reaction in this situation (Table 9.1).

A comparison of the turn-off time of the failed autopilot shows us another picture (Table 9.2). Here the ability to quickly perform voluntarily actions explicitly depends on the pilot's awareness of the situation and his or her past experience.

Thus, physically powerful noninstrumental signals (angular acceleration) attract the pilot's attention to a fault timely and impel the pilot to take control of the plane to prevent plane rotation. Nevertheless, at a later stage, the reliability of the pilot's actions depends not only on the characteristics of a signal but also on the dynamic mental model of the flight, which defines how judiciously and correctly a pilot identifies the cause of a sudden plane rotation.

TABLE 9.1

Comparison of the Latent Periods of Motion Reactions during Fault Detection

Confidence Interval Boundary of Reaction Time (Seconds)		Probability (p) of Differences between Pilots' Reaction Time on First and Repeated Faults (Wilkinson Criterion)
For unexpected fault (first) refusals	For anticipated (repeated) faults	$p = .402$ (i.e., the difference is statically insignificant)
0.28 ± 0.028	0.27 ± 0.013	

Influence is an acceleration of $30°/s^2$ at the first and repeated faults.

TABLE 9.2

Time Characteristics of Actions of Pilots with Different Awareness Levels

Participants	Maximum Time of Autopilot Turnoff (Seconds)	Cases of Instant (Less Than 2 Seconds) Autopilot Turnoff	Cases of Tardy Autopilot Turnoff (More Than 40 Seconds)
(Instructors) informed about faults	36	16%	–
Unaware test pilots	58	33%	11%
Unaware ordinary pilots	108	5.5%	33%
Ordinary pilots after repeated faults	42	65%	4%

The leading role in taking appropriate actions in this case plays the function block of self-regulation that is responsible for creation of a dynamic mental model of self-regulation (see Figure 9.1). The speed and adequacy of the performance of a conscious action depends on the efficiency of the interaction of two sub-blocks ("operative image" and "situation awareness"). It was discovered that the pilot can be aware of only an operative image component of the flight, which overlaps with the subblock SA. From the latter, it follows that the pilot cannot verbally describe all the components of the dynamic model and his or her strategies of performance in an emergency situation. This implies that we need to analyze the pilot's dynamic model by means of collation of recorded information about flight characteristics, the pilot's motion, visual and voice reactions through the interview about typical characteristics of an emergency situation, and signs of faults. Only a comparative analysis of the objectively obtained data in the experiment with the data obtained during interviews can help us to understand the pilot's dynamic model of the flight and his or her strategies for performance in the emergency situation.

The dynamic model of the flight is affected by the operator's goal, set, and those aspects of flight situation that are subjectively more significant for the pilot.

Application of the suggested complex approach has shown that noninstrumental signals do not present information sufficiently clearly for most pilots, although they detect the signal instantly. Moreover, pilots interpret the same signals differently. Different interpretations—and therefore, different identification efficiency—are defined as the pilot's ability, based on his or her previous experience, to form a correct dynamic model of the fault situation. The operative image of the fault situation is especially important if limited possibility exists for the verbal description of the fault signs.

Different identification and interpretation strategies are observed and they depend on the characteristics of the dynamic mental model of a flight, in general, and the operative image, in particular (see functional block "subjectively relevant task conditions," block 9 in Figure 9.1). The executive actions of pilots (program performance) also depend on the dynamic model of a flight. We describe five identification and interpretation strategies of fault situation next.

The *first strategy* is instantaneous identification, when the pilot's dynamic mental model of the situation coincides with the content of the current situation. Signals are compared with the mental model already developed in the past, and the *situation is identified instantly*. The pilot experiences a feeling of familiarity and perceives noninstrumental signals as definite information. Analysis of the situation flows unconsciously. Hence, in such a situation, an operative image of the flight is particularly important (Figure 9.1, subblock "operative image").

The *second strategy* is identification and interpretation of the situation after the mental search between alternatives is evaluated based on operative thinking. Operative thinking provides the development of a mental model that unfolds over time in the internal mental plane without addressing the external stimulus. The pilot has an adequate dynamic model and an operative image of the flight. However, within the model, similar signals are not differentiated sufficiently.

The *third strategy* arises when the pilot needs additional information for evaluation of the content of incoming information and development of a mental model of the flight. The pilot can identify information only after paying attention to the equipment, that is, identification and interpretation relies on additional signals and perceptual and thinking actions. The dynamic model and image of situation derived from experience are not completely adequate and, therefore, identification and interpretation of information cannot occur in the internal mental plane.

The *fourth strategy* is the identification and interpretation based on not only perceptual and thinking actions but also motor actions. The pilot cannot develop an adequate dynamic model and operative image of a flight based on the previous experience and mental analysis of situation. In addition, he or she needs to use the trial-and-error method, which includes motor actions and analysis of their consequences. Motor actions perform explorative functions.

The *fifth strategy* can be viewed as identification and interpretation, but very conditionally. It adjoins the fourth mode, but the difference is that it does not lead to a correct interpretation of the situation.
Our experimental findings are summarized in Table 9.3. It shows that the reliability of a pilot's actions depends on the identification and interpretation strategies of activity. Let us look at some features of a pilot's behavior in the process of fault identification and decision making.

All pilots receive identical signals—the abrupt change of the aircraft position. However, the uncertainty of information depends not only on its objective characteristics, but also on the strategies of orienting activity and is derived from the subsequent subjective interpretation of information. The effectiveness of a pilot's actions depends on the speed of translation of a vague signal into a definite signal, that is, the pilot has to compensate the deficiency of the information model and the safety of the human–machine system completely depends on human capability.

In the first strategy, information identification and interpretation is performed instantly. The pilot has to make no mental efforts to identify the situation. In the second strategy, an analysis of the situation is performed in the internal mental plane, without resorting to any external supporting means of activity. The analysis of the situation is performed in an abbreviated manner—a mental search and comparison of alternatives last from 2 to 20 seconds. Information becomes subjectively definite, and in a survey, the pilots assure that signs of such failure "cannot be confused." This strategy is close to the first one.

In other strategies, the pilot needs an external supporting means of activity. This requires additional explorative actions. Because there is no objectively given distinctive feature of failure in the external environment in a ready

TABLE 9.3

Efficiency of the Identification and Interpretation Strategies of Activity

Identification Mode	Identification and Interpretation Time	Estimation of Action Reliability	Number of Pilots Using This Strategy
I. Instantaneous identification and interpretation	2	Actions are reliable	11
II. Identification and interpretation after a mental comparison of signals	20	Actions are reliable	9
III. Identification and interpretation after a search of the additional visual information	60	Actions are insufficiently reliable because of a time lag	10
IV. Identification and interpretation after utilizing explorative motion actions	108	Actions are unreliable because of a time lag and possible harmful influences on the plane	11
V. Chaotic explorative motor actions	–	Actions are unreliable, fault is not identified	4

form, a pilot utilizes mental actions (third strategy)—and in more complex situations, external explorative motor actions (fourth strategy). This is the trial-and-error method. These actions do not always help in obtaining the required information. Thus, all other identification and interpretation strategies do not exclude erroneous actions in the process of transformation of vague uncertain signals into definite certain ones. Explorative strategies are primarily associated with the functional block 8. We further study the role of visual signals in the process of identification and interpretation of a situation. Hence, the filming of the sight process during a flight becomes important. The study demonstrates changes in strategies of gathering visual information. For instance, a pilot's visual fixation at high altitudes becomes longer (Tables 9.4 and 9.5).

Changes in the structure of information gaining are indirect indicators of the following unfavorable changes in a pilot's activity in an emergency. Emotional-motivational mechanisms influence the quality of interpretation of information (blocks 6 and 7, Figure 9.1). First, emotional tension increases because of a physically strong and psychologically uncertain stimulus. Second, the search for information that is necessary for decision making is intensified. Instruments cannot give the necessary information; therefore, it results in increasing of the emotional tension. Third, the significance of different displays is altered. As a result, the pilot pays increased attention to

TABLE 9.4

Total Duration of Sight Fixation on the Instruments (% of Sight Fixation on Each Instrument)

Flight Conditions	Attitude	Course Direction Indicator	Variometer	Speed Indicator	Altimeter	Other Instruments
Without faults	38	15	17	7	11	12
Roll fault	54	10	25	5	5	–
Pitch fault	39	7	38	5	8	–

TABLE 9.5

Average Duration of One Instance of Sight Fixation on Instruments (Seconds)

Flight Conditions	Attitude Indicator	Course Direction Indicator	Variometer	Speed Indicator	Altimeter
Without faults	0.7	0.7	0.6	0.6	0.7
With faults	1.5	0.5	0.6	0.5	0.5

instruments that, according to the pilot's experience, become more significant. Increased attention is caused by the need to make unusual efforts to control the handle and, as a result, "aircraft feeling" deteriorates.

All these unfavorable changes are the results of insufficient information that the pilot receives about the fault. This means that there is no absolute assurance of success in the pilot's actions during such decision making.

9.7 Pilot's Strategies of Activity while Receiving Visual Information

Identification of errors is the most important point for our research. In this section, we analyze information representation by means of an alphanumeric display with an illuminated board (these indicators *use* alphanumeric and symbolic characters for representation of information) and pointer indicators.

Table 9.6 clearly shows that the pointer indicators have a low attraction effect, which in turn reduces a pilot's chances to detect a signal and act rationally. Moreover, this is irrespective of a pilot's training, the level of the pilot's operative thinking, and the content of his or her dynamic model and operative image, in particular. Pilots can act purposefully only after the signal-detection stage. Hence, the task of ergonomists and psychologists is to find ways to help pilots detect the required signal, that is, to adapt the technology to a human

TABLE 9.6

Quality Characteristics of Pilots' Actions

Type of Informational Model	Probabilistic Characteristic (p)			Time Characteristics (seconds)		
	Detection	Identification	Errors	Detection From–to (Mean)	Identification From–to (Mean)	Action From–to (Mean)
Flight experiment						
Illuminated board	0.98	0.93	0.14	1–58 (17.3)	1–45 (13.5)	1–52 (15.2)
Pointer indicators	0.53	0.53	0	4–186 (25.8)	3–5 (3.5)	4–32 (11.1)
Ground experiment						
Illuminated board	0.94	0.88	0.18	1–109 (8.2)	4.0–32 (9.1)	1–68 (12.3)
Pointer indicators	0.51	0.51	0	5–405 (127.5)	4–22 (12.3)	0.6–36 (14.2)

being. For example, a correctly designed illuminated board helps pilots not only to detect the signals on time, but also to interpret the meaning of the input information adequately (Figure 9.1, functional block 1). In this case, an active thinking process is not required; the pilot has to memorize the instructions only. After signal detection, the pilot's actions are determined by the informational model and the memorized scheme of actions.

Verbal–visual representation of information facilitates identification. In most cases, information processing lasts from 0.5 to 5.5 seconds, on an average. Verbal representation of information removes the information uncertainty and shortens the process of decision making. Our data show that the time of receiving information from such instruments is shortened by 30% in the case of verbal representation of fault. Our findings also show that verbal information does not interfere with a flight's control. It makes possible the combining of piloting and fault-elimination actions. If fault is represented by a text on the displays, the length of the decision-making process is determined by the following two factors: (1) sufficient attraction effect and (2) semantic correctness of a text, which can provide a meaningful interpretation of the input information.

Verbal information is always interpreted in accordance with a goal that is accepted or formulated by the pilot (see the relationship between blocks 1 and 2, Figure 9.1). In addition, block 6 is also critical for the interpretation of verbal information in stressful situations (see the relationship between blocks 7, 6, and 2; Figure 9.1). From this statement, it follows that for the correct interpretation of information, it is important to take into account the goal and emotional-motivational components of activity. Interaction of these components of activity is often overlooked in cognitive psychology. A verbal message is a type of signal coding that satisfies the requirement of meaningful interpretation of input information in emergencies. The possibility of missing a verbal emergency signal by a pilot is virtually ruled out. The pilot is sure of timely receipt of the emergency signal and the complexity of meaningful interpretation of information is reduced. This factor helps the pilot to concentrate on the basic task of piloting without being distracted by tracking of plane systems.

Representation of emergency signals with a human voice is a new type of signaling and it is different from other types, even from written messages. It is difficult for a pilot to relate to voice messages, a signal lamp, or a title on the display. Verbal signals are not only easier to interpret, but they become more significant for humans (blocks 6 and 7). We are concerned about the possibility of increasing the pilot's suggestibility with voice messages that can lead to inadequate reaction to the voice message. There also are other problems with verbal signals. For example, it is difficult to ensure the reliable identification of two verbal signals at the same time, which can happen if a radio exchange and emergency signals are overlapped. To increase the reliability of signal identification in this type of a "voice mix," it is necessary to follow the recommendations developed by psychologists. The first recommendation is to use voices with different timbres and pitches. For example, a female voice report reporting about the emergency with the background of a male voice from the ground control. The second recommendation is as follows: during training, pilots should be informed that emergencies are reported in a female voice.

The findings show that it is reasonable to use voice signalization to increase the reliability of signal identification (Ponomarenko and Lapa 1985). However, many problems of voice usage under extreme flight conditions are still unsolved. We have solved some of these problems in experiments using a flight simulator. We have conducted an experiment with a single-seated aircraft simulator equipped with a prototype of a signal system.* The fault simulator changes the indications of the corresponding instruments to imitate a specific type of fault (for example, in a fire, the temperature of exhausts was increased) and the pilot hears a voice message through the headphone.

In the experiment, the pilot has two tasks—complete a low-height flight, and while enroute the flight, react to faults correctly and on time. Sixteen pilots participated in the experiment (4 test pilots and 12 regular plane pilots, ages 22–40). The pilots made 52-hour-long flights. The 248 faults were imitated during all the flights. Each pilot made two or four flights with two types of signaling systems, which were used alternately. In the first type of system, messages were in the form of ascertain; in the second type, messages were in the form of step-by-step instructions. According to the experimental program, voice messages were introduced as follows: alone or simultaneously with light indications; against the background of the pilot's radio exchange with the experimenter (voice noise); and a verbal message with a delay of 5–10 seconds after the light indications.

---

* The experiment was led by Zavalova and Lapa.

The goal of the experiment was to reveal the pilot's activity characteristics on reception of the emergency voice signals. We introduced the voice signals at stages of the flight where accurate piloting or looking outside the pilot's cabin for ground signal observation was required. The findings proved the high efficiency of the voice signals. We were interested in getting answers to the following two questions:

1. Do long voice signals (up to 15 seconds) interfere in the pilot's actions? Analysis of oscillograms, which registered the duration of a voice signal, showed that actions start and end during the receiving of a voice signal, and the signal did not interfere with action. If a voice signal is delayed after the visual signal, then the beginning of speech did not interfere with actions either. This can be explained by the fact that standardized verbal signals are simpler for a pilot and therefore processing of these signals can be more automatic and require less concentration.

2. Do the characteristics of the actions depend on message formulation? It turned out that the messages in the form of ascertains and step-by-step instructions influence the structure of a pilot's actions differently. For example, messages about fire, with the instruction "reduce plane velocity" (prescribe the order of actions) and simply with a statement that "there is fire" (did not prescribe the order of actions), produce a different order of pilot's actions, even for skilled pilots who knew the manual very well. The significance of the different indicators for a pilot had changed after receiving the verbal information. As a result, the order of the pilot's action for gathering the required information was also changed after receiving the instruction: the pilot tracked the velocity indicator more frequently. Therefore, functional blocks 6 and 7 (evaluative and inducing components of motivation) can influence the cognitive functions of a pilot. The type of verbal message influences the structure of a pilot's actions. Instructional forms of messages impose on the pilot a formally correct but extrinsic order of actions. Nevertheless, knowledge of the instruction does not compel a pilot to follow it. An exact following of instructions is a characteristic of apprenticeship, not of mastery. Such an obligation to instructions may reduce skilled pilots' responsibility and their active involvement.

An assessment of the pilots' subjective judgments and attitude revealed that most of the pilots prefer ascertain type of messages because they provide switching of attention to an event and do not deflect from actions and their independent performance. However, participants approve the usefulness of brief step-by-step instructions in case the unskilled pilot forgets the order of actions. In any case, instructional step-by-step commands should be used in events where a pilot has no reserve of height and does not know what actions to choose. If a pilot has time to think, then step-by-step instructions should be avoided.

Due to the high subjective significance of voice messages, wrong voice messages may lead to a much greater adverse result than the wrong messages of other types of indicators. The probability of erroneous actions after receiving a wrong voice message is high. For example, in response to a wrong message "turn off engines," in 11 of 27 cases, the pilots turned the engines off although there was no necessity to do it. In general, pilots performed unnecessary actions in response to incorrect verbal commands in 12 of 22 cases. Such an influence of voice messages demands enhanced requirements to ensure the technical reliability of a voice-signal system. Thus, the factor "significance of information" influences its interpretation.

9.8 Discussion

Activity is a complex multidimensional system, which should be studied from different perspectives by utilizing different methods. All methods of analysis in systemic-structural activity theory are organized into four iterative stages: the qualitative stage, the algorithmic stage, the time-structure analysis stage, and the quantitative stage. The qualitative stage includes diverse methods. The most important of them is the functional analysis of activity, which utilizes a systemic, qualitative analysis of activity. This method considers activity as a complex goal-directed self-regulative system. Currently, the concept of self-regulation is understood differently in various areas of psychology. Theoretical models of self-regulation that lie outside of activity theory describe human behavior as a reactive self-regulative process. The proposed models reduce self-regulation to elimination of errors committed at the behavioral level. Such models do not explain how behavior or activity is regulated in the absence of errors made at the behavioral level. For example, the pilot can control flight based on analysis of anticipated events even when there are no disturbances and errors at the time. Such mechanisms of self-regulation as conscious goal-setting, the formation of mental models of situations, forecasting, formation hypotheses, planning, and so on are not discussed in the primitive homeostatic models of self-regulation.

In contrast to the functional analysis of activity, the other qualitative methods of analysis concentrate their efforts on separate parameters of activity. For example, cognitive analysis is a parametric method of study because this method does not use systematic principles for the analysis of human behavior or activity. Therefore, cognitive analysis should be combined with functional analysis for studying activity. In this chapter, we concentrated our efforts on the functional analysis of pilots' activity in emergency situations.

Functional analysis becomes possible when we develop models of self-regulation of activity. The major units of analysis in this situation are functional mechanisms or blocks. Each functional block can be considered a stage of information processing with a particular purpose in activity regulation. Each functional block can include different cognitive processes depending on the specificity of task performance. These stages have a nonlinear looped structure organization, with multiple feed-forward and feedback influences. Basic concepts, such as a goal, conceptual and dynamic mental models, meaning, evaluative and inducing components of activity, and subjective criteria of evaluation of performance results, are considered functional mechanisms or blocks. For the application of the self-regulative model of activity, each functional block is considered a different window through which we consider the same activity during task performance. Hence, we have multiple representations of the same activity, and these representations can be compared with each other. During any particular study, the most important parameters for the given functional blocks can be considered.

In this study we did not use the general model of self-regulation. Rather, we utilized the self-regulative model of orienting activity when the process of self-regulation is carried out mainly in the mental plane. External motor actions can be performed only periodically. The main purpose of external motor actions is not just actual transformation of the external situation but rather analysis of these actions' consequences. The major purpose of motor actions is explorative. It is natural that such actions cannot be used in all cases. That is why in orienting activity explorative actions or operations are carried out mainly in the mental plane. Thus, we wanted to demonstrate that existing models of self-regulation outside of activity theory are not adequate, because they cannot function in the absence of external motor actions and errors.

Usually, functional analysis utilizes experimental procedures in combination with observation and subjective judgments of experimental events by different subjects involved in this experiment. Discrepancies between these data are an important source of information for the analysis and discussion of obtained data. Individual differences in the interpretation of data by subjects are also an important source of information for further analysis. Sometimes, additional experimentation is required after such an analysis. Experiments can be conducted not only by utilizing simulators, but also in real flights. The simulator cannot always reproduce a real, psychologically adequate work environment. Sometimes, a combination of these methods is needed from a safety standpoint.

Particular attention should be paid to the cognitive strategies of activity performance. Any motor response should be analyzed in unity with cognitive regulative mechanisms based on the important principle of unity of cognition and behavior in activity theory. Selection of the most efficient strategies of performance with the subsequent acquisition of these strategies by pilots is an important practical application of the obtained data. The obtained data can also be used for the enhancement of equipment designs. This can be explained by the fact that changes in the configuration of equipments can change the strategies of performance in a probabilistic manner.

At present, experimental procedures and their interpretation are performed from different, insufficiently developed, and integrated theoretical perspectives.

An experiment and the interpretation of the obtained result depend on the intuition of the researcher to a significant degree. Functional analysis provides integration of diverse data, which unifies all available information and helps to interpret it from the unitary theoretical position. The experimental data are analyzed and interpreted based on standardized principles of analysis. This method helps us to use data obtained in cognitive psychology more efficiently. Analysis of the obtained data gives us an opportunity to formulate conclusions that are more specific.

Experimental data demonstrate that a signal that arises in a given situation can be instrumental or noninstrumental, relevant or irrelevant. Pilots have to know how to select relevant information and distinguish it from the irrelevant one. Furthermore, noninstrumental signals are particularly important in emergencies. Special training procedures should be developed that help pilots transfer noninstrumental signals into pragmatic signals that can be used for interpretation of emergency situations. Noninstrumental signals can significantly change the strategies of interpretation and transformation of information in emergency situations.

In emergency situations, erroneous perception and interpretation of correct information significantly increases because an incorrect formation of both the goal and a dynamic model of the situation provoke a false hypothesis about the situation, which blocks the receipt of adequate information. This is particularly relevant to an emergency, when a task should be completed under time-restricted conditions. Under such conditions, the significance of different instruments changes and, therefore, the strategies for gathering such information change as well. Indirect, noninstrumental signals that help to quickly detect and interpret the failure require quick confirmation by instrumental signals. In the absence of such information, a collision of different hypotheses can happen, which makes it more difficult to make the right decision. Finally, a pilot must be able to predict the course of the developing events and determine their causes. Therefore, anticipation of events in an emergency is critical. The dynamic mental model of the events is particularly important in this situation.

Training to act correctly in an emergency must provide the knowledge of not only how to use different motor actions for the manipulation of controls but also help to correctly interpret the emergency situation. Therefore, identification strategies should be developed for various emergencies. Special rules and heuristics for the identification of such situations can be developed. Therefore, an algorithmic description of activity can be useful in such situations. Two types of algorithms can be used here. One is called the algorithm of identification, and the other is called the algorithm of transformation. This is the second stage of activity analysis. This stage is called "algorithmic description of activity" and is performed after the qualitative stage of analysis, particularly after the functional analysis of activity.

References

Anokhin, P. K. 1969. Cybernetic and integrative activity of the brain. In *A Handbook of Contemporary Soviet Psychology*, eds. M. Cole, and I. Maltzman, 830–57, New York: Basic Books.

Bandura, A. 1997. *Self-Efficacy: The Exercise of Control*. New York: W. H. Freeman.

Bedny, G. Z., and W. Karwowski. 2006. The self-regulation concept of motivation at work. *Theor Issues Ergon Sci* 7(4):413–36.

Bedny, G. Z., and W. Karwowski. 2007. *A Systemic-Structural Theory of Activity. Application to Human Performance and Work Design*. Boca Raton, FL: Taylor & Francis.

Bedny, G., and D. Meister. 1997. *The Russian Theory of Activity: Current Application to Design and Learning*. Mahwah, NJ: Lawrence Erlbaum Associates.

Bernshtein, N. A. 1966. *The Physiology of Movement and Activity*. Moscow: Medical Publishers.

Boekaerts, M., P. R. Pintrich, and M. Zeidner., eds. 2005 *Hand-Book of Self-Regulation*. San Diego, CA: Academic Press.

Carver, C. S., and M. F. Scheier. 2005. On the structure of behavioral regulation. In *Handbook of Self-Regulation*, eds. M. Boekaerts, P. R. Pintrich, and M. Zeidner, 255–74. San Diego, CA: Academic Press.

Endsley, M. R. 2000. Theoretical underpinnings of situation awareness: A critical review. In *Situation Awareness Analysis and Measurement*, ed. M. R. Endsley and M. Garland, 3–32. Mahwah, NJ: Lawrenece Erlbaum Associates.

Konopkin, O. A. 1980. *Psychological Mechanisms of Regulation of Activity*. Moscow: Science Publishers.

Leont'ev, A. N. 1978. *Activity, Consciousness and Personality*. Englewood Cliffs, NJ: Prentice Hall.

Locke, E. A., and G. P. Latham. 1990. Work motivation: The high performance cycle. In *Work Motivation*, eds. U. Kleinbeck, H.-H. Quast, H. Thierry, and H. Hacker, 3–26. Mahwah, NJ: Lawrence Erlbaum Associates.

Ponomarenko, V. A. 2006. *Psychology of Human Factor in Dangerous Professions*. Krasnoyarsk, Russia: International Academy of Human problems in Aviation and Astronautics, and Krasnoyarsk Aviation Technical College.

Ponomarenko, V. A., and V. Lapa. 1985. *Profession-Pilot*. Moscow: Military Publisher.

Ponomarenko, V., G. Z. Bedny, and R. N. Makarov. 2006. Activity pilot in flight and its imaginative components. *Triennial International Congress*, Maastricht, the Netherlands.

Preece, J., Y. Rogers, Y. Sharp, H. Benyon, D. Holland and S. Carey. 1998. *Human-Computer Interaction*. Harlow, UK: Addison-Wesley.

Pushkin, V. N. 1978. Situational concept of thinking in activity structure. In *Problems of General and Educational Psychology*, ed. A. A. Smirnov, 106–20.

Rubinshtein, S. L. 1959. *Principles and Directions of Developing Psychology*. Moscow: Academic Science.

Sokolov, E. N. 1963. *Perception and Condition Reflex*. New York: Macmillan.

Tikhomirov, O. K. 1984. *Psychology of Thinking*. Moscow: Moscow University Publisher.

Uznadze, D. N. 1961. *The Psychology of Set*. New York: Consultants Bureau.

Vancouver, J. B. 2005. Self-regulation in organizational settings: A tale of two paradigms. In *Handbook of Self-Regulation*, eds. M. Boekaerts, P. R. Pintrich, and M. Zeidner, 303–42. San Diego, CA: Academic Press.

Wiener, N. 1948. *Cybernetics: Or Control and Communication in the Animal and the Machine*. Cambridge, MA: The MIT Press.

Zarakovsky, G. M., and V. V. Pavlov. 1987. *Laws of Functioning Man-Machine Systems*. Moscow: Soviet Radio.

Zavalova, N. D., and V. A. Ponomarenko. 1980. The structure and content of image as mechanism of regulation of actions. *Psychol J* 1(2):37–50.

Zavalova, N. D., B. F. Lomov, and V. A. Ponomarenko. 1986. *Image in Regulation of Activity*. Moscow: Science Publishers.

Zimmerman, B. J. 2005. Attaining self-regulation: A social cognitive perspective. In *Handbook of Self-Regulation*, ed. M. Boekaerts, P. R. Pintrich, and M. Zeidner, 13–41. New York: Academic Press.

10

Functional Analysis of Pilot Activity: A Method of Investigation of Flight Safety

V. Ponomarenko and W. Karwowski

CONTENTS

10.1 Introduction

Records regarding aviation accidents indicate that 60–80% of accidents result from erroneous actions of flight crews. The main types of erroneous actions of the aircrew are deviations from established rules and procedures for flight operations, overconfidence, incorrect operation of equipment, and other related factors. Surveys show that the violation of rules and flight procedures are due to factors such as lack of discipline, insufficient professional preparation for operation in atypical circumstances, the absence of an individual approach to training, a tendency to simplify the content of training methods, and overvaluation of professional qualifications. This chapter discusses these factors.

One of the most important detractors to flight safety is the disconnection between equipment and human cognitive requirements and capabilities for operating that equipment. Other safety concerns stem from deficiencies in the protection of the body against ideally bound conditions such as temperature, noise, and vibration levels.

Different types of aircrafts require varying lengths of roll on takeoff. The roll requires suitable control actions. A rapid transition from one aircraft type to another may cause premature takeoff or abortion of the takeoff. Studies of aborted takeoffs show that the pilot often incorrectly concludes that there is insufficient runway left, insufficient thrust, or decreasing thrust levels. Accommodating the previous type of aircraft takeoff roll causes the development of conflicting information in the pilot. Prior experience tells the pilot to begin rotation, while musculoskeletal signals deem that the speed is not sufficient. As a result of this conflict, the pilot begins to mistrust or ignore the aircraft instruments and relies on faulty data to make critical operational errors. Therefore, the pilot aborts the takeoff, or worse, rotates prematurely and takes off with an insufficient airspeed.

Flight safety depends in part on the effectiveness of the pilot's professional training. The effectiveness of the flight simulators is insufficient. Scores illustrating the effectiveness of simulators for aircraft landing show a rating of 0.7; for special situations during flight, 0.5; for visual flight, 0.3; and for takeoff, 0.2. We must consider the benefits of the retention of training, as breaks in training may lead to loss of skills and knowledge.

Humans have limitations regarding spatial orientation during flight. Of particular importance are factors such as orientation in space, changing direction of flight, vestibular and visual illusions, and the perception of three to five simultaneous flows of information. In the case of receiving information from simultaneous flows, the pilot changes the point of visual fixation up to 200 times a minute. Even with effective technology and a well-trained crew, the demands of the situation may overwhelm human psychophysiological possibilities. Such situations might lead to catastrophe. Social factors could also have a significant influence on flight safety.

Therefore, flight accident prevention requires a joint study of human and technical aspects of each incident, that is, a comprehensive study of the properties of the individual, the specificity of particular activity, and the pilot's work conditions. Ponomarenko and Zavalova (1994), Ponomarenko et al. (2003), Bodrov (1989), and Kotik (1978) made substantial contributions to the development of these complex approaches. The functional analysis of the pilots' activity is especially important (Bedny and Karwowski 2007). In this chapter, we describe some of the concepts of this approach.

10.2 Pilot–Aircraft–Environment as a System

A pilot's work activity is performed within the "pilot–aircraft–environment" system. The framework of this system is presented in Figure 10.1. A central element of the system is the pilot, who operates the aircraft and ensures the achievement of the flight goal.

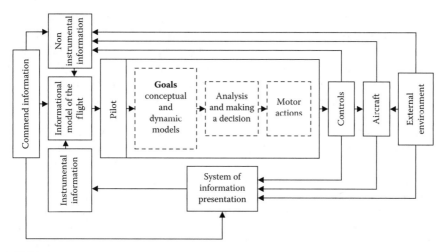

FIGURE 10.1
"Pilot–aircraft–environment" system.

During aircraft operation, the pilot interacts with the aircraft and its equipment by means of controls and communication with the external air-traffic controller. The accumulation of all information that a pilot receives for successful flight operations is called the "informational model of the flight." This informational model consists of three components: instrumental information, noninstrumental information, and command information. Instrumental data are focused on the dashboard and are passed on to the pilot through visual devices such as attitude indicator, vertical speed indicator, altimeter, and course and airspeed indicators.

The attitude indicator or artificial horizon shows the pitch and roll angle (i.e., angular coordinates defining the orientation of the aircraft relative to earth). The vertical speed indicator informs the pilot of the instantaneous rate of descent or climb of the aircraft. Flight and navigational information, as well as information on the aircraft systems status, is presented on control displays as shown in Figure 10.1.

The primary means of information display is the visual control panel. The noninstrumental information includes an overview of cockpit space, angular and linear acceleration of the aircraft, vibration, noises, haptic feedback of the controls, and even smells. Additional information is also provided by the pilot's perception of control devices and the applied force, which depends on the flight mode and stage of the mission (takeoff, landing, and cruise).

The command or operational information about the flight stage is provided in the form of commands from the air-traffic controller or from the flight leader. The command may come from the radio directly to the pilot, and in this case, it should be considered noninstrumental information. In other cases, these commands can be transferred through devices in the form of specified parameters of flight.

Environmental factors include gravity, magnetic field of the earth, weather conditions, and pilot dead reckoning. Devices that transmit the information include radios, radiolocation devices, radionavigation devices, and global positioning systems. Factors in the external environment include decreased atmospheric and partial pressure of oxygen, sharp fluctuation in barometric pressure, and extreme temperature ranges. These characteristics affect the pilot's condition and ability to maintain ideal levels of aircraft control.

The process of aircraft operation can be described using a model of self-regulation of activity (Figure 9.1, see Chapter 9). The goal of the piloting activity is to transfer the aircraft from one area to another while keeping the aircraft in a safe state and overcoming the external perturbations that upset that safety state. Based on available information from various sources, the pilot builds a mental model of the current situation, as well as forecasts the aircraft position in the absence and presence of control actions. This is called "staying ahead of the airplane." Based on the mental model and formulated goal, the pilot creates a series of response actions and then performs these actions. These pilot actions have an impact on the operational devices, which cause a change in the aircraft position or mode of equipment operation.

Information about the effect of the pilot's action on the operation controls, aircraft's position, and environmental conditions are provided to the pilot as feedback. Such information may come through displays in the form of instrumental information or directly to the pilot as noninstrumental information. The pilot evaluates this information in light of achieving the goal. This way, the pilot can evaluate the result of the overall goal and the results of individual actions as well. We want to draw attention to the fact that the executive and the evaluative stages of self-regulation of activity are presented in the general model of self-regulation (Bedny and Karwowski 2007). Depending on this internal evaluation, the pilot may adjust specific inputs or carry out large-scale changes to his or her actions. The proposed model "pilot–aircraft–environment" (Figure 10.1) could be useful for the analysis of this pilot activity.

Currently, many of the functions of information preparation for decisions are measured and compiled by air-borne and surface-computing devices. Most of the pilotage functions (minor control inputs in autoflight) are also automated. Due to automation eliminating the need for constant attention, surveillance and control activity are important in maintaining safe flight, especially if there are problems. This in turn increases the importance of functions such as vigilance, which can be defined as the readiness to act in an unexpected situation. Cognitive functions, such as vigilance have to be distinguished from others, such as the reaction time of pilots, as these two properties may have no bearing on each other. Vigilance is crucial in addressing the automation of the flight. The pilot's role in the automated system should not make him or her passive and unable to respond appropriately. The role of the pilot in such situations should be viewed as "hot standby." Certain functions of the manual flight should be allocated to the pilot (Zavalova et al. 1986).

Thus, the system "pilot–aircraft–environment" is the focus of this study. The components of this system have their own inherent qualities. However, the elements of this system possess systemic characteristics, which manifest themselves through the interaction of its components. The simplest example is the pilot's seat. If we consider the chair only in terms facilitating the ease of control manipulation, we can make no claims regarding the chair's comfort level. The convenience of viewing different displays should also be included in the consideration. Furthermore, it is important to include the chair's effectiveness for ejection (when appropriate) in the analysis. Finally, we also have to take into account the extent to which the chair is appropriate in terms of spatial characteristics of cockpit, design, weight, cost, ease of maintenance, and a host of other design characteristics.

According to systemic-structural activity theory (SSAT), it is necessary to distinguish between the "man–machine interface" and human activity in a system's standpoint (Bedny and Karwowski 2007). Examination of the activity as a system requires the selection of standardized units of analysis. This approach allows the comparison of structural characteristics of the equipment with regard to the activity.

10.3 Study of Pilot Activity from the Functional Analysis Perspectives

In a complex activity, it is often impossible to distinguish between the mental processes of sensation, perception, memory, and reasoning. These processes are all interrelated in complex ways. For example, perception is related to reasoning. The process of perception involves memory. The process of identification includes image recognition, and so on. That is why a functional analysis of the activity suggests that the activity should be analyzed not only in terms of psychological processes, but also in terms of functional blocks, which are the main mechanisms of regulation of activity.

Functional and cognitive analyses are interrelated. Each functional block integrates cognitive processes in a different way depending on the functional specialty of the block and the nature of the problem. The system moves from analysis to cognitive function and vice versa. Functional analysis is a qualitative systematic analysis of the activity in which the activity is considered a self-regulating system.

The model of self-regulation of orienting activity, which is presented in Section 9.4 does not describe the executive aspects of self-regulation activity. For this purpose, we need to utilize a general model of activity and self-regulation (Bedny and Karwowski 2007). This model includes functional mechanisms "formation of a program of task performance," "making a decision," "program of performance," "subjective standard of successful result," and so on. Each functional mechanism or functional block determines the range of issues that need to be examined in the analysis of activity. The choice of functional block for the analysis depends on the specificity of the tasks. Models of self-regulation activity also allow the examination of the relationships between the mechanisms or blocks themselves. Activity models of self-regulation allow possibilities of analyzing the strategies employed during the performance of various tasks from a common, theoretical point of view. Fundamental aspects of work such as goal, motives, significance and arousal level, planning of activity, decision making, execution, and self-assessment are better considered from the functional analysis perspective, where activity is studied as self-regulative system.

The functional analysis of activity dates back to the work of Anokhin (1962) and Bernshtein (1966). A more complicated concept of a reflex system, which included feedback, was introduced, replacing the theory of a reflex arc of "stimulus-response" reaction. This was performed long before the emergence of cybernetics. According to the concept of a reflex circuit, the body evaluates its own activity and corrects it. According to Bernshtein (1966), the body interacts and adapts to the environment and also alters the local environment. The basis of such behavior is not limited to a feedback mechanism, but a mechanism capable of predicting future events based on which expectations about a desired outcome can be developed. These predictive activity mean an organism can plan his or her behavior to achieve these desired outcomes.

Bernshtein's studies focused on motor actions. Anokhin concentrated his efforts on developing a description of conditioned reflexes operating as self-regulative systems. A conditioned reflex is described as a dynamic organization developed for a particular purpose. This organization selectively integrates different central and peripheral neural mechanisms. Any conditional reflex also includes feedback. Anokhin and Bernshtein introduced their concepts of self-regulation before the concept of feedback was introduced by Wiener (1958). A detailed presentation of these concepts is beyond the scope of this chapter. Their description can be found in a book edited by Cole and Maltzman (1969) and in Bedny and Meister (1997). We briefly mentioned the work of these scientists here as their work is important for understanding the principle of functional analysis and why it arose.

The physiological self-regulation model did not take into account the limitations of the application of physiological models to the explanation of psychological processes. There was a need to develop a model that described the self-regulation process at the psychological level. Self-regulation at the psychological level was first described by Konopkin (1980) and Kotik (1974). The self-regulation model described by Konopkin received the greatest recognition and was eventually accepted as a theoretical framework. The proposed model describes self-regulation of sensorimotor activity. The model contains seven functional blocks of self-regulation mechanisms. Bedny (Bedny 1987; Bedny and Zelenin 1988) also developed concepts regarding psychological aspects of self-regulation. Currently, the concept of activity self-regulation proposed by Bedny and Karwowski (2007) is the most detailed and developed. The overall model of self-regulation proposed by these authors includes 20 functional blocks. This model describes not only the cognitive but also the emotional and motivational aspects of self-regulation. It also includes conscious and unconscious levels of self-regulation. The concept of self-regulation proposed by these authors gives the clearest description of the functional blocks and their use in the analysis of specific occupations. The model of self-regulation of orientation activity is presented in our joint work with Bedny in this book (see Section 9.4 of this book). As can be seen from the model, unconscious levels of self-regulation include functional mechanisms such as the "mechanism of orienting reflex" introduced by Pavlov (1927), "afferent synthesis" suggested by Anokhin (1962), and "set" described by Uznadze (1961). These mechanisms are considered in the framework of psychological self-regulation of activity. The role of these mechanisms in the process of unconscious psychological self-regulation of activity was presented in Bedny and Karwowski (2007). However, the leading role in this model belongs to the conscious level of self-regulation. The advantage of the described models is the fact that they reveal the relationship between the conscious and unconscious levels of self-regulation.

10.4 Examples of Functional Analysis of a Pilot's Activity

Pilot activity can be described in general terms as a process of perception and goal formation followed by information processing, which leads to the creation of a mental model of the flight, execution of control actions, and evaluation of the outcomes. Analysis of the pilot's eye movement patterns shows that the strategy of the movements of the pilot's gaze depends on the requirements of tasks and the subjective importance of selected information. An active focus of attention depends on the relevance attributed to the object of that focus, which is linked to the functional mechanism "assessment of sense of task" (see Figure 9.1). The change of attention from one focus to another cannot be always equated with the movement of gaze to the desired focus. There are cases in which the pilot is looking at the device yet does not perceive the appropriate information. This can be explained by an information significance factor. The functional block "assessment of sense of task" demonstrates the importance of such factor. Subjective significance is related to an increased activation of the neural centers of the brain involved in certain activity. The higher activation of these centers causes inhibition of other neural centers, and thus the conscious attention switching ability to new stages of activity is hampered. The gaze movement can be conducted on an automated, stereotypical level. However, the process of conscious switching of attention and the related gaze movements are the most important.

The pilot can subjectively evaluate the same task or its components as more or less significant. Tasks deemed more important by the pilot elicit higher activations of neural centers involved in regulation of activity. This activation directly affects the process of attention switching. The functional block "sense" is responsible for evaluating the significance of a task or situation. It is the evaluative, cognitive-emotional component of activity. The pilot may under- or overestimate the objective significance of a task or situation. This estimation error also influences the process of attention switching (Bedny and Karwowski 2007).

For example, experiments support that during the rapid eye movement from observation of information outside the cabin to inside the cabin instrument, such as the altitude indicator, the pilot will fail to notice a change of 5–7° in 30–60% of the time. This is explained by the fact that the differentiation of outside-cabin reference points is characterized by a higher subjective significance and requires greater attention. From the standpoint of functional analysis of activity, the significance of the factor increases the concentration of attention. The higher the concentration of attention to one focus, the more difficult it is to switch attention to a new focus.

The dynamic model of the situation (functional block "subjectively relevant task conditions," Figure 9.1) includes information on the current state and near-future state. It may also include information on previous events. This means that a dynamic mental model of the situation also includes information about the possible future states of the spatial location of the aircraft and causality of developing events.

This information allows pilots to rapidly process the information. The pilot, referring to an instrument, does not read the indication as an entirely new visual idea or concept, but compares the current reading with a dynamic mental model of the situation. Therefore, he or she can rather precisely perceive the gauge indication in a short time of eye fixation. In such situations, the pilot is not looking for an unknown event, but for the confirmation of predicted and expected information. An experienced pilot has a device reading time 1.5 to 2 times shorter than that of less-experienced pilots. However, it was found that the speed of perception of younger and less-experienced pilots was faster than the experienced, but older pilots. The foresight of gauge readings frees attention resources for the performance of other elements of the task, but in some cases, this can lead to perception errors, especially when actual indications do not match expected readings. Analysis of self-regulation mechanisms allowed the establishment and clarification of the relationship between psychological and physiological mechanisms of activity.

The significance and activation of neural centers are interconnected. A subjective evaluation of complexity (functional block "assessment of task difficulty") has an impact on the significance assessment. If a pilot underestimates the difficulty of an objectively complex problem, then the concentration of attention on the task may be insufficient. Consider the function of "conceptual (stable) model" block, which was described in Chapter 9. This functional mechanism contains information about the entire upcoming flight. The dynamic models (functional block "subjectively relevant task conditions") of flight evolved against the background of this mental model. Consider the example of an experimental analysis of this functional mechanism in conjunction with some other functional mechanisms or blocks of the pilot's activity regulation. In practice, the preparation for flight raises the question of the subjective assessment of the degree of complexity and significance of the upcoming flight and its stages. Not all possible situations can be predicted. The pilot often cannot objectively assess the difficulty and significance of future flight or its individual stages, even if he or she is familiar with them in advance. Verbal responses do not always reflect the correct situation (functional block "assessment of task difficulty" is also described in the general model of activity self-regulation; Bedny and Karwowski 2007).

To perform an objective assessment of the pilot's attitude toward future flight events, physiological indexes can be used. Sports science investigates this problem as a prelaunch condition of an athlete. In flight conditions, such predictions occur as a result of the construction of the conceptual model of the flight by the pilot, which assesses its significance and difficulty. Emotional memory has special importance during this predicting process.

This problem occurs as a result of the construction of a conceptual flight model and assessment of its significance and difficulty. Pilots familiarize themselves with the upcoming flight, thinking about the main flight stages and creating a conceptual model of the flight. To support the existence of a relationship between flight difficulty and its subjective impression of significance, the experiment discussed next was conducted.

10.4.1 Method

The study involved experienced test pilots with an average age of 30–40 years and 15 years of job experience. Participants were asked to perform test flights. All participants had a clear understanding of the difficulty and dangers that may arise during future flight. The flights in the experiment were divided into four groups of complexity, (rated from 1 to 4, where 1 was the least complex and 4 was the most complex). Physiological functions were measured before the flights by using special indexes.

10.4.2 Results

The results of these measurements for the 20 subjects in this group are presented in Table 10.1.

The results support that the highest complexity level (level 4) caused the largest changes in physiological indexes. The content of task 4 consisted of new equipment testing without natural reference points during close to the land flight.

The differences in physiological measures between groups 1 and 4 were statistically significant at 0.05 ($p < .05$). Increased heart rate, breathing rate,

TABLE 10.1

Effect of Difficulty Level and Significance of Upcoming Flight on Physiological Measures

Difficulty Level	Physiological Indexes					
	Heart Rate (bpm)		Breath Frequency (cycle/min)		Concentration of Sugar in Blood (mg %)	
	Baseline	In Cabin before Flight	Baseline	In Cabin before Flight	Baseline	In Cabin before Flight
1	64 ± 6	75 ± 5	12 ± 2	12 ± 4	65 ± 3	58 ± 2
2	64 ± 6	84 ± 2	12 ± 2	12 ± 6	65 ± 3	76 ± 4
3	64 ± 6	84 ± 5	12 ± 2	12 ± 4	65 ± 3	92 ± 6
4	64 ± 6	92 ± 4	12 ± 2	14 ± 5	65 ± 3	120 ± 12

and an increase in the blood glucose level indicated an increase in the sympathetic nervous system activation. This activation allowed the body to function at increased efficiencies and rates, affording the performance of more complex and demanding tasks. There was a marked interaction of psychological and physiological levels of activity regulation. Experienced pilots could clearly identify the complexity and importance of the upcoming flight. At the same time, it was possible that inexperienced pilots could underestimate the complexity of the upcoming flight of a level 4 flight and could overestimate the complexity of a level 1 or 2 test flight. Underestimation could result in different physiological measures. The data obtained can be explained through the analysis of mechanisms or functional blocks such as "conceptual model," "assessment of sense of task," and "assessment of task difficulty" (see the model of self-regulation, Figure 9.1).

Preparedness for emergency situations depends primarily on psychological mechanisms of activity regulation, which in turn affect the physiological mechanisms of self-regulation. Interaction of these mechanisms causes effects such as increase in muscle tone, increase in circulation, and rapid scanning and shifting of attention. This predisposes the pilot's body to unexpected situations, possibly preventing a situation where the pilot may be unprepared to respond to an emergency situation. Blocks such as "assessment of task difficulty" and "assessment of sense of task" provide preparedness to such situations. The ability to operate effectively in an emergency situation depends on the operating agent's condition—both physiological and psychophysiological factors. The efficiency of operations is also dependent on psychological stability (emotional and strong-willed stability, properties of the psyche) and confidence levels, as well as on the levels of professional training received and assimilated. From the functional analysis standpoint, professional training is linked primarily to the functional blocks such as "past experience" and "conceptual model" (see Figure 9.1).

The ability to operate effectively in an emergency situation also depends on the emergency alerts' ability to attract attention and transfer appropriate meaningful information. These issues were discussed in Chapter 9. The objective complexity of an emergency is transformed into subjective difficulty. An objectively complex situation can be easy for a well-trained pilot (with a low level of difficulty due to training and preparedness).

After the orientation stage of self-regulation (formation of task goal, formation of conceptual and dynamic mental models, assessment of difficulty, and significance; assessment of sense of task), the pilot evaluates the situation, and such evaluation does not always require any action to change the situation at hand. In such cases, evaluation of the situation is completed without taking any action. Evaluation of the situation is the main purpose of orienting activity. However, very often, a pilot formulates a new goal and performs some executive actions for goal achievement based on the orienting stage of activity. The pilot creates a program of specified action, makes a decision, and performs established tasks hopefully associated with the completion of the desired goal. At this stage (executive stage of self-regulation), the following blocks are important: "formation of program task performance," "making a decision," and "program performance." After the pilot makes a decision and performs the response actions, the results of these actions are evaluated (evaluative stage of self-regulation). Based on the feedback, action corrections are applied if deemed necessary (see the general model of self-regulation, Bedny and Karwowski 2007). Sometimes, orienting activity includes decision and behavior actions. However, the major purpose of such actions is not to change the situation, but rather to evaluate the consequences of actions for better understanding of the situation.

Feedback may materialize from several varying sources. The question arises as to how a pilot builds single or integrated motor actions when there are multiple sources of feedback? Does the pilot drastically alter a response already in progress when faced with new information? Studies have shown that this is not the case. Studies revealed that motor actions that are under visual control can be continued without the visual control of the device. Motor actions based on the sensorimotor regulation principle consist of two cycles. The first cycle starts from the moment of visual fixation on the device, suggesting the need to start a goal-oriented motor action.

After the pilot initiates the goal-oriented motor action or several goal-oriented actions (possibly lasting from 0.5 to 20 seconds) and receives positive information about performance, the pilot switches his or her gaze and attention to other instruments. However, during this period, those actions initiated are continued. A single motor action or multiple simultaneous actions are carried out without visual control. However, they are completed only after additional visual information provides confirmation that these actions achieved the desired goal.

If the additional fixation determines that the goal is not achieved, second fixation gives rise to additional actions. Each motor action, or several tightly interdependent actions, consists of two cycles: the first cycle under the visual control and the second without it. This becomes possible because the program formation stage of action regulations is completed. This example demonstrates the way in which we can provide an explanation of the pilot activity in terms of functional blocks or mechanisms of self-regulation. Each functional block can also be described in terms of cognitive processes. We have described briefly how functional analysis can be used to explain pilot activity and how from a functional analysis the approach can be transferred to the cognitive analysis or to the traditional qualitative analysis. Functional analysis describes activity as a self-regulative system that includes a number of mechanisms or function blocks. These mechanisms or blocks are the main units of analysis.

10.5 Functional Analysis of Emergency Situations

Analysis of an emergency situation should start with a qualitative analysis of the activity. From the SSAT perspective, the following qualitative methods of analysis can be used: objectively logical analysis, cultural–historical analysis (Vygotsky 1971), individual psychological analysis, and functional analysis (Bedny and Karwowski 2007). In this section, we utilize objectively logical analysis and then transfer to functional analysis.

Investigation of accidents must start with an analysis of emergency situations. The emergency is a situation associated with adverse deviations in an otherwise controlled process (manufacturing process, equipment operation, and so on). The variability of these uncontrolled processes may become the root cause for trouble. Emergency situations make the implementation of activity as planned impossible. Emergency actions also demand alterations in the modes of equipment operation. The causes of emergencies range from incorrect human actions, equipment failure to a combination of these and other factors.

Human error as a cause of accidents can only be determined after the identification of specific erroneous human actions. However, erroneous human actions do not immediately condemn the person or persons associated with the task. For example, a person can perform some erroneous actions leading to an accident due to equipment design lacking ergonomic guidelines for design or online interaction. The degree of culpability is determined by the nature of the erroneous actions. Erroneous actions must be appropriately distinguished from malicious actions. This suggests that accident investigation should be conducted from the systems approach point of view, which examines the interaction between humans and technology.

Emergency situations change quickly over time during and in the absence of specific human counteractions. During stressful conditions, automatic changes of the mode of the equipment functions often results in accident

(Bedny and Meister 1997). The need for analysis of accidents in the aviation industry from a system's point of view was established in the works of Zavalovoy and Ponomarenko in the 1970s (Ponomarenko and Zavalova 1994). Particular attention in this approach was focused on the notion that errors can be caused only by the individual negative characteristics of the pilot; in other words, the pilot's error can be caused by negative characteristics of the technology or the interaction of both factors.

Personnel response to emergencies has a very limited time for assessment and planning for those emergency situations. Therefore, characteristics such as reserved time become especially important. Reserved time is defined as the surplus of time over the minimum that is required for an operator to complete the required task, ranging correct deviation of a system's parameters from the allowable limit to bringing the system back under control. Functional analysis perspectives posit that when activity is considered self-regulating systems, there is a profound difference between the objectively existing reserve time and the operator's subjective evaluation of this remaining time (Bedny and Meister 1997). If the characteristics do not match the expectations, problems may arise, including inadequate or incomplete evaluation of the situation and inadequate preparation of activity strategies mitigating emergency situations. The appropriateness of behavior during an emergency situation is directly related to early detection, correct interpretation of the situation, decision making, choosing the appropriate plan of action, and other aspects associated with preparedness. However, the short time for responses coupled with serious consequences for erroneous actions make the situation personally significant and emotionally charged. Analysis of accidents can be conducted from the functional analysis perspective, where the activity is considered from the self-regulation point of view. The unit of analysis for a functional mechanism or functional block is used.

The orientation in a given situation requires block mechanisms such as "goal," "conceptual (stable) model," "subjectively relevant task conditions (dynamic model)," "assessment of task difficulty," and "assessment of sense of task (significance)." Dynamic models of the situation include two subblocks that are partially overlapped. One subblock is responsible for a verbal–logical reflection of the situation, and the other for an imaginative reflection of the situation (Bedny and Karwowski 2007; see also Chapter 9). Imaginative components are particularly important for pilots. We conducted multiple studies describing the role of operative (dynamic) images of flight in pilot activity (Ponomarenko and Zavalova 1994; Ponomarenko 2006).

An emergency situation during flight is usually accompanied by emotional tension associated with the following factors: the possibility of serious consequences, difficulty determining the nature of the emergency, and difficulty finding an appropriate solution.

The more complex an emergency situation, the more severe the consequences, and the more significantly it is perceived, thus placing increased demands on the pilot. The complexity and significance of the situation are not only determined by objective factors. These objective factors are based on the subjective assessment of the situation. Hence, the general model of self-regulation of activity and the self-regulative model of orienting activity include not just the functional unit "significance" but also the functional block "assessment of difficulty." These units are responsible for the subjective assessment of the significance of the situation and the subjective assessment of difficulty (Bedny and Karwowski 2007). For example, the objective complexity of the situation may be underestimated by the pilot. As a result, serious errors may be committed. There are differences in the assessment of task difficulty between experienced and novice pilots. These differences may affect the assessment of task significance. A highly significant task for the pilot can cause high emotional tension, perhaps leading to decreased capacity for good decision making.

Excessive emotional tension in an emergency situation causes the following two types of reactions:

1. Constraint and inhibition expressed in delayed actions or omission of actions
2. A sharp increase in excitability, leading to the impulsiveness of action or hasty decisions

The effects of stress also alter the processes of perception and operative thinking. One of the common perceptual difficulties that arise under stress is decreased capacity for attending to individual instruments and an incorrect holistic interpretation of the situation. This has been observed as a violation of attention functions. A typical violation of the processing of information is a shift from quantitative to qualitative reading of devices.

Emergency situations can be classified based on the following criteria: time and accuracy of detection, complexity of interpretation and decision making, level of emotional tension, and complexity of performed actions in response to emergency. Emergency situations can be considered diagnostic tasks performed during stress. Studies show that an emergency situation can be analyzed from the standpoint of functional analysis, while data interpretation is made in terms of functional blocks.

10.6 Experimental Study of an Emergency Situation in a Real Flight

Research was conducted on an AN-12 aircraft, which was equipped with a special failure-simulating panel. The experimenter used this panel to simulate an engine failure. Twelve pilots took part in the experiments. The pilots' tasks were to find the simulated deviation. Failure of the engine was included in the test during flight. However, the pilots did not know about the possibility of engine failure. One of the four engines was unexpectedly stopped by the instructor using a handle.

As these experiments were relatively dangerous, statistical data are limited. Table 10.1 presents the pilot's performance after the engine failure. The pilot's speech was recorded during the task and served as the pilot's report on what happened and was his radio exchange with the experimenter. The conversation with the pilot regarding the information used in the emergency conditions was also registered after detecting a breakdown and restarting the engine. Information about the pilot's emotional tension was obtained by recording some of their physiological reactions during the flight. The first objective characteristic of the efficiency of the pilot's actions during the breakdown is the time he or she takes to detect the emergency situation. Two criteria for the evaluation of the pilot's performance were selected: detection time of faults in the engine function and time needed to complete the diagnosis of the situation. Detection time includes the period between the actual engine failure and the first vocal reaction by the pilot. During detection time, the pilot can only detect that something is wrong and hypothesize that it may be an engine failure. Diagnosis time includes the time from the moment of failure until the pilot's report about the stop of an engine. Therefore, the diagnosis time includes the detection time and the correct interpretation of engine failure. Table 10.2 presents the experimental data.

The obtained data demonstrated that it takes the pilot a substantial amount of time to perform the required actions. This could cause problems during autoflight, as vigilance over instruments is reduced. The delay in the performance of the diagnostic task was considered to be an erroneous action (see, for example, pilots 4–8 and 11). Physiological data demonstrated that the task performance

TABLE 10.2

Pilot's Detection and Diagnosis Time on Engine Breakdown in Flight (Seconds)

Pilot	Detections	Diagnosis
1	8	53
2	2	55
3	2	80
4	7	140
5	2	145
6	3	200
7	6	260
8	6	230
9	5	40
10	2	15
11	6	280
12	2	5

was accompanied by high emotional tension manifested through increased heart rate. Heart rate increased, on average, by 20 beats per minute. An increase in the pressure of the control column grasp was also observed. This meant that unexpected emergency situations were perceived as highly significant by the pilots. In order to obtain information regarding the cause of delay in diagnosing the engine breakdown, it is necessary to consider the specificity of information gathered by the pilot. The study showed that the pilot used three types of information about the engine breakdown. The first type of information was gathered by looking at different instruments that showed the parameters of the engines of the aircraft, such as the number of revolutions per minute, fuel and oil pressure, exhaust gas temperature, and revolving moment. The second type of information was obtained from bulb indicators. Two bulbs are located on the ceiling of the cockpit. One of them alerts the pilot about the "engine breakdown," and the other one shows that the "pump is working." A third bulb is located on the panel near the left pilot, and it alarms about the state of the "propeller." A fourth is on the panel near the right pilot and signals "icing formation." The study also demonstrated that the pilot utilized noninstrumental information as well. The pilots obtained this information via subjective feelings. Such information is associated with forces perceived by the body, feedback from control surfaces manifested in flight controls, cockpit sounds, and other information gleaned by the senses. The sequence of different types of incoming information was determined based on the radio reports and is presented in Table 10.3.

The data support that the strategies of information gathering can be described as follows. The first signal that the pilot receives does not originate from flight instruments. These signals include information regarding angular acceleration and a feeling of decreased reactive force from one side resulting from windmilling of one of the engines. At the same time, the pilot starts feeling changes in the resistance in the control column, pedals, and hears changes in sounds, and so on. Based on these data, the pilot concludes that "something is wrong with the engine." Based on this information, the pilot formulates very general goals and tasks suited for achieving these goals.

When the pilot asks himself "What is wrong?" this stage can be determined to be task formation, a diagnostic task. The pilot attempts to interpret the meaning of input information. "Meaning of input information" is the first function block of the conscious level of activity self-regulation. This is "object meaning" associated with networks of feelings and experiences with a particular object or situation. Only at the next stage can it be transferred into categorical or verbal meaning (see Figure 9.1). It is very important to define the identifying features of the situation (Bedny and Karwowski 2004). Based on the analysis of such features, the pilot can understand the meaning of the situation. The gathered information and results of its integration guide the pilot in developing dynamic mental models of the situation. The meaning of the situation and its dynamic model define the nature of information gathering at the next step of task performance (integrative stage of situation interpretation). Our analysis demonstrates that the pilot starts searching for

TABLE 10.3

Sequence of Gathering Different Types of Information (Percentage of Cases)

Source of Information	Information Features	Sequence of Information							Did Not Use
		1	2	3	4	5	6	7	
Personal feeling	Turning moment, force on Pedals and control column, changing of body position, changing of sound	100							
Data on indicators	Decreasing RPM	70	20	10					
	Turning moment	30	60	10					
	Fuel pressure			10	70	20			
	Oil pressure			10	10	30		20	80
	Index TOG			10	10	30			50
Alarm	Engine breakdown							10	90
	Pump working						10		90
	Propeller						20		80
	Ice formation						10		90

instrument information associated with the engine state. The pilot pays attention to the control panel first and to the instruments that show revolutions and turning moment (torque). As a result of such a strategy, in most cases, the index-emergency alarm is out of the pilot's attention area. These data can be explained by the fact that the four bulbs are located far away from each other. Therefore, the signaling system is not included in the information gathering strategy. This makes it difficult to develop an adequate mental model of the emergency situation. At the first stage of engine breakdown, the pilot only guesses that something is wrong with the engine but does not have a clear understanding of the source of the malfunction. This is why the pilot's attention is directed not at the alarm signals (light bulbs), but to the indicators.

Therefore, alarm signals are not included in the mental model of the situation, which leads to the deviation of the information gathering sequence from the one prescribed by the instruction. In most cases, the pilot detects and diagnoses the engine failure without looking at the alarm signals. Lack of very important data during information gathering and mental model creation of a flight situation can become a reason for errors and delayed decisions, especially in the case of conflict between the actual and expected events.

Table 10.4 presents the sequence of information gathering according to the standard procedural manual and the real sequence of information received in the case of engine breakdown.

There is a contradiction between the recommended methods of information gathering based on standard manuals and the actual strategies of gathering information during engine failure.

To change the diagnostic strategy of the pilot, a new design for the panel has been suggested. According to the new design, the master alarm signal (signal "DANGER") is now located directly over the group of instruments that reflect the parameters of the engine function. Noninstrumental information attracts the pilot's involuntary attention to this area. As a result, the pilot cannot miss the signal.

Ambiguous imaginative feelings regarding engine failure that reflects noninstrumental signals are transformed into verbally consistent logical forms. The pilot makes a rapid preliminary conscious conclusion about the breakdown of the engine. Information from the instruments used by the pilot supports his or her conclusion. Information about the parameters of the failed engine can be seen on the instruments and is confirmed by the light indicator that illuminates when the engine breakdown occurs. Therefore, the signaling system attracts the pilot's attention at the correct time, and the method of information gathering becomes more reasonable. Thus, mental models of the situation are developed more quickly. (A mental model integrates not only systems of meanings but also systems of images.) Detecting and diagnosing the breakdown at the right time depends not only on the ideal design of instrumental information, but also on the pilot's experience. Therefore, a special training program was developed.

TABLE 10.4

Prescribed and Real Strategies of Information Gathering during Engine Failure

Sequence of Information Gathering According to Standard Procedural Manual	Real Sequence of Information Gathering in the Case of Engine Breakdown
Progressive pitch. Tuning of aircraft	Feeling of acceleration, resistance on control column and pedals, changing position of body, sound changes
Light indicator "ENGINE BREAKDOWN" is turned on	Decreasing RPM
Decrease in the turning moment	Decrease in the turning moment
Pressure drop in the gas system	Pressure drop in the gas system
Decreasing RPM	Decreasing exhaust gases temperature
Decreasing exhaust gases temperature	Lighting of signal bulb (alarm signal) about engine failure
Pressure drop in the oil system	Pressure drop in the oil system

On the other hand, the pilot's performance depends on whether the goal of the task is correctly formulated. The overall goal of the flight is usually clear. However, the goal of the intermediate tasks can be incorrectly formulated by the pilot depending on the circumstances. Moreover, an objectively presented goal can be interpreted subjectively in many different ways. Let us consider the role of subjective formulation of a goal for flight safety. During flight, the directory control of the aircraft and signals are automatically processed by a computer. The pilot performs a compensatory tracking task and follows the perpendicular indexes. The pilot's goal is to maintain the point of the index-crossing at the indicator center, which ensures the precise maintenance of the flight path. Experiments and practice demonstrate that in such conditions, the pilot's goal is often distorted. The real goal of the flight, "to maintain the correct position of the aircraft in space," is replaced by the goal of "maintaining the index-crossing at the indicator center." This elevates the piloting quality, but lowers the flight reliability. The reliability drops because of the degradation in the pilot's readiness to notice a violation of the flight path because he or she may not know the aircraft's real position in space. The adequate dynamic model of the space position is formulated slowly. The pilot stops noticing danger in the selected flight parameters. Although these parameters are in his or her field of view, the pilot's attention is concentrated on the command indexes. Here, the intermediate goal of the activity has a negative effect on information selection and the formation of the dynamic model of the situation.

10.7 Conclusion

SSAT utilizes systematic principles of analysis and allows the development of more efficient methods of studying pilot performance. Functional analysis when activity is considered self-regulative system is especially useful for the qualitative systemic analysis of the pilot's activity. This approach gives unified and standardized principles of analysis and description of experimental data. A description of the pilot's activity is performed in terms of functional blocks utilized in the models of activity self-regulation. Depending on the exact nature of the task performed by the pilots, we can pay more attention to certain functional blocks of activity self-regulation than to others.

In this chapter, we have demonstrated the importance of such functional blocks as "goal," "conceptual (stable) model," "subjectively relevant task conditions (dynamic model)," "assessment of task difficulty," "assessment of the sense of task (significance)," "formation of the level of work motivation," and so on. The presented material demonstrates that the dynamic model of the flight and its important mechanisms such as the operative image are critical for task performance in emergency conditions. Therefore, not only

verbalized, but also nonverbalized activity components are critically important in the creation of a dynamic mental model of the situation. The image of
the flight performs two functions in the regulation of the pilot's activity: one
is associated with the orientation in situation (cognitive functions) and the
other is involved in the regulation of motor actions (executive functions).

The dynamic mental model and its imaginative components are very
important in understanding the principles of activity self-regulation and
for the development of adequate performance strategies in emergency conditions. The study of strategies regarding task performance is the main
purpose of the functional analysis of the activity. We have discovered the
differences between the strategies prescribed by formal instructions and
the real strategies of pilot activity. Strategies of activity are developed based
on mechanisms of self-regulation. The most important mechanism for the
development of the pilot's strategy is the mechanism responsible for the creation of a dynamic mental model of the flight. This mechanism or functional
block includes the subblocks "operative image" and "situation awareness."
Based on a dynamic mental model, the pilot develops a real strategy, which
deviates from the strategy prescribed by formal procedures and training. We
discovered that an inefficient design of the instrument panel and existing
instructions caused an erroneous formation of a mental model describing
the situation. This resulted in inefficient strategies of information gathering
in an emergency situation. Our study demonstrates that functional analysis
is a very useful tool for studying pilots' activity in emergency situations.

References

Anokhin, P. K. 1962. *The Theory of Functional Systems as a Prerequisite for the Construction
      of Physiological Cybernetics*. Moscow: Academy Science of the USSR.
Bedny, G., and D. Meister. 1997. *The Russian Theory of Activity: Current Application to
      Design and Learning*. Mahwah, NJ: Lawrence Erlbaum Associates.
Bedny, G. Z. 1987. *The Psychological Foundations of Analyzing and Designing Work
      Processes*. Kiev: Higher Education Publishers
Bedny, G. Z., and M. P. Zelenin. 1988. *Ergonomic Analysis of Work Activity and the
      Problem of Safety in Merchant Marine Transportation*. Moscow: Merchant Marine
      Publishers.
Bedny, G. Z., and W. Karwowski. 2004. Meaning and sense in activity theory and their
      role in study of human performance. *Ergonomia* 26(2):121–40.
Bedny, G. Z., and W. Karwowski. 2007. *A Systemic-Structural Theory of Activity.
      Application to Human Performance and Work Design*. London: Taylor and Francis.
Bernshtein, N. A. 1966. *The Physiology of Movement and Activity*. Moscow: Medical
      Publishers.
Bodrov, V. A. 1989. Problems of professional and functional reliability of operator.
      Psychol J 10(4):142–9.

Cole, M., and I. Maltzman, eds. 1969. *A Handbook of Contemporary Soviet Psychology*. New York: Basic Book Publishers.

Konopkin, O. A. 1980. *Psychological Mechanisms of Regulation of Activity*. Moscow: Science Publishers.

Kotik, M. A. 1974. *Self-Regulation and Reliability of Operator*. Tallinn: Valgus.

Kotik, M. A. 1978. *Textbook of Engineering Psychology*. Tallinn: Valgus.

Pavlov, I. P. 1927. *Conditioned Reflex*. London: Oxford University Press.

Ponomarenko, V. A. 2006. *Psychological Factor in Dangerous Profession*. Krasnoyarsk: Political Publisher.

Ponomarenko, V. A., and N. D. Zavalova. 1994. *Applied Psychology. Problems of Safety in Aviation*. Moscow: Science Publishers.

Ponomarenko, V. A., V. V. Lapa, and A. V. Chyntyl. 2003. *Activity of Crew in Aviation*. Moscow: Aviation Safety Union Publishers.

Uznadze, D. N. 1961. *The Psychology of Set*. New York: Consultants Bureau.

Vygotsky, L. S. 1971. *The Psychology of Arts*. Cambridge, MA: MIT Press.

Wiener, N. 1958. *Cybernetics*. Moscow: Soviet Radio. (Russian Translation).

Zavalova, N. D., B. F. Lomov, and V. A. Ponomarenko. 1986. *Image in Regulation of Activity*. Moscow: Science Publishers.

11

Methodology for Teaching Flight-Specific English to Nonnative English–Speaking Air-Traffic Controllers

R. Makarov and F. Voskoboynikov

CONTENTS

11.1 Introduction

The English language is recognized all over the world as a communication tool for pilots and air-traffic controllers of international flights. Development of effective communication skills for nonnative English–speaking air-traffic controllers and pilots is of great importance to provide safety of international flights. The communication process for air-traffic controllers and pilots is not a simple way to convey the meaning of messages; these messages must be immediately transformed into actions. The work environment for air-traffic controllers is always stressful due to their responsibility for peoples' lives and the time limits for making decisions and responses. Therefore, a training system must be designed for them in such a way that it will allow the development of psychophysiological features of personality such as emotional stability, a high degree of concentration, the ability to memorize and keep information in working memory, creativity, and decisiveness. The training must also take into the account the fact that nonnative English-speaking people communicate differently than native English-speaking people do. Thus, the training of future air-traffic controllers must not be a simple summarization of separate skills but rather an integrated system.

11.2 Psychological Analysis of Air-Traffic Controllers' Activity

11.2.1 General Characteristics of Air-Traffic Controllers' Activity

Air-traffic controllers are the main people involved in controlling airspace and monitoring flights; they are responsible for starts, takeoffs, launching and landing of aircrafts, and so on. About 65% of dangerous situations in airspace are the result of inadequate actions by the air-traffic controllers. Air-traffic controllers' duties include evaluation of air traffic, locating aircrafts in the airspace, following their movements, overseeing aircrafts' routes, and following the rules and regulations of the flights. They perform all their actions by analyzing the information directed to them in the form of signals on the screens and information from radio channels, telephone communication, or secondary sources of information (symbols, rules, and regulations). Based on the analysis of all those sources of information, air-traffic controllers form their decisions and formulate and send commands to the aircrafts' crews.

Psychological analysis of air-traffic controllers' activity is important to develop effective methods of training. People communicate with one another in ways that guide the interpretation of the meaning conveyed by the language they use. Sentences are constructed using phrases, and phrases and words convey only the part of information about their intended composition. In addition, prepositions are important clues to the listener about the phrase that is to be interpreted. People of all languages face problems in conveying the meanings of solutions in a different language. For example, in English, adjectives and modifiers appear after the noun. In many other languages, adjectives and modifiers appear before the noun. In English, the word order adds importance to the information. In other languages, the word order may be important only for emphasis, and the roles of the different parts of the sentence are signified by changing the form of the word to indicate their case.

The main characteristics of air-traffic controllers' activity is the interpretation of visual information on the screen, radio communication with pilots, and communication with other air-traffic controllers. They must analyze and act upon constantly changing characteristics of information—its capacity, time, and meaning. They usually oversee several flights simultaneously by following the information on the screen and by communicating with other air-traffic controllers, who are responsible for different sectors of airspace, through the radio. They must be able to evaluate their own actions and realize how their actions are understood and interpreted by others. This means that an air-traffic controller must be able to coordinate his or her actions with the actions of others and at the same time must understand and predict the sequence of events. The only means that air-traffic controllers have to control and monitor aircraft movements is their verbal communication through radio. Therefore, their radio commands can be viewed as verbal actions, which are perceived and interpreted by pilots. They must be able to use the language effectively to perform different operative tasks under time limits and stress. Their speech constitutes a major part of their thinking process. Thus, the verbal aspects of air-traffic controllers are of significance.

Air-traffic controllers and pilots communicate without looking each other, which presents additional difficulties for nonnative English-speaking people. It is a well-known fact that the body language adds value to information in the process of communication. When compared to the work of the operators in many other fields, air-traffic controllers do not have visual contact with the objects they monitor. This is a very important characteristic of their work. Moreover, even indications of aircrafts' positions on the screen are episodic and discrete. It is a necessity for an air-traffic controller to switch attention to other important elements of the entire situation.

To perform his or her activity correctly, an air-traffic controller must forestall adequate mental models of a situation and formation of verbal control actions. Substantial peculiarity of such tasks is that an air-traffic controller builds a system of commands and actions in his or her mind, which is based on the evaluation of the constantly changing situation. An important characteristic of successfully implemented commands is the ability to foresee the way a situation may possibly change. Without the ability to create mental models of a situation, such foreseeing is impossible. The dynamic model of a situation is based on the verbal information received from the pilots and the visual information on the screen. One of the main characteristics of the air-traffic controllers' activities is that they relate to operative tasks and functions of operative thinking (Bedny and Karwowski 2000; also see Chapter 4 of this book).

11.2.2 Operative Thinking and Working Memory

By analyzing the air-traffic controllers' activities, we can identify two regimens that influence task performance. The first regimen is a normal flight condition, and the second one is a significant deviation from a normal flight course. In the latter case, the process of operative thinking gets much more complicated. Operative thinking is intimately connected to the characteristics of a person's memory and particularly with the characteristics of his or her working memory. Air-traffic controllers must keep some general rules and principles of flight monitoring in a long-term memory to resolve typical problems. They also need to keep information on all dynamic units of situations in the working memory until the operative tasks are completed. Only after the completion of the task can this information be forgotten. Active manipulation of information in the working memory is an important component of operative thinking (Bedny and Karwowski 2007). To improve air-traffic controllers' performance, we must look for all possible ways to decrease their memory load.

Another important aspect of communication is operative thinking under stress conditions, which usually takes place in task performances in dynamic situations under time constraints (Bedny and Meister 1997; Bedny and Karwowski 2007). The most important features of operative thinking were originally described by Pushkin (1965). He pointed out that the major features of operative thinking are practical problem solving and performance under emotional stress and time constraints. Operative thinking has three functions: planning, control, and regulation of cognitive and behavioral actions. (It is described in depth by Bedny and Karwowski in Chapter 4 of this book.)

Operative thinking is an essential cognitive process responsible for the construction of dynamic mental models of a situation. The analysis of operative thinking process demonstrated that for future air-traffic controllers, studying the English language only by means of traditional methods is not sufficient. The most effective way they can learn the language is in the conditions of resolving operative tasks, which could be modeled in role games. Analysis of operative thinking also demonstrates that the most productive way to develop communication skills is in problem-solving situations, which could be created on the basis of studying past critical situations and accidents in the airspace.

11.2.3 Characteristics of Attention in Air-Traffic Controllers' Activity

The work of air-traffic controllers requires extraordinary qualities of attention. We will describe two important characteristics of attention here: (1) distribution of attention and (2) switching of attention. Distribution of attention demonstrates the ability to perform several actions connected to each other. For example, while talking to a pilot, an air-traffic controller perceives information on the screen. When the situation becomes more complicated, distribution of attention could transform to switching of attention. Analysis of communication problems between the air-traffic controller and the aircraft crew, such as incorrect interpretation of commands, mixing aircrafts' codes ("hear-back" or errors of feedback), shows that the study of attention is necessary in designing effective methods for teaching professional English.

There are different kinds of attention. *Involuntary attention*, which is almost automatic, manifests itself without effort by the organism. It is drawn to environmental phenomena. *Voluntary attention* is a conscious and effortful switching of concentration on different events in the environment. The concepts of voluntary and involuntary attention are similar to the concepts "data-driven" and "conceptually driven" processing in cognitive psychology. From the functional analysis perspective, attention is considered a complex self-regulative system. This system integrates the informational and energetic mechanisms. The existence of forward and backward interconnections among the functional mechanisms of attention allows the formation of attention strategies in task performance. The model of attention is described by Bedny and Karwowski in this book (see Chapter 12). Formation of attention strategies is an important aspect of training for air-traffic controllers in the process of developing communication skills.

11.2.4 Emotional Factors in Air-Traffic Controllers' Activity

Air-traffic controllers' work requires not only specific knowledge and specific skills but also developed psychophysiological qualities such as emotional stability, a high degree of concentration, an excellent memory, creativity, and decisiveness. Therefore, training of future air-traffic controllers must not be a simple summarization of separate skills, but rather a development of interdependent system of knowledge, skills, and personal features. The systemic-functional approach in professional training must be based on the human factor because the malfunctioning of a system has a definite influence on biosocial being of humans.

Emotional stability is one of most important characteristics of personality that has a significant influence on the performance of any activity. In stressful situations, an air-traffic controller must maintain the required level of thinking, memory, and attention. In the process of professional selection of candidates, a psychologist or an ergonomist must analyze the above-described functions. For people who conduct radio communication in a foreign language, the psychic tension significantly increases. In activity theory, psychic tension can be divided into operational tension and emotional tension. Operational tension is determined by a combination of task difficulties and lack of available task time. Emotional tension is determined by the personal significance of a task to the operator (Nayenko 1976). These two types of tension are closely interrelated, and under certain conditions, one type of tension causes the other. For example, the lack of available time for a significant task causes more tension than the lack of time for an insignificant task.

In our research, we dedicate special attention to professional selection of air-traffic controllers. Specialists in the foreign language department of Kirovogradskiy University of Aviation work closely with other departments' specialists in order for them to develop the ability to use English efficiently in stressful situations.

11.2.5 Fatigue in Air-Traffic Controllers' Activity

Fatigue is a dynamic component of an activity, which changes during work shifts, and it has its definite effect on air-traffic controllers' performance. Verbal components are a prevailing characteristic of air-traffic controllers' work, and are substantially different from the work characteristics of many other professions. Air-traffic controllers' speech consists of standardized and regulated sets of phrases by their structures and contents. These phrases present an exteriorization of their mental activity. Close to the end of working shifts, the structure of phrases and commands changes, the number of errors increases, phrases get shorter, more prophylactic kind of commands take place, and more errors of verbal inconsistencies and unclear commands appear at the end or in the middle of phrases (pronouncing commands in syllables, not responding on requests, sending additional requests, repeating commands, changing the tone of voice, and so on). Another symptom of decreased work effectiveness manifests itself in the form of changes in the semantics of commands (commands without clear directions), syntactic structure of phrases, mixing of identification codes of aircrafts, and incorrect direction of commands. Most of the time, such errors take place in nonstandard situations: under stress, when tired, and in conditions of time limits for making decisions.

11.3 Functional Analysis of the Air-Traffic Controllers' Activity

11.3.1 Dynamic Mental Model

The functional analysis of activity considers activity a self-regulatory system. Major units of functional analysis are functional blocks or functional mechanisms (see Chapters 1 and 9 of this book). One of the most important aspects of air-traffic controllers' activity is the creation of a dynamic mental model of the situations. According to Bedny and Karwowski (2007), the functional mechanism (functional block) of self-regulation—subjectively relevant task conditions—is responsible for the creation of such a model. The creation of a dynamic mental model allows us to integrate the received pieces of information and to structure them in accordance with a task's goal. The dynamic mental model of a situation improves the ability to search and select the necessary information, its analysis, synthesis, and interpretation. The dynamic model includes imaginative and verbally logical components. The following two subblocks or mechanisms are responsible for an adequate formation of the above-described components: operative image and situation awareness; they partially overlap each other. An important source of information for the creation of a dynamic mental model is the visual means to display the information. Thus, air-traffic controllers must constantly transform the visual information in their minds into verbal information and vice versa.

The creation of an adequate mental model depends on the ability of air-traffic controllers to combine verbal and imaginative information. Imaginative information includes visual as well as acoustical and other modalities. The main difficulty in creating such a mental model is that not all imaginative components of it could be transformed into adequate verbal components. This, in turn, creates some additional mental load on the air-traffic controller's state of mind. Transformation of visual information into verbal information requires the development of a language to communicate with aircraft crews, as well as a language to communicate with the computer. To perform the latter, certain computer skills are necessary. For that purpose, a specific keyboard and symbolic indicators are used. The process of exchanging information with the computer in a dialog regimen is significantly different from the verbal exchange of commands with aircraft crews. This requires a concentration of attention to enter the information into a computer and time for responses due to possible pauses, which differ from typical pauses in verbal exchange. The possibility of getting rejections and errors from the computer forces the air-traffic controllers to recheck the important information. For the same reason, the air-traffic controllers may be forced to switch the flight monitoring process by using the nonautomatic method. That is, there might be a necessity for an extreme re-creation of the mental model of the air-traffic situation.

11.3.2 Objective Meaning and Subjective Sense

Time constraints for decision making and stress are unavoidable in any air-traffic controller's work, which is a specific peculiarity of the mental process. The air-traffic controllers' mental stage is attributed to their responsibility for peoples' lives. In such cases, an objective real situation may be interpreted incorrectly by an air-traffic controller, that is, subjectively, as a human being. Analysis of such situations presupposes the differentiation of the objective meaning of a situation and the subjective sense. The functional block "assessment of sense of a task" is involved in this transformation. It is an emotionally evaluative mechanism responsible for the evaluation of significance of a situation or its elements (Bedny and Karwowski 2004). The factor of subjective significance influences the selection of information and its interpretation; therefore, the objective meaning of a task and its subjective interpretation might not coincide. This in turn influences the development of a dynamic model of a situation and the subjectively relevant task conditions. It can lead to nonadequate decision making and incorrect executive actions. That is, the emotional-motivational component of activity influences the cognitive components (Bedny and Meister 1997; Bedny and Karwowski 2006). Cognitive psychologists do not consider these aspects of human performances in their studies sufficiently.

There is also a functional mechanism such as "assessment of task difficulty." The objective complexity of a task and subjective evaluation of its difficulty may not be equal (Bedny and Karwowski 2007). The more complex the task, the greater the chance that it is more difficult. Along with the increased intensity of air traffic as an objective characteristic of complexity, the air-traffic controller's subjective characteristics, such as difficulty and mental efforts associated with it, increase as well. The significance of a task also depends on the evaluation of its difficulty. Assessment of task difficulty depends on personal features such as self-efficiency and self-esteem. An increase in the task difficulty influences the increase of the task significance. Hence, emotional tension changes as well. However, the significance of a task can also be independent of the difficulty factor. For example, a task may be relatively not complex but still subjectively important. This is explained by the fact that even a relatively easy task does not eliminate the feeling of responsibility for people's lives. Some research has shown that in such work conditions, air-traffic controllers' pulse rate increased for up to 32%.

11.4 Activity Approach to Training

11.4.1 General Principles of the Training Process

In the former Soviet Union, there were two major vocational training systems, which are used in today's Russia as well. One was called the "operationally complex system," which was developed in the beginning of twentieth century. The other one was known as the "problem-analysis system" (Batishev 1977). The first is usually applied to traditional blue-collar workers. The major notions of this system are the operational stage and the complex stage of training. In the former stage, students acquire specific skills for performing typical elements of the production operations. They are called "operational skills," where students learn how to perform separate operations using particular tools. In the latter stage, students perform a holistic production operation that integrates a number of separate operational skills.

The problem-analysis system is used in the training of operators, such as chemical plant operators, pilots, air-traffic controllers, and so on, who are involved in the control and regulation of automated processes. The main elements of this kind of training system are not the production operations but the resolution of task problems. The major functions performed by these operators are the following: monitoring technical system parameters, analyzing the system malfunctioning, controlling the system state, and so on. At the preliminary stage, we should outline the basic task problem correctly, which is required for a particular profession; this is a difficult stage of the problem-analysis system of training. In the second stage, teaching methods are used for the development of skills for solving different isolated problems. In Section 11.4.2 we will describe the problem-analysis method of training.

In the Soviet Union's educational system, particular attention was paid to the development of the ability of students to independently formulate and solve different problems. That is why in the Soviet Union, multiple-choice questionnaires were not used, and are not often used in the Russian school system even today. Students only need to recognize or reproduce some information, whereas other information should be actively used by them in practice. Special attention was paid to mental development, which was not a simple absorption of knowledge but rather the ability to acquire knowledge independently, apply and transfer it to different situations, and generate new knowledge based on the existing knowledge. For the same reason, much effort was devoted to developing "discovery" learning in the education process, and particularly in vocational training (Kudryavsev 1975; Matyushkin 1972).

In discovery learning, the student is confronted with problems that require a search for unknown laws, methods of actions, and applicable rules. Whichever training method is used, there are at least two stages: introductory (theoretical) followed by a task performance. Discovery learning is used in an introductory stage in which a student, under supervision, discovers different methods of a task performance, develops theses, and selects the preferred one. After learning these methods, the student must follow them strictly (Bedny 1981). Obviously, feedback is essential to the learning process. The final stage of learning involves the evaluation of the outcome of a task performance, either by the instructor or by a machine.

The behavioral approach has not been used in the development of teaching methods in the former Soviet Union and in today's Russia because it is contrary to activity theory. Activity theory emphasizes that learning must include goal formation, consciousness, speech, motivation, and social interaction. During the learning and training processes, students acquire knowledge and skills. Knowledge can be declarative and procedural (Anderson 1985; Landa 1984). Declarative knowledge is stored in the memory in the form of images, concepts, and propositions, and it is a necessary basic knowledge; however, it is not sufficient for practical application. Procedural knowledge is much more important in the training process because language is used in resolving real practical tasks-problems (Bedny and Meister 1997). Procedural knowledge is knowledge about motor and cognitive actions. Thinking is not knowledge but what we do with the knowledge. Its structure is organized as a set of mental actions or operations carried out to solve the problems (Landa 1982). A student can have a large repertoire of images, concepts, and propositions in his or her mind, but a small repertoire of mental actions with him or her. As a result, the student cannot know which mental actions to apply to which knowledge.

11.4.2 Self-Regulation Theory of Learning

To provide an efficient method of learning and training, we should consider learning a self-regulation process. It is also necessary to define the goal of learning activity, discover the initial state of the students' knowledge and skills, define effective strategies of task performance at different stages of the training process, and provide students with the required feedback. Bedny developed a self-regulation theory of learning (Bedny and Meister 1997), in which an activity of a learner is considered a self-regulative system. Concepts such as functional blocks and learner strategies are critical in such learning. For the analysis of the learning process, the model of self-regulation of activity should be used (Bedny and Karwowski 2007). This model includes 20 functional blocks or mechanisms of self-regulation. This model helps to describe the strategies of activity performance at different stages of the learning or training process. This model describes the interaction of conscious and unconscious levels of self-regulation, which could be transferred into each other to some extent. According to the self-regulative learning theory perspective, learning can be voluntary and involuntary, as well as conscious and unconscious (Bedny and Meister 1997). The self-regulation concept of learning is a systemic method of analysis of the learning process.

Considering learning from the self-regulation perspective is also useful in describing the concept of functional system developed by Anokhin (1955). Anokhin described association as a process of formation of a self-regulatory system. Association cannot be reduced to a simple reactive system, and is a result of functioning cyclic system in which feedback is a critical mechanism.

These data also demonstrate that learning should be described as a self-regulative process.

Analysis of these factors brings us to further discussion, in which we will utilize the problem-analysis system of learning and training and some important ideas derived from the self-regulation concept of learning. First, task analysis allows us to uncover typical wrong strategies of task performance in air-traffic controllers' actions. For this purpose, we performed an analysis of past accidents and incidents in airspace, which helped us to create our method of training for developing the required verbal skills. The main goal of our research was creating practical methods for teaching specific English. By studying and analyzing the air accidents and incidents, we concluded that the training in technical English is an important part of the entire training system. However, concentrating the training on exercises for specific English is not sufficient for the purpose. Learning and applying standardized commands and standardized skills in standardized conditions may be helpful to a certain degree.

In air-traffic controllers' work, there are many situations in which they must demonstrate creativity and logical thinking. Hence, there is a need for the development of flexible verbal skills that are to be applied in stressful conditions. Studying and analyzing the air accidents is the best method for developing operative thinking and the ability to break the problems into categories by the degree of extremity, as well as the ability to direct necessary commands to the pilots in English. Therefore, the traditional method to study English, including memorization of standardized commands, is important only for providing the students with basic knowledge. Learning how to apply the knowledge to practical use can be achieved by a systematic analysis of problem situations and by searching for strategies for solutions. The main purpose of the analysis of past radio communications in the training process is to prevent air accidents in the future. The ability to conduct radio communication can be developed by two important elements of the study: multiple repetitions of commands in standardized situations and the practice of using the English language in nonstandard task problem situations.

11.4.3 Analysis of Typical Errors in Radio Communication in the Training Process

The analysis of typical errors is an important source of information for the simulation of radio communication in the training process. We will describe some of them below. One of the problems in radio communication between pilots and air-traffic controllers is errors of feedback (read-back or hear-back problems). Pilots repeat air-traffic controllers' commands with errors. These kinds of errors are so-to-say mutual errors, that is, the air-traffic controllers do not catch the pilots' errors, and the pilots do not notice the air-traffic controllers' errors. Each of them hears what they want to hear. For pilots, it takes place mostly while they read numeric indications on their dashboards. By studying the activity of the pilots and air-traffic controllers, scientists concluded that there are four main reasons these problems occur:

1. Due to the similarity of the aircrafts' signal codes (examples will be described in more detail in Section 11.6).
2. Only one pilot conducts radio communication on the frequency of air traffic; the second pilot conducts communication with the people who are in charge of automatic transformation of information in an airport area, or with an airline.
3. There may be a discrepancy in what is in the mind and what the mouth says, for example, confusion between "10,000" and "11,000" in an echelon of a flight; confusing "right" and "left" during a parallel takeoff of two aircrafts; accepting information about an airplane in a different echelon of a flight and receiving permission to enter into the echelon; confusing the speed limit with the altitude of a flight.
4. Pilots hear what they expect to hear when commands are not very clear.

To minimize these communication errors in the future, we offer exercises, which are based on the investigation of the air accident on September 10, 1976, in Zagreb airspace. The following seven procedures can reduce communication errors:

1. *Errors of feedback*: The pilot of an aircraft repeats the air-traffic controller's command about the frequency or about the echelon of a flight with a mistake. The air-traffic controller's task is to identify and liquidate the error.
2. *Transposition of numbers, that is, repeating them in a wrong order*: The air-traffic controller's task is to identify and liquidate the error.
3. *Unclear command of an air-traffic controller*: The air-traffic controller's task is to identify the discrepancy and to perform the instructions correctly.
4. *The use of "affirmative" and "roger"*: In cases when the pilot must repeat an air-traffic controller's command, the air-traffic controller's task is to request a confirmation of the command.
5. *The use of "affirm" instead of "affirmative"*: When the pilot confirms the air-traffic controller's command by saying "affirmative," possible background noise in the microphone may cover the first part of the word, so the pilot may hear only the "tive," which may be understood as "negative." Under new rules and regulations, confirmation

must be "affirm," instead of "affirmative." In this exercise, the air-traffic controller's task is to request clarification.

6. *Errors as a result of identically sounding signaling codes of aircrafts*: If signaling codes are similar, they need to be changed to avoid errors. The air-traffic controller's task is to request clarification.

7. *Accepting information about an airplane in a different echelon of a flight and receiving permission to enter into the echelon*: The air-traffic controller's task is to identify the problem and liquidate the error.

The minimum time interval between the passages of aircrafts must not be less than 10 minutes. However, in role games it may be reduced for training purposes. Solid skills in identifying and liquidating the errors in radio communication will allow future air-traffic controllers to dedicate their full attention to monitoring air-traffic for the best possible safety of the flights in international airspace.

11.5 Games as a Method of Training

11.5.1 Requirements of Game Scenarios

Games are a specific kind of human activity involving situations imitating the real environment. In professional relations, role games are based on certain problem situations. For air-traffic controllers, games have special meaning; they imitate a complex kind of human activity including the process of conducting radio communication in English in extreme situations of the international airspaces. Many individual traits manifest themselves in games. Role games also serve as a reliable tool for evaluating the students' preparedness for the future work. Games allow creating artificial technical and social aspects of the air-traffic controllers' professional activity, which presents an opportunity for developing the required verbal skills.

To develop role games, we used the documents of investigations of accidents and incidents by the International Organization of Civic Aviation (IOCA). In the Addendum to the Standards and Recommended Practice in Attachments 1, 6, 10, and 11 of the IOCA, there is a description of language use in radio communication, and in addition to Attachment 1, there are requirements to the knowledge of languages, which include a number of criteria and the IOCA evaluation scale of knowledge in languages. This scale differs from the evaluating criteria existed in the past. The most important element of the scale is that it reflects new requirements of the IOCA, with more emphasis on the ability of verbal communication versus the ability to read and write. The basics of the training should be active participation in the process of obtaining information as well as active thinking versus memorization of the given information. In other words, while studying and analyzing air accidents, the students must themselves find the problems and solutions in a particular situation. The ability to absorb language skills on their own is a positive factor resulting from the students' participation in the role games.

Role games must be flexible to allow a change in the scenario and to develop different scenarios. This will help the students learn how to react and use the professional language effectively in different kinds of extreme problematic situations. In the creation of game scenarios, two factors must be taken into account: the characteristics of the students (age, sex, intellectual level, previous profession, general and specific skills, and the degree of motivation in learning) and the physical characteristics of the environment (means of information, means of communication, technological means, computerization, and all the other means for providing information to the air-traffic controllers).

11.5.2 General Characteristics of a Game Process

There are two kinds of active training methods: (1) imitation of professional activity and (2) nonimitation. Imitation consists of two subgroups: games (role games, business games, didactic games) and nongames (analysis of real situations). Games and discussions are most often used for training. The role game method is based on the psychological analysis of the air-traffic controllers' activity. Thus, the training must be based on two reciprocal stages: traditional methods of teaching and specific professional English in combination with the participation in the role games. The sequence of those stages does not need to be in a certain order.

In role games, there are simulated activities of people in small groups. Role games are based on real situations that require the ability to implement decisions in the conditions of their professional activity. Business games simulate the conditions of a group activity to solve the problems related to the involvement of different specialists who finalize the result of the game. Didactical games reflect the future air-traffic controllers' activity by the analysis and perception of study materials for finding decisions in the problem situations. Didactical games have one specific element that distinguishes them from other kinds of games: they have a specific goal in the study and have the respective results.

Role games take place not only in a class environment but also in the environment of re-created pilots' cabins and air-traffic controllers' operation rooms. Such imitating environments allow the modeling of real flight conditions, including elements of critical situations. The instructor monitors the entire game from a separate room by communicating with the pilots and the air-traffic controllers through a microphone. The students, who are not playing roles at the time, listen to the entire communication process; they participate mentally in evaluating the parameters of a flight and make their own decisions. This way of learning is called "learning by observation" and was described in the work of Bandura (1977). Bedny (1978) in his research on the analysis of production operations had discovered two factors that influence learning by observation. The first factor is that the acquisition of some components of work is easier to observe. The second factor is that when the task is more difficult to observe, it might be discovered only during the actual performance. For example, some aspects of performance such as switching or distribution of attention, coordination of muscle efforts, and so on could be developed by the actual performance. In learning by observation, a person cannot imitate the emotional-motivational aspects of a work, which is important for the work of air-traffic controllers. This means that although learning by observation is an important part of the training, it is not sufficient for developing the necessary skills to conduct an effective radio communication in the real-work environment. Important psychological qualities are developed by active participation in games and playing roles under time constraints and stress.

For that purpose, several different scenarios of the games are created. Students are divided into several groups; each student has a chance to play his or her role in different game scenarios, knowing that his or her performance will be evaluated. Such conditions allow not only the modeling of cognitive qualities but also the emotional-motivational aspects of the activity. Participation in the games is the best way to develop practical verbal skills in a professional language, and it also helps develop operative thinking and the ability to change the strategy whenever the situation requires. Under stress and time constraints, role games develop the ability to act creatively. In role games, students use their language skills differently, which creates a positive emotional-psychological climate and in turn a positive atmosphere in the process of learning English.

Participation in role games is the most effective method for evaluating the language preparedness of the future air-traffic controllers for their work in international airspace. It allows instructors to model real situations not only for training in English but also for forming the students' moral and psychophysiological qualities, which are important for conducting a safe radio communication in extreme situations. This also allows instructors to test the students' ability to combine radio communication in English in nonstandard conditions with their future professional activity. Successful performance of the language tasks could be achieved either by high-quality professionalism or by an ability to endure an excessive psychophysiological load.

Specifics of the games are characterized by mental processes having its outcome in verbal actions. The information that an air-traffic controller obtains from the screen must be in tact with his verbal actions. The verbal information may be inadequate or delayed, may come from different sources, and may be interpreted by the crew members or by the other air-traffic controllers. In some cases, the information can be duplicated, and in other cases, the game situation presupposes that an air-traffic controller might try to slow down the upcoming stream of information by rephrasing the questions or commands, by asking to repeat or confirm the command, or by requesting additional information. Sometimes, the air-traffic controllers switch from English to Russian and break the commands into parts.

For nongames, we have identified a set of materials that includes the all possible sources of knowledge: printed materials (reports of investigations of air accidents, sketches, tables, newspaper and magazine articles), photographs of the accidents, video materials (television news and documentary films), and most important, audio recordings of last-minute radio communications between the air-traffic controllers and the aircrafts' crews.

In the beginning stages of training, performing radio communication even in the native language is quite difficult. Obviously, it is much more difficult to do it in a nonnative language, especially in extreme situations (the level of emotional tension increases, which lessens the ability to understand the nonnative language). Air-traffic controllers do not have a face-to-face communication with pilots, in which a huge chunk of information could be provided by means of body language. Bryan Day (2004) showed that, in people's interactions, about 56% of information is transferred by means of body language (nonverbal communication), where only 7% by verbal communication (words), and about 38% by the tone of voice. The latter factor is usually eliminated by electronic voice modulation. Thus, radio communication is the only contact the air-traffic controllers have. As long as no visual memory is involved in the communication between the pilots and the air-traffic controllers, special attention should be paid to the verbal-imaginary sources in the training, such as the analysis of real radio communication records. Mathews (2004) called them "Grail's Cup." The use of verbal-imaginary sources was very limited until recently because of the Civic Aviation Convention regulations (Chapter 5, paragraph 13: "Non-divulging of records").

Analysis of the air accidents and incidents is not limited to the identification of errors in radio communication. The record of every air accident has a variety of valuable information in it: an air-traffic controller's work environment, sequence of events, consequences of an accident, technical characteristic of a given aircraft, description of an airport, conclusions, and recommendations. In analysis of air accidents, students are actively involved in discussions, form their own opinions about games, and form their own conclusions. In the role game method, we use task problems with different levels of complexity. Upon the completion of the classes, the students will prepare reports in which they describe the analysis of the previously used game situations. Systematic writing of reports is good practice in English and helps store the necessary knowledge in the memory for future use. The instructor should ignore some errors in the students' communication if those errors do not distort the meaning of the exchanged information, but the instructor should help them overcome the language barrier and reduce emotional tension.

11.5.3 Basic Psychological Characteristics of the Games

The basic characteristics of the role game method are described next (Makarov and Gerasimenko 1997).

11.5.3.1 Strict Regulations and Time Constraints for Testing and Evaluation

In conditions of high emotional tension, there are strict regulations and time constraints for the perception of information and making decisions. Air-traffic controllers receive information from the electronic devices in the form of signals and numerals and transform them in their mind into verbal commands in a nonnative language. Implementing this principle is needed not only to develop their professional ability to conduct the radio communication but also to help them keep the optimum level of physical and mental health.

11.5.3.2 Extra Psychological Load

Extra psychological load is based on the optimization of psychological and physiological efforts of humans in the process of modeling extreme situations. Normally, a person would need to make less and less psychophysiological effort to perform, as his or her professional qualities are developed with experience. Analysis of the past air accidents leads to developing the important skills such as breaking the data into prioritized elements of occurred situation depending on the degree of extremity. It helps the students overcome the language barrier and reduce emotional and physiological efforts in nonstandard radio communication. It also helps them release their mental reserves in making appropriate decisions.

11.5.3.3 Gradual Increase in Psychological Load

This requirement is materialized by implementing the following in the training process: After conducting a difficult task, students should have a break in the form of an easier task; then, their psychological load should be increased to a certain level, higher than the load they had before. This can be achieved by increasing the volume of the study material, by presenting new study material, by higher intensity and complication of tasks, and so on. Adaptation to constant mental and emotional tension is a very important part of the air-traffic controllers' professional training. This will widen their adaptation ability to be prepared for future work.

11.5.4 Components of the Games

Participants: The participants are investigators, specialist(s) in recoding the parameters of flight information, specialist(s) in recoding the information of the communication between the crew members, representatives of a rescue crew, an air-traffic controller, an airline representative, and an airport representative. They gather at a round table for a discussion about an aviation accident, to identify possible reasons of a catastrophe, and to compose a conclusion report and make recommendations.

Purpose: The purpose is to strengthen the knowledge and skills obtained in the language training class; to develop the cognitive and searching activity; to advance the motivational level and ability to analyze the possible reasons of events; to advance the skills in self-learning and in ability for self-teaching; to develop a skill in planning the work, in finding and analyzing the information, in identifying a prevailing problem, to learn to compose, rephrase, and state the study material; to learn to transfer theoretical knowledge into practical skills when analyzing extreme situations; to form qualities such as responsibility, discipline in performance, self-confidence, self-critique; to develop mental qualities as such logic thinking, concentration and distribution of attention, adequate self-evaluation; to collect and store the knowledge about reasons for extreme situations and for events; to form the skill for evaluating the results of self-activity and the activities of others; and to learn how to compose reports.

Conditions: Under the training program schedule, a time limit is allocated for the game. Therefore, a simplified scenario with the main points of existing information about an accident in the report is used. Specific attention is paid to the analysis of the radio communication between a pilot and an air-traffic controller. An instructor oversees the entire game and takes notes for future analysis of the students' mistakes, further discussions, and distribution of the tasks to the students.

Steps: Before the game begins, the instructor introduces the rules of the game to the students, assigns roles, and gives the students specific tasks and time for consultation. The participants are given a plan of a future report in which they can mark their thoughts and take notes.

Procedure: The leading investigator opens the discussion and lets the specialists state their thoughts and opinion about the situation. Specialists in recoding present the information they retrieved from the instruments of the aircraft; then, they let the participants listen to a record of last-minute radio communication, answer the participants' questions, and let them state their opinions. Only after that do the specialists in recoding give their conclusion. The game participants analyze the information, make a conclusion as to the presuppositions and reasons for the accident, and form and compose recommendations.

Composing and preparing a report: In the reports, the students present the information according to their roles: about the accident, about the aircraft crew, and about the aircraft's condition at the time of the flight and damages to it. They describe the errors in radio communication between the air-traffic controller and the pilot; describe the airport, navigation devices, weather conditions, landing, and monitoring airspace; and describe the rescue and fire crews' operations and the passengers' conditions.

Discussion: This takes place after all the reports are submitted. The instructor analyzes the students' mistakes without their presence and works out exercises using the necessary training material for the students. The instructor should begin the discussion in a positive manner, that is, state the correct points in the students' reports. Thus, students will have to concentrate on refreshing and storing the necessary knowledge.

Criteria of evaluation: The criteria for conducting and discussing the role games are as follows: degree of motivation in performing a task; emotional stability in performing the difficult tasks in conditions of destructions and time limit; the ability to visualize a flight upon information received in English and to store it in memory; the ability to break the information by the degree of extremity; originality in making decisions; the ability to recognize and liquidate the mistakes in the communication by requesting additional information; the ability to apply the English knowledge; pronunciation; creativity in performing role games tasks; the ability to evaluate the student's own decisions as well as decisions of other participants; the ability to state an opinion in the discussion and to defend a point of view in English; and the degree of activity in participating in the game discussion on a given topic.

11.5.5 Designing Problem Situations

Taking into account that knowledge and skills realize themselves best in the condition where they were first formed, we paid special attention to the problem-solving methods of teaching. We understand this method as a way to present specially developed ill-defined situation to the students. It requires the students not simply to solve the problem, but, based on the given ill-defined situation, to develop or formulate task problems. The main element of game scenarios must be problem-solving situations, because such situations usually take place in the emergency conditions. Game scenarios must be designed in such a way that the air-traffic controllers would be able to find an adequate solution of the operative task in their verbal communication. In modeling game scenarios, there can be a verbal exchange of information between the instructor and the students and among the students. All communications must be recorded and analyzed later. To create a problem situation, we should identify the conflicts. The content of every problem situations present itself as a learning problem. Makhmutov (1975) pointed out that a task or a question could be seen as a problem when there is a conflict between the requirements of a task and a student's background or means for reaching the goal. The problem-solving situation takes place when there are several possible options in making a decision in conditions of limited information.

Problem situations must be designed as models of normal and dangerous situations. The latter can be achieved by including additional elements in the situations, such as redundant or irrelevant information, by gradually increasing the responsibility for incorrect and untimely solutions, by decreasing the time limit for the responses, and so forth. Moreover, it might be helpful to purposely include some unclear information as well as some verbal information that is difficult to understand. The students' involvement in resolving typical problems of flight monitoring while imitating real situations is one of the most important aspects of training. We recommend the use of special training devices and role games, where verbal communication between the pilots and the air-traffic controllers is the center part of the training. Recording of communications in the games can be easily registered, and listening to the recordings can be an excellent tool for analyzing the mistakes. It will have a definite positive influence on the preparedness of nonnative English-speaking students for future work.

11.6 Requirements of Knowledge in English

According to Document 9835 of the IOCA, pilots and air-traffic controllers should have a required level of knowledge of English language. However, analysis of air accidents demonstrated that a good knowledge of English does not necessarily lead to reliable radio communication in extreme situations. The air-traffic controllers' ability to use English in real-life situations is much more important. The knowledge and skills that the air-traffic controllers need to possess and a set of qualities that they need to develop are as follows:

- They need to know the standard phraseology of radio communication, basic structure and models of sentences, and aviation terminology; should be fluent in everyday English and in the topics necessary for their work; and need to possess knowledge of problems in conducting radio communication in extreme situations on the international air space.
- They need to have the ability to understand and keep space–time characteristics of flights in memory; the ability to analyze the alternative models in perception and realization of nonstandard solutions in a form of commands; the ability to conduct reliable radio communication under time constraints, high emotional tension, high intensity of mental and perceptive functions, overloaded short

memory, and frequent turning into long memory; the ability to react immediately on the information received; and the ability to initiate and carry out radio communication during sudden changes of a situation and in destructions.

- They need to be able to analyze specific and critical situations in the process of radio communication in managing air traffic; apply the knowledge and skills for realizing and understanding the nature of problems; project the progress of a situation and anticipate possible results; use vocabulary effectively; state thoughts concisely and without double meaning; keep the speech pace when changing standard phraseology to a nonstandard one; use the paraphrases in unusual and sudden situations; perceive the information from an aircraft crew by ear; minimize misunderstanding by means of rechecking, confirmation, and/or correction; specify the meaning of what was being said in the case of unclear understanding by an aircraft crew; reduce the pace of communication; and speak clearly in the event that an aircraft crew is having difficulties understanding the situation.

- They need to possess a high level of motivation to perform the professional activity; high emotional stability in extreme situations; clear pronunciation; and professional qualities in implementing their knowledge of English and other skills into practice.

To develop these qualities, we offer the following exercises: exercises for proper pronunciation in English; exercises for learning fundamental structures and models of sentences for proper communication; exercises to increase vocabulary and its effective use in practice (e.g., learning synonyms); exercises that are directed toward forming the ability in paraphrasing; exercises for identifying errors and to react by commands at a regular pace; exercises for memorization of aviation terminology by studying the reports of past air accidents and available printed materials and media materials; exercises for developing the ability to re-create the flight characteristics and keep them in memory; exercises that help to increase the pace and speed of speech in performing tasks. The following additional exercises are also offered: working with materials of mass media (audio and video recordings); participating in discussions of flights' safety issues; and participating in role games.

11.6.1 Fragment of Radio Communication

A fragment of a real radio communication between the aircraft crew and the air-traffic controller (on June 17, 1997), which we used in a role game for training, is shown in Table 11.1.

TABLE 11.1

Fragment of a Real Radio Communication

Time	Participant	Content of Radio Communication	Notes
20:48:40	AC	Kr-sk control BAW028 good morning	
	ATC	BAW028 Kr-sk control good morning, go ahead	
	AC	BAW028 we are abeam KRS at 47, maintaining 10.600 m, standard, estimating CZ at 56	
	ATC	BAW028 maintain FL 10.600 m call me over CZ	
	AC	Will call you over CZ BAW 028	
	AC	Kr-sk control ECL 9893 good night	
	ATC	ECL 9893 Kr-sk control good night	
	AC	9893 reached 10100 m abeam Kr-sk 08 to Irkutsk	
	ATC	Maintain 10100 m outbound 23, report abeam	
	AC	10100 Maintaining abeam	
	ATC	BAW028 Kr-sk control passing CZ, maintain FL 10.600 m, report FIR boundary	
	AC	BAW028 Roger will do	
	ATC	BAW028 we have been climbing to avoid another aircraft, we had a TCAS warning at 10.700 m and we are now descending back to 10.600 m standard	
	ATC	Say it again, please	
	AC	BAW028 we just had a TCAS climb warning due to opposite traffic at nearly 10600 m, we are now descending back to 10600 m	
	ATC	Roger	
	Opposite traffic	10700 takes by TCAS, opposite traffic	
	ATC	Opposite traffic at FL 10100 m	
	AC	BAW028 that is confirmed, there was at that height, we had a deviation to my usual height warning in this case, that is indicated at 10100 m we are now at 10600 m climbing to 10700	
	ATC	Roger	
	ATC	BAW028 Control	
	ATC	BAW028 Kr-sk Control	
	AC	BAW028, go ahead	
	ATC	BAW028 roger maintain level 10.600 m, passing FIR boundary contact Kemerovo Control 129,3 have a nice flight	
	AC	129,3 BAW028 thanks very much, good-bye 129,3	

AC = aircraft crew; ATC = air-traffic controller; TCAS = traffic collision avoiding system; CZ = point of flight; FIR = flight information region; Kr-sk = Krasnoyarsk.

11.7 Methods of Evaluation

We offer the following methods of evaluation: (1) tests for evaluating the level of communication skills in English, (2) problem tasks, and (3) analysis of concrete situations.

Tests for evaluating English skills should include grammar structure, vocabulary, and aviation terminology. The criteria should be the number of correct answers in multiple choice questions.

Problem tasks allow identifying the following: the depth and capacity of knowledge regarding the errors in radio communication; the ability to think logically and the ability to analyze the events and to find possible consequences of the events; flexibility in using English in real nonstandard situations; speed and originality of thinking, that is, the ability to find and materialize correct and decisive commands to the aircraft's crew in extreme situations; and the ability to state thoughts and define viewpoints in English in conditions of high concentration in resolving complicated tasks. The criteria for evaluating performance in problem tasks are the ability to use the obtained knowledge for making correct decisions and to formulate it in English.

Below is an example of a problem task in which there was a contradiction between the visual information and the verbal information given in English.

> Aircraft B767 (course 100) and aircraft MD80 (course 217) of the echelon FL290 are on crossed lanes. B767 must go through by about 15 miles behind MD80. When the aircrafts were about 80 miles away from each other, an air-traffic controller gave a command to both the aircrafts to stay on the course. A minute before their courses would cross, he gave a command to B767: "An approaching aircraft is under 30° on your left, from left to right, the same echelon, an aircraft MD80, at this time 25 miles away, approaching you." The pilot of B767 began concentrating on the image on display, which was on the left of the TCAS display and asked the air-traffic controller: "Where is the aircraft going to?" The air-traffic controller responded to his question with information.
>
> However, the pilot of B767 transmitted back: "We are going to take a course on 120°" and began to make the right turn. Horizontal separation was promptly reduced, and the TCAS on both aircrafts went on. Starting to go lower, the pilot of B767 transmitted: "We would like to go lower to echelon 270." Later, to justify his decision to change course, the pilot of B767 said to the air-traffic controller: "The aircraft was coming towards us; therefore we made the right turn trying to avoid collision." The pilot reduced the separation to 2 miles. Still, why did the pilot of B767 make the decision to make a turn in spite of the air-traffic controller's command? And why did he turn right after all?

Analysis of concrete situations has to do with the use of language in air-traffic controllers' work at international airspace and allows an evaluation of the following: the depth and broad knowledge about errors in radio communications, which could have caused air catastrophes, as well as the possible reasons for them; and the ability to recognize and to eliminate the errors in radio communications, which would lead to the safety of air traffic.

EXAMPLE:

01 33:11 *Tower*: Air Cal 336, you are cleared to land.
01 33:33 *Tower*: Air Cal 931, taxi into the takeoff position; hold it and be ready.
01 33:37 *Air Cal 931*: 931 is ready.
01 33:52 *Tower*: Air Cal 931, traffic is clearing at the end of the lane, you are clear for takeoff, Boeing 737 a mile and a half away of landing lane.
01 33:57 *Air Cal 931*: I see, we are diverging.
01 34:13 *Tower*: Air Cal 336, go around, 336, go around!
01 34:16 *Air Cal 336*: Can we go for landing? Ask him if he can hold on.
01 34:18 *Tower*: Air Cal 931, continue to hold on if you can.
01 34:21 *AC 336 to Tower*: Can we land?
01 34:22 *Tower*: Air Cal 931, aircraft is behind you! Abort the takeoff!
01 34:25 *Tower*: Air Cal 336, go around, 931 will abort the takeoff!
01 34:27 *AC 336*: Captain: Gears up!
01 34:36 IMPACT!

Aircraft lands with gears retracted. Thirty-four people were wounded, four of them seriously.

11.8 Examples of Past Radio Communications for Analysis

Boeing B-747-121 of Pan American Airlines did not clear a takeoff lane. The pilot of the aircraft transmitted information of his takeoff to an air-traffic controller: "We are at the takeoff now." However, the air-traffic controller did not inform the aircraft crew of Boeing B-747-206B of KLM airline about it. As a result, there was a collision on the takeoff lane and 583 people died.

Instruction to the students: Explain the air-traffic controller's action and identify the reason for the error.

Why did the air-traffic controller not inform the aircraft crew of Boeing B-747-206B of KLM, which was ready for the takeoff, that the takeoff lane was taken? He either did not hear or did not understand the pilot's command for takeoff. If he did not understand, the command may have had a mistake. In the English language, the preposition "at" means the *position* of a subject, not an *action*. That is, this phrase may only mean that an aircraft is on the takeoff lane ("at the takeoff position" or "ready to roll"). The action of an actual takeoff may sound like this: "We are taking off now" (present continuous tense in English grammar).

On March 5, 1973, an aircraft Boeing 747-249F of the Flying Tiger Line crashed into a hill on the way to the Kuala Lumpur Subang airport (KUL).

Instruction to the students: Listen and read a fragment of the radio communication record and try to find the reason for the catastrophe.

This accident happened due to the similarity in pronunciation of numeral "2" and the preposition "to" in English language. An air-traffic controller gave a command to the aircraft crew: "Get into echelon (go down) 2400" ("two-four-zero-zero"). The pilot confirmed the command: "OK! To 400" (it sounded "to-four-zero-zero"—similar to "two-four-zero-zero"). The air-traffic controller did not catch the mistake. These kinds of similarities, in which words with different meanings sound the same in pronunciation, exist in English. To avoid these kinds of unfortunate situations in the future, the students must perform a series of exercises. For example, listen to the sentences and short dialogs. They are all correct; however, every sentence has a word in it that may have different spelling and meaning if taken separately. Translate the sentences. Identify which word may create a problem. Describe the different meanings of a particular word.

In another case, incorrect use of the word "hold" led to a catastrophe in which 34 people were wounded and 4 were seriously wounded.

Instruction to the students: Read a fragment of the radio communication record and explain the meaning of the word "hold" in all the cases where the word was used. Change the command in a way that would prevent misunderstanding in a radio exchange between the air-traffic controller and the pilot.

Students' knowledge in English is not up to the task, and therefore they cannot find a solution. They would need to use a dictionary for that. However, a regular dictionary will not help much for this particular task. Only in an aviation dictionary can they find different meanings of the word "hold" in everyday English and in specific aviation English. The instructor gives the students only factual parts of the situation without analyzing a pilot and an air-traffic controller's actions. Therefore, the students are forced to activate their cognitive skills in forming a problem, analyzing the problem questions, constructing an algorithm in a search of an answer, and finding and implementing a solution. Thus, the students realize their actions in the following sequence: analyzing the situation → formulating the problem → establishing the hypothesis of wrong actions of a crew or an air-traffic controller. This approach enables the students to realize all their knowledge for this particular situation.

On July 2, 2002, at 23:43 hours Europe time in German airspace, a collision of a passenger aircraft TY–154M and a cargo airplane Boeing 757 resulted in the death of 71 people, including 52 children.

Instruction to the students: Listen to the last 50 seconds of radio communication between the crews of TY-154 and Boeing 757 and the air-traffic controller, analyze the situation, find the errors, and discuss it in English.

EXAMPLE

Time until collision		
50 seconds	Ty-154, TCAS	Aircraft is approaching! Aircraft is approaching!
45 seconds	Air-traffic controller	BTC2937 (code of Ty-154, go lower, echelon 350, descend, an aircraft is approaching).
40 seconds	Captain of Ty-154	We are descending.
38 seconds	Boeing 757, TCAS Boeing 757 began descending	Descend, descend!
35 seconds	Ty-154, TCAS	Climb up! Climb up!
34 seconds	Second pilot of Ty-154	(System) says "Climb up"!
32 seconds	Captain of Ty-154	Descending!
30 seconds	Air-traffic controller	BTC2937, descend, echelon 350, speed up your descending.
25 seconds	Captain of Ty-154	I am descending to echelon 350, BTC2937.
20 seconds	Air-traffic controller	Yes, an aircraft is approaching, under two o'clock, now is on 360°.
13.3 seconds	Captain of Boeing 757	Begin descending by the TCAS command.
11 seconds	Crew of Ty-154	Where is it?
9.5 seconds	Ty-154, TCAS	Climb up!
5.3 seconds	Second pilot of Ty-154	(System) says: "Climb up!'
3.8 seconds	Crew of Boeing 757 saw the Ty-154 on their right.	Pilot jerks his steering wheel.
1.8 seconds	Crew of Ty-154 saw Boeing 757 on their left at 300°.	Pilot jerks his steering wheel.
00 seconds	COLLISION	

In addition to the air-traffic controller's conflicted commands to the TCAS, students have to discover one more contradiction. For possible visual orientation in the airspace, the air-traffic controller commands an angle between the nose of the aircraft and the direction of the approaching object to the Ty-154 crew. According to the rules, the pilot must visualize his aircraft on the face of a clock and relate the object to the numbers on it. For example, "under two o'clock" means there is an object on your right at the angle of 60° from the nose of your aircraft. By giving this particular command, an air-traffic controller oriented the pilot of TY-154 to look to the right. However, Boeing 757 was actually at his left—under 300° or "under 10 o'clock." Apparently, the air-traffic controller mistakenly transferred the command to the wrong aircraft's crew.

11.9 Conclusion

In this chapter, we reviewed a training method for air-traffic controllers that is based on the principles of activity theory. The purpose of the method is to develop the communication skills of air-traffic controllers in English. The main concepts in the learning process are activity, goal, cognitive and behavioral actions, self-regulation, strategy, sociocultural context, and so on. The major units of analysis in the training process are cognitive and behavioral actions (including verbal actions) and functional blocks. The trials and errors in learning are considered in line with the self-regulatory process. One important prerequisite for learning is the unity of external and internal activities.

For pilots and air-traffic controllers, communication is not just a simple process of information exchange but an important tool for performing practical actions under time constraints and stress. For developing an effective method for English teaching, we used the problem-analysis system of learning and training and some important ideas that were derived from the self-regulation concept of learning. Particular attention was paid to the development of the ability to use English in the process of operative thinking in performing operative tasks. In the learning process, we should develop the ability to form appropriate goals and adequate dynamic mental models of the situation on the basis of verbal information. The method of teaching the English language to the air-traffic controllers was divided into two stages. The first stage involved the traditional method of acquiring a foreign language. The second involved the development of communication skills in problem-solving situations, which could be created on the basis of studying past critical situations and accidents in the airspace. This method combines the acquisition of the English language in game conditions, which imitate real professional situations. Our experimental method prepares students to communicate effectively in English in their future work as air-traffic controllers in extreme situations of international airspace.

References

Anderson, J. R. 1985. *Cognitive Psychology and Its Application.* 2nd ed. New York: Freeman.
Anokhin, P. K. 1955. Features of the afferent apparatus of the conditioned reflex and their importance in psychology. *Prob Psychol* 6:16–38.
Bandura, A. 1977. *Social Learning Theory.* Englewood Cliffs, NJ: Prentice Hall.
Batishev, S. Y. 1977. Teaching workers in industry. In *The Fundamentals of Vocational Pedagogy*, eds. S. Y. Batishev, and S. A. Shaporinsky, 412–503. Moscow: Higher Education Publishers.

Bedny, G. Z. 1978. Some psychological aspects of production operations acquisition process and time study. In *Questions of Work Psychology and Professional Training*, ed. V. G. Asseev, 34–56. Irkutsk, Russia: Higher Education Publishers.

Bedny, G. Z. 1981. *The Psychological Aspects of a Timed Study during Vocational Training*. Moscow: Higher Education Publishers.

Bedny, G. Z., and W. Karwowski. 2004. A functional model of human orienting activity. In Special Issue, *Theoretical Issues in Ergonomics Science*, ed. G. Z. Bedny 5(4):255–74.

Bedny, G. Z., and W. Karwowski. 2006. The self-regulative concept of motivation. *Theor Issues Ergon Sci* 7(4):413–36.

Bedny, G. Z., and W. Karwowski. 2007. *A Systemic-Structural Theory of Activity. Application to Human Performance and Work Design*. London: Taylor & Francis.

Bedny, G., and D. Meister. 1997. *The Russian Theory of Activity: Current Application to Design and Learning*. Mahwah, NJ: Lawrence Erlbaum Associates.

Bedny, G., and W. Karwowski. 2000. Theoretical and experimental approaches in ergonomics design: Towards a unified theory of ergonomics design. In *Proceedings of the Fourteenth Triennial Congress of the International Ergonomics Association and Human Factors and Ergonomics Society*, 197–200. San Diego, CA: Human Factors and Ergonomics Society.

Day, B. 2004. Language testing in aviation: The stakes are high. In Proceedings of the ICAO Aviation Language Symposium, Montreal, Canada.

Kudryavsev, T. V. 1975. *Psychology of Technical Thinking*. Moscow: Pedagogical Publishers.

Landa, L. N. 1982. Descriptive and prescriptive theory of learning and instruction. In *Instructional Design, Theories and Models. An Overview of Their Current State*, ed. C. M. Reigeluth, 55–74. Hillsdale, NJ: Lawrence Erlbaum Associates.

Landa, L. N. 1984. Algo-hueristic theory of performance, learning and instruction: Subject, problems, principles. *Contemp Educ Psychol* 9:235–45.

Makarov, R. N., and L. V. Gerasimenko. 1997. *Theory and Practice of Designing Goal-Directed Models of Operators in Complex Control Systems*. Moscow: International Academy of Human in Astronautics' Systems.

Makhmutov, M. I. 1975. *Problem-Solving Approach to Teaching: Basic Theoretical Questions*, 367. Moscow: Pedagogika.

Mathews, E. 2004. *Aviation Language Training: a Summary of Best Practice*, 1–3. Montreal: ICAO Aviation Language Symposium.

Matyushkin, A. M. 1972. *Problem Situation in Thought and Instructions*. Moscow: Pedagogical Publishers.

Nayenko, N. I. 1976. *Psychic Tension*. Moscow: Moscow University.

Pushkin, V. N. 1965. *Operative Thinking in Large Systems*. Moscow: Science Publishers.

Section V

Special Topics in the Study of Human Work from the Activity Theory Perspective

12

Functional Analysis of Attention

G. Bedny and W. Karwowski

CONTENTS

12.1 Introduction

In this chapter, we focus on the mechanisms of attention from the systemic-structural activity theory (SSAT) perspective (Bedny and Karwowski 2007). SSAT is specifically adapted to the study of human work (Bedny and Meister 1997; Bedny and Karwowski 2007; Chebykin et al. eds. 2008; Bedny et al. 2008). SSAT considers activity a multidimensional system that is studied from various perspectives, and it includes multiple interdependent procedures, steps, and levels. These can be grouped into three major approaches: the cognitive approach, in which the concept of process is central; the morphological approach, in which the concept of mental and motor actions is most important; and the functional approach, in which the concept of self-regulation and functional blocks is important. The main approach we consider here is functional analysis, which is used to develop a model of attention. This method is similar to the cognitive approach. We show that not only the activity process but also separate cognitive processes can be described as complex self-regulative systems.

The concept of attention plays a central role not only in cognitive psychology but also in activity theory. People do not analyze all the information that is available to them at any given time. They can select particular information and be in a more or less attentive state. The study of attention is important from both theoretical and applied perspectives. It can be useful to study the factors such as automation, prediction of multitask performance, cognitive efforts, and task complexity. Task complexity is a major factor in creating a challenge for an operator's performance. A complex task requires greater cognitive effort (Bedny and Karwowski 2007). The more complex the task, the more the subject concentrates during task performance. With the concept of task complexity, the concept of task difficulty arises. These two characteristics of the task should be differentiated. Task complexity is an objective characteristic of the task, whereas task difficulty is the performer's subjective evaluation of the task complexity. Depending on the individual features of a subject, the same complex task will be evaluated as relatively more or less difficult. An increase in the task complexity increases the performer's mobilization of mental efforts and increases the concentration of attention. The subject can perform time-sharing tasks differently depending on the complexity of each task.

At present, two different types of attention models are used. One type considers attention a mental effort (Kahneman 1973). The other treats attention as an information-processing system. The mental effort models can be related to either "single-resource theory" or "multiple-resource theory" (Wickens and McCarley (2008). Kahneman's model can be related to the single-resource theory. Kahneman suggests that there is a single, undifferentiated pool of resources available to all the tasks and mental activity. If the task difficulty increases or the person performs two tasks simultaneously, more resources are required. Also, allocation of these resources is required. The more difficult a particular task is, the fewer the resources that are available for the second task. One limitation of the single-resource theory is in interpreting the well-established empirical finding that when concurrent tasks are in different modalities or use different codes (spatial, verbal), allocation of the resources becomes much easier. Single-resource theory is only able to predict variation in the task difficulty.

Multiple-resource theory argues that instead of a single, undifferentiated resource, people have several differentiated capacities with distinct resource properties. For example, it is easier to perform two different tasks that require different modalities than the two tasks that require the same modality. In this situation, time-sharing tasks will be more efficient (Wickens and McCarley 2008). Regarding the cognitive information aspects, Norman's model is of interest (Norman 1976). In this model, two mechanisms are important: data-driven and conceptually driven processing. Data-driven processing is an automatic process and depends on the input information. When this process predominates, we will use the term "involuntary attention" as per activity theory (Dobrinin 1958). Expectation and generation of hypotheses about the nature of sensory signals, conceptualization, and past experiences are also important. Information from memory is combined with information from sensory data. This is conceptually driven processing, which we will call "voluntary attention" as in activity theory. Voluntary attention includes not only cognitive components but also energetic components; Norman's model ignores the energetic components.

The relationship between peripheral and internal processes is also important in the study of attention. One of the drawbacks of multiple-resource theory is its inability to determine whether the advantages of cross-modal tasks over intramodal tasks are attributable to the central or peripheral processes. For example, time-sharing may not be the result of central resources, but rather the result of peripheral factors that constitute the two intramodal tasks. Two visual tasks may pose confusion and masking, just as two auditory messages may mask each other (i.e., this exhibits peripheral over central effects). Wickens and Hollands (2000) wrote that the degree to which the peripheral factors, rather than central factors, are responsible for the cross-modal interference or better than the cross-modal time sharing remains uncertain. In one part of the Wickens and Hollands study, it was shown that when visual scanning is carefully controlled, cross-modal displays do not always produce superior performance. This can be explained by the fact that attention can be considered a self-regulative adaptive and adjustable system and attention can be characterized not only by attention limitations but also by the ability of the subject to use the attention features efficiently in any particular task or in a required period of time. The ability to adapt and tune different features of attention to certain task requirements is provided by the mechanisms of self-regulation of activity. The self-regulation mechanisms are responsible not only for tuning, sustaining, and regulating attention but also for all other cognitive processes. The self-regulative process we discuss here is not homeostatic but rather goal-directed (Bedny and Karwowski 2004a, 2004b). The self-regulated mechanisms of attention as a goal-directed process are not sufficiently studied.

Many single-channel theories of attention are based on the research of the psychological refractory period (PRP; Meyer and Kieras 1997; Pashler and Johnston 1998). However, the PRP is not the only mechanism of attention. Attention also depends on consciously regulated strategies. For example, in our earlier studies (Bedny and Karwowski 2008), we found that the subject can program for the second motor action when he or she performs the first motor action. Such a strategy depends on the complexity of the first and second motor actions. For example, if the first action is very complex or is associated with danger, the subject can use sequential strategy. As we will attempt to demonstrate in this chapter, the ability to perform elements of activity in parallel depends not only on the mechanisms of the PRP but also on the strategies of self-regulation of activity in general. Attention is not the "performer" of two tasks. The subject, with his or her past experience, motivation, conscious goal and strategies, and so on, is the performer of these tasks.

Thus, at present, no theory of attention completely explains the phenomenon of attention. The goal of this chapter is to present some new data that can be useful in understanding attention mechanisms. We consider the attention processes when the subject performs sequential tasks of various complexities, and as a result, models of attention are developed.

12.2 Experimental Method

12.2.1 Review of the Existing Experimental Methods to Study Attention

One of the most important procedures for studying attention is the use of two simultaneous or time-sharing tasks. The time-sharing method refers to situations when an individual simultaneously receives two messages (Navon and Gopher 1979). For example, each ear is used as a separate input channel. Typically, the instructions introduce the goal of tracking either one or both of the channels. In the first situation, we talk about switching attention, and in the other situation, about the allocation of attention (Norman and Bobrow 1975). The other widely accepted method of studying attention requires the performance of two tasks in a sequence. The interval between the two tasks (stimulus-onset-asynchrony [SOA]) can be changed during the experiment. The performance time for the second task is increased for the short SOA. However, the complexity of choice-reaction tasks on which the difficulty of task performance depends is also a critical factor in dual-task performance. Usually, this characteristic of task does not change during the experiment. A more complex choice-reaction task has been described by Pashler and Johnston (1998) in their experimental study. The first task included two acoustic stimuli and the second one three visual stimuli. The interval between these two tasks was changed from 50 to 450 milliseconds. The less the time interval between these two tasks, the longer is the response time for the second task.

We approached this issue from a different perspective. We asked the subjects to perform two tasks in sequence. The second task was given immediately after completing the first task. The complexity of the tasks varied. We analyzed how the complexity of the first task influenced the performance of the second task. Moreover, we asked how this influence changed when the complexity of the second task also changed. The choice-reaction tasks used in this study include a number of alternatives for the first and the second tasks in different sets of experiments. The left-hand reaction is a response to an auditory stimulus. The right-hand reaction is a response to a visual stimulus. The number of auditory stimuli varies from one to four. The number of visual stimuli varies from one to eight. Therefore, the complexity of both tasks can be changed by changing the number of presented stimuli. The right-hand reaction is performed immediately after the left-hand reaction, and the number of possible choices for the right-hand varies from one series of experiments to another. The interaction of complex choice reactions performed sequentially is analyzed from the viewpoint of the concept of self-regulation of activity. In conclusion, we note that the complexity of the reactions determined not only the feature of the presented stimuli but also the nature of the required responses. These two factors are taken into account in our experiment. The complexity of responses in all the experiments was one and the same. Such complexity was determined by the distance of hand motions and the standard resistance of push switches when the subject clicks on them with his fingers.

12.2.2 Procedure

For studying the interaction of the complex choice reactions, we developed an experimental bench that had a subject panel on one side and a researcher panel on the other. The subject panel had a digital indicator with numerals that light up and a sound device to create clear-tone sound signals. There were two start positions on the panel in front of the subjects: the left-hand start position and the right-hand start position. Push switches were located at different radial directions from the start position. Four switches were used for the left-hand position and eight for the right-hand position. Two reaction-time meters were located at the left-hand and right-hand positions on the researcher's panel. The reaction time of the left hand was registered by the left meter and that of the right hand was registered by the right meter. The researcher created the program using the keyboard on the panel for the subjects. The task consisted of the following sequence of events:

1. Subjects kept their sight on the digital indicator. When one of four sound signals was given through a headphone on the left ear, they had to react by pushing the corresponding switch with the left hand. During the execution of this action, subjects had to keep looking at the digital indicator.

2. Immediately after the sound signal and the subject's response, one of the eight numbers lit up and the subject had to react with the right hand by pushing the corresponding switch. The time for each reaction was measured separately. The number of alternatives in each set of the experiments was the same.

3. A warning signal was turned on to notify the start of the next sequence of signals.

Four male university students were selected for this experiment. Prior to the experiment, they were trained for 2 days (1-hour session per day) to work with the panel. The experiment was conducted for 3 days.

12.2.3 Design

In the first set of experiments, only visual stimuli were presented to the subjects, and the subjects had to react only with the right hand. The number of stimuli was increased from one to eight. This set was marked 0–8, with 0 indicating there was no previous reaction with the left hand to the sound stimulus and 8 indicating a reaction with the right hand to the visual stimulus, with eight alternatives. The average result of all measures was calculated on the basis of 40 reactions of each subject on the corresponding signals (10 preliminary reactions were not considered for the calculation of average reaction time). Erroneous reactions were not considered. The next set of experiments consisted of a measure of simple reaction time when the reaction was executed with the right hand after the left-hand reactions. The number of sound stimuli was gradually increased from one to four. (The given sequence was 1–1, 2–1, 3–1, and 4–1, where 1–1 means one sound stimulus for the left-hand reaction and one visual stimulus for the right-hand reaction, 2–1 means two sound stimuli for the left-hand reaction and one visual stimulus for the right hand reaction, and so on.) The subjects were instructed that after receiving the sound signal they had to react using their left hand. Immediately after that, a "1" would appear on the visual indicator. The subjects would have to react to the sound stimuli using the left hand and to the visual stimuli using the right hand as soon as they could. The same instructions were given for each set of the experiments.

In the third set, after the simple reaction to the sound signals, the visual signal was given and the subjects pushed the corresponding switch. In this set of experiments, the following programs or signals were used: 1–2, 1–4, 1–6, and 1–8. For all the trials, only one sound signal was used, and the number of visual stimuli was varied from two to eight.

On the second and third days, three sets of experiments were conducted (sets 4–6). In the fourth set, the following programs or signals were used: 2–2, 2–4, 2–6, and 2–8. The subjects had to execute the left-hand choice reactions with two alternatives, and the number of stimuli for the right hand was increased from two to eight. In the fifth set, the following programs were used: 3–2, 3–4, 3–6, and 3–8, and in the sixth set, 4–2, 4–4, 4–6, and 4–8 were used. We used a partial counterbalance schema of the experiment. Two subjects started to perform from the simple combination of tasks and finished with the more complicated combination of tasks. The other two subjects started with the more complicated combination of tasks and finished with the more simple one (4–2, 4–4, 4–6, 4–8, and finished with 2–2, 2–4, 2–6, 2–8).

An additional set of experiments was performed on the fourth day, when we offered two signals with such similar tones the subjective discrimination was evaluated as difficult. The following programs of signals were used: 2′–2, 2′–4, 2′–6, 2′–8. The result of this set was compared with that of the previous sets for the same subjects. Table 12.1 represents the general plan of the experiments. The subjects received information about the reaction time for visual and acoustic stimuli after each trail. The significance of the experimental results was checked by two-way analysis of variance (ANOVA).

TABLE 12.1

General Plan of the Experiment

Day of the Experiment	The Number of the Set of Experiment	Program (Relationship between Sound and Visual Signals)
Day 1	1	0–1; 0–2; 0–4; 0–6; 0–8
	2	1–1; 2–1; 3–1; 4–1
	3	1–2; 1–4; 1–6; 1–8
Days 2 and 3	4	2–2; 2–4; 2–6; 2–8
	5	3–2; 3–4; 3–6; 3–8
	6	4–2; 4–4; 4–6; 4–8
Day 4	7	2′–2; 2′–4; 2′–6; 2′–8

12.3 Results

Let us consider the results obtained on the first day. In the preliminary experiment, we measured simple right-hand reaction for visual stimulus. The average time for a simple right-hand reaction, when only one visual stimulus was presented, was equivalent to 0.23 seconds. Then, we conducted the major sets of the experiments. In the first set of experiments, the number of visual signals was changed from two to eight (sequence 0–8). The average time for right-hand reactions to the visual signals (without reaction to the sound stimulus by the left hand) for each subject when the number of stimuli was changed from two to eight is shown in Table 12.2. If the number of stimuli increases, reaction time also increases. The interdependence of reaction time and the number of stimuli can be depicted by a logarithmic curve (Figure 12.1 curve with the symbol ▲ —no sound signals). This function is called Hick's law (Hick 1952).

The result of the second set of experiments, when the number of sound signals for the left-hand reaction was changed from one to four and only one visual stimulus was given, shows that the average right-hand reaction time for a visual stimulus was 0.12 seconds, sufficiently less than the simple reaction time for the right hand on the digit signals only (compare 0.23 seconds and 0.12 seconds). The difference between the means according to the t test was statistically significant ($p < .05$).

The result of the third set of experiments, when only one sound signal was used and a number of visual signals for the right-hand reaction was changed from two to eight, is shown in Table 12.2. Based on these data, a curve with the symbol □ (Figure 12.1) was drawn. The reaction time for the right-hand reaction after a simple left-hand reaction to the sound stimulus decreased compared with the reaction time of the right hand without previous reaction of the left hand. This difference is not statistically significant. Nevertheless, this

TABLE 12.2

Average Time for Left- and Right-Hand Reactions to the Sound and Visual Signals

Left Hand		Right Hand			
		Reaction Time			
Number of Sound Signals	Reaction Time	Number of Visual Stimuli			
		2	4	6	8
0	–	0.37	0.52	0.6	0.62
1	0.33	0.34	0.5	0.6	0.61
2	0.62	0.41	0.59	0.69	0.71
3	0.75	0.43	0.63	0.75	0.77
4	0.78	0.46	0.66	0.82	0.82
2'	0.74	0.43	0.65	0.77	0.8

result is interesting because the curves with the symbol ▲ and the symbol □ do not intersect.

On the second day, the fourth, fifth, and sixth sets of the experiments were conducted. The results of the fourth set of experiments, when the number of sound stimuli was two and the number of visual signals was changed from two to eight, is shown in Table 12.2. The results of the fifth and sixth sets of the experiments, when the number of sound signals was three and four, respectively, and the number of visual stimuli was changed from two to eight, also are shown in Table 12.2. Curves with symbols ◊, Δ, and x were drawn on the basis of the data from Table 12.2 (Figure 12.1).

The results of the fourth day of testing (when subjective discrimination of two sound signals were evaluated as difficult) are shown in Table 12.2 (the last row, 2'). Curve ■ illustrates the right-hand reaction time to visual stimuli when the subjects preliminarily reacted to two poorly distinguishable sound signals. In this case, we can see that the discriminative features of a sound stimulus have approximately the same effect as the number of sound stimuli. We can see that curve ■ is positioned between curves Δ and x. The position of this curve is much higher than that of curve ◊, when the subjects preliminarily reacted to two well-distinguished sound stimuli.

The data demonstrate that the more complicated the previous reaction to the sound signal is, the higher is the curve position, which describes the reaction time of the visual stimulus. We compared the significance of the differences between the reaction times for when the reaction was performed only by the right hand (curve ▲) and by the right hand with previous left-hand reactions (curves ◊, Δ, ■, and x).

The within-subjects two-way ANOVA has been conducted to test the effects of the number of visual stimuli, sound stimuli, and their interaction on reaction time to the visual stimulus.

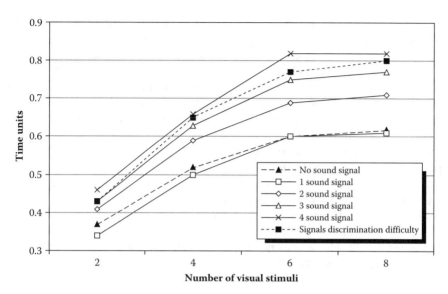

FIGURE 12.1
Reaction time for visual stimuli performed by the right hand.

The within-subjects ANOVA revealed that as the number of visual stimuli increased the, reaction time (RT) increased F (3, 45) = 54.51, $p < .001$. Furthermore, as the number of preceding sound stimuli increased, the RT to the visual stimulus increased as well [$F(3, 45) = 6.11$, $p < .01$]. The effect of the interaction between the sound and visual stimulus difficulty was also significant [$F(9, 45) = 3.02$, $p < .01$]. These results show that as the number of the preceding (acoustic) stimuli increases, the acoustic stimuli have progressively more interference with the reaction time to the following visual stimuli. Furthermore, there is an interaction between the difficulty of the sound stimulus and that of the visual stimulus. This suggests that as the number of the visual stimuli increases, the preceding stimuli have a progressively greater effect on the performance of the second task.

12.4 Interaction of Complex Reactions

Let us analyze the results. Curve ▲ illustrates how the right-hand reaction time changes when only visual stimuli (from two to eight) are given. This curve is a logarithmic function. If we scale axis x as log N, we get Hick's law. It describes the speed of the human information processing when the choice reaction is accomplished. Such research is the foundation of the implementation of the information theory in psychology. All of the following results will be compared with curve ▲ in Figure 12.1.

In the second set of experiments, two sequential reactions were performed. One reaction was performed with the left hand as a response to the sound stimulus when the number of stimuli varied from one to four. The second simple reaction was performed with the right hand as a response to the visual stimulus. A prior reaction to the sound signal reduced the second simple reaction time. This was proved statistically and did not depend on whether the left-hand reaction is a complex or a simple one. This was caused by the partial time overlap between the first and second responses. Observation showed that it was very difficult for the subjects to act quickly with the right and left hands without partial overlapping of the reactions. The reactions took place despite that the subjects were instructed to use the visual stimulus as their start signal for the right-hand reaction. In other words, the start signal for the right hand was not actually an external stimulus; rather, it was the start of the left-hand movement. However, the subjects did not realize the overlapping factor.

For the third set of experiments, the overlap between the first and second responses was eliminated (program 1–2, 1–4, 1–6, 1–8). The reason was that decision making (what switch to press by the right hand) could be made only after the digital gauge was lighted up. We assumed that when a subject switched his attention from the first reaction to the second one, the reaction time would be increased for a selected right-hand reaction. We also assumed that when the left-hand reaction to a sound signal was simple, the increase in the right-hand reaction time to the visual signal would be insignificant. However, the results of the experiment showed the opposite. The right-hand response time with previous simple left-hand reaction decreased compared with the single right-hand reaction (see curves ▲ and □ in Figure 12.1). To understand this result, we analyzed the subjects' behavior. The observation of strategies of the subjects and their debriefing showed that all the subjects considered a single right-hand reaction to be more difficult than a right-hand reaction with a previous simple left-hand movement.

We noticed the following two factors that influence the time of the second reaction. One factor is an elimination of the time uncertainty of presenting the second (visual) signal after the previous left-hand reaction. The second factor is the shift of attention from the first reaction to the second one. The first factor increased the speed of the response to the second signal, but the second factor decreased the speed of this response. In this set of experiments, a left-hand movement is a simple reaction to the sound signal and the performance of such a reaction required a low-level of attention. Due to this, the second factor had a weak influence and practically did not increase the response time of the second reaction. At the same time, the elimination of the time uncertainty of appearance of the second signal reduced the response time of the second reaction. This means that in this case the factor of elimination of time uncertainty of presentation of visual signal overrides the factor of the shift of attention from the first reaction to the second. As a result, we can see an insignificant reduction in the response time in the case of a visual signal (compare curves ▲ and □). However, these differences are not statistically significant.

On the second and third days of the experiment, programs 2–2, 2–4, 2–6, 2–8 were used in the fourth set; 3–2, 3–4, 3–6, 3–8 in the fifth set; and 4–2, 4–4, 4–6, 4–8 in the sixth set. In fact, the shape of the curves did not change (Figure 12.1). However, as a result of the increasing complexity of the previous reactions, the location of the curves is higher in this case. Therefore, increasing the complexity of the previous reaction influences the performance of the following one. The level of the second reaction complexity is very important too. The more complex the previous and the following reactions, the more they influence each other. This outcome can be interpreted on the basis of attention theory. During time-sharing tasks performance, we have two streams of information. One source of information was presented to the right ear and the other one was presented to the left ear. In our experiment, two streams of information have been presented sequentially so that with each time period a subject dealt with one source of information only. The more complicated the previous and the following portions of information and response selection processes in both tasks were, the more difficult it was to shift active attention from one source of information to another. The result of the conducted experiments proves this conclusion.

The increase in difficulty by shifting attention from one reaction to the other overrides the elimination of the time uncertainty of the onset of the second reaction. As a result, the time of the second reaction increases. During the debriefing of the subjects, the following interesting fact was discovered: during the performance of the most complicated task in which four sound signals were presented, there was a subjectively noticeable break between identifying the digit and making a decision about the performance of the second reaction. (One subject said, "I see the digit, but cannot make a decision and move my hand. My hand sticks to the start position.")

12.5 Mechanisms of Attention and Strategies of Information Processing

The obtained data explained the influence of the mechanisms of attention on the strategies of information processing. Recognition of a stimulus is made by a passive automatic process using a low-level of attention, but decision making is linked with active processes using a high level of attention. Active processes reorganize slower than passive automatic processes. The reorganization of attention mechanisms, when sequential unconnected portions of information are presented, is the same as when the subject shifts attention from one portion to another portion of the simultaneously presented information. In our experiment, the subjects cannot keep all of the information about two reactions in the short-term memory. It becomes necessary to use information from long-term memory. The search for information in long-term memory using a scanning device can start at any node point in a structure of the information in the long-term memory (Norman 1976). Hence, during the extraction of the information from long-term memory, the alphabet used by the subject constantly changes. This alphabet is dynamic. As a result, the speed of information processing changed as well.

Let us analyze the result of the experiments conducted on the fourth day when two subjectively sound signals that were difficult to differentiate were presented. The similarity of curve Δ (three sound signals were presented) and curve ■ (sound signals that were difficult to differentiate were used) in Figure 12.1 allows us to conclude that the deterioration in differentiation of the sound signals influences the reaction time of the visual signal in the same way the increase in the number of alternative sound signals does. We discovered the reaction time of the visual stimulus performed by the right hand increased as a result of the deterioration in differentiation of the sound signal for the first reaction performed by the left hand. In this situation, it is more difficult to decide which sound signal was presented. The first task (performed by the left hand to a sound stimulus) became more complex and therefore more difficult for the subjects. As a result, the process of adjusting the mechanisms of attention from the first to the second task became more complicated. Hence, the reaction time for the second task depends not only on the amount of information presented for the first reaction, but also on the differences between the signals. This shows that the complexity of the two considered subtasks determines the strategies for shifting attention from one task to another.

The outcome is that two tasks performed sequentially cannot be considered independent. When subjects repeated the experiment with two reactions, they combined these reactions into a holistic structure and complex activity strategies were developed. These strategies can be conscious or unconscious. For instance, when the programs 1–1, 2–1, 3–1, and 4–1 were performed, a premature right-hand reaction was performed unconsciously. This contradicted the instructions. The subjects developed their strategies to optimize their activity. Therefore, none of the given instructions can strictly predetermine the possible strategies of the actual activity (Bedny and Seglin 1999). In our experiment, a significant increase in the speed of the reaction to a sound signal led to a delay in the reaction to the visual signal. The subjects tried to choose the optimum speed of the reactions, which allowed them to respond quickly to both signals. The strategies of activity were optimized through the coordination of the external conditions and the internal capabilities of the subject. As seen, the subjects did not react to various independent stimuli but rather developed distinct strategies to achieve the specific goal of the unitary task "react with maximum speed by left and right hands." The conscious goal influences the strategies of activity, and such strategies can be conscious or unconscious.

Comparing the results of the observation, debriefing the subjects, and analyzing the experimental data show that the individuals do not react to various stimuli but actively select and interact with the information. Depending on the obtained results of the activity, individuals reformulate the goals and strategies of the activity. This results in the transformation of conscious contents of the activity into unconscious and vice versa. The voluntary attention of an individual is the mechanism through which consciousness is attained. Our observations showed that during the experiment, subjects shifted their attention between the two tasks, allocating their attention and efforts in an attempt to perform one task quicker, slowing down the other task, or vice versa, correcting errors, and attempting to improve the strategy of activity. As a result, individual actions are integrated into a holistic structure based on the self-regulative mechanisms. Thus, describing an individual behavior in terms of stimulus-response is crude and inaccurate. Individual behavior cannot be explained as the sum of independent reactions to a series of independent stimuli. When we develop a model of attention, we should consider the self-regulation process and voluntary and involuntary attention mechanisms. The experimental data show that the model of attention should incorporate the self-regulation process and various voluntary and involuntary attention mechanisms should be considered.

In cognitive psychology, a delay in the second task reaction is called PRP. The processing of each task can be divided into three stages: (1) the early stimulus processing stage, (2) the central processing stage, and (3) the late processing stage (Pashler and Johnston 1998). The first and third processing stages can be performed simultaneously for both tasks. However, the central processing stages of the two tasks cannot be combined or performed in parallel.

In SSAT, these two tasks are considered a single task, which integrates the interdependent and voluntary regulated actions to a significant degree. Each task (more precisely subtask) has its own subgoal. A high-order goal of a general task is to perform each subtask as quickly as possible. During the experiment, the subjects realized that the first and the second tasks were not independent and developed complicated subjectively suitable strategies of the holistic task performance. Limitations in the information processing are one of the factors that influence the PRP. This factor always interacts with the conscious and unconscious mechanisms of activity regulation. In SSAT, these stages are associated with various cognitive and motor actions. The early stimulus processing stage in SSAT can be described as a simultaneous perceptual action; the central processing stage is a group of actions called decision-making actions at the sensory-perceptual level or explorative thinking actions, which are performed based on the sensory-perceptual information; the late processing stage can be considered a motor action that usually requires a low or average level of concentration.

Simple perceptual actions and average complex motor actions can be performed in combination with other similar actions. At the same time, decision-making actions at the sensory-perceptual level, decision-making actions at the verbal logical level, explorative-thinking actions, and so on cannot be performed simultaneously (Bedny and Karwowski 2007). There are also complicated perceptual actions that consist of a chain of subsequent perceptual actions that are involved in the recognition of unfamiliar stimuli or in complicated motor actions that require a high precision or that are performed under stress. Sometimes, motor actions also can require a high level of concentration. Such perceptual or motor actions cannot be performed simultaneously. Thus, per activity theory, we do not discuss the possibility of one or several bottlenecks in the attention mechanism but rather distinguish between automatic and voluntary conscious processing mechanisms of attention. Automatic processing mechanisms of attention facilitate the performance of various elements of task (actions or operations) simultaneously, while mechanisms of attention involved in voluntary and conscious processing of information very often provide a possibility of sequential processing. As we can see, the concept of "stages processing analysis" introduced by Sternberg (1969) and the data that describe an opportunity to perform these stages in sequence or simultaneously are in agreement with the data that are presented by SSAT. At the same time, in contrast to the data obtained in cognitive psychology, in SSAT rules that describe the possibility of combining mental and motor components of an activity (actions or operations) during the task performance are developed (Bedny and Karwowski 2007). This information provides a link between the theoretical data and the applied research.

12.6 Model of Attention

According to SSAT, attention can be described as a goal-directed self-regulative system. Consideration of psychological functions and activity from perspectives of self-regulation is regarded as functional analysis of activity (Bedny and Karwowski 2007; Bedny and Karwowski 2004a,b). In the functional analysis of activity, informational and energetic components are closely connected (Bedny and Karwowski 2006). Therefore, the described model of attention includes not only cognitive but also emotional-motivational mechanisms. The basic units of analysis of self-regulative systems are functional mechanisms or functional blocks. The functional blocks can be considered specific stages of information processing, which have feed-forward and feedback connections with some other stages. Each stage has a particular purpose in activity regulation. Depending on the task specificity, cognitive processes can be combined differently at different stages of attention. Hence, the content of each stage or block depends on the task specificity. During the development of the cognitive model of attention, we pursued both theoretical and practical goals. The model should not be overloaded with insignificant details. At the same time, it should be able to explain and predict the behavior or activity of a person performing time-sharing tasks and tasks of various complexities.

In our experiment, two factors can reduce the mutual interference of the considered tasks. The first factor is the use of two information-processing channels, namely, two modality-defined resources. One of them is auditory and the other is visual. According to the multiple-resource theory, this should alleviate the simultaneous performance of the two considered tasks (Wickens and McCarley 2008). The second factor is the performance of the two tasks in sequence by the subjects. The subjects should perform the second subtask only after completing the first. However, our research showed that an increase in complexity of each task increased the effect of its interference even when the tasks were performed in sequence and the subjects used two modality information processing channels.

This contradicts with multiple-resource theory, in which people have several different capacities in terms of the resource properties. In this theory, the two tasks in consideration are independent and should not influence one another, but our research data show that it is not the case. The interference of the two tasks, especially when their complexity has been increased, allows us to conclude that resources are shared. Complicated tasks make access to resources more difficult and reduce the ability to allocate the resources; that is, there are undifferentiated, limited resources and there is a mechanism that regulates the allocation of these resources. The mechanism that is responsible for the investment of required resources of attention in performance is called the "available level of arousal." The second mechanism or block that is responsible for voluntary allocation of resources and their coordinated usage is designated the "regulator integrator." Physiological studies consider specific and nonspecific arousal. When we evaluate the task difficulty, nonspecific arousal is especially important (Aladjanova et al. 1979). The more difficult the task is for the subject, the more mental effort it takes. These efforts demand nonspecific arousal. Therefore, the difficulty of the executed task is a critical element of time-sharing tasks. The more complex the task is, the higher the probability that this task will be difficult for the subject. Hence, the higher the task complexity, the higher the degree of limited energy resources it requires.

Each information-processing task should have an appropriate level of energy support not only at the physiological but also at the psychological level. Hence, there is another block called the "evaluative and inducing level of motivation," which is responsible for energy supply at the psychological level (Figure 12.2). The first one (the available level of arousal) is considered to be a physiological mechanism and the second one as a psychological mechanism. The last one regulates the emotional and motivational states of a person during the task performance. This block is involved in creation of the vector "motives → goal," which gives activity its goal-directed flavor (Bedny and Karwowski 2007).

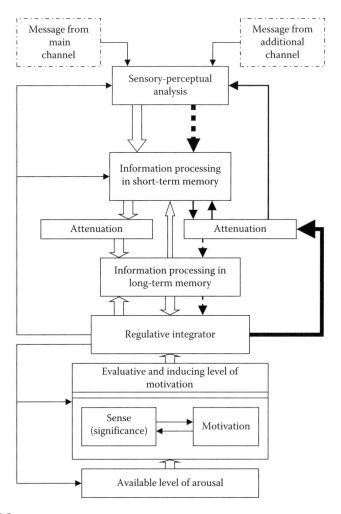

FIGURE 12.2
Model of attention.

The evaluative and inducing level of the motivation block consists of two subblocks, which mutually influence each other. One subblock is called "sense" (significance) and the other one "motivation." The subblock "sense" is responsible for evaluation of personal significance of a goal and various components of an activity. The subblock "motivation" refers to the inducing components of an activity (Bedny and Karwowski 2004a,b; Bedny and Karwowski 2007). The factor of significance influences the method of information interpretation and creates motivational forces associated with the main channel of information processing. The functional block "regulative integrator" includes mechanisms that are responsible not only for unconscious but also for conscious information processing and for the development of conscious goal of activity.

Hence, the interaction of the "regulative integrator" block with the "evaluative and inducing level of motivation" block provides voluntary regulation of attention and selection of the more significant goal of activity and switching attention from one task to another.

Through feedback, the regulative integrator activates and regulates the functional blocks "available level of arousal" and "motivational level"; these three interrelated blocks are shown in Figure 12.2. We can create the model of attention using single-resource theory without considering specific arousal. According to the functional analysis of activity, any self-regulative model includes not only the energetic but also the cognitive mechanisms (informational mechanisms).

We will start our discussion with an analysis of the relationship between the peripheral and central processes of attention and their effect on our model of attention. Sometimes, cross-modal time-sharing is better than intramodal (Wickens and Hollands 2000). This can be due to the central or peripheral processes in which the central processes are associated with separate perceptual resources and peripheral processes with visual scanning, imposing confusion, masking, and so on. In our experiment, the subject was given only one task per a certain period of time, which involved visual or auditory processing. In this case, interference of the peripheral processes was eliminated. The process of switching attention became complicated only because central processing is a critical factor in the performance of the two tasks. However, our model of attention does not ignore peripheral processing. Peripheral and central processes interact with each other. This relationship is shown in our model by feedback from "regulative integrator" to various information-processing blocks and to the "sensory-perceptual analysis" block in particular.

The model of attention (Figure 12.2) has two channels of information processing. One of them is the main channel, and the other one is an additional channel. Each channel of updating information has a limited capacity. The complexity of updating information by each channel depends on the quality and quantity of information and on the speed of its update. There must be a mechanism that is responsible for the coordination of information processing between these two channels and switching of attention from one channel to other. The more complex the listed characteristics of information and the loading of channels, the more difficult it is for the regulative integrator to allocate the informational resources. Hence, the regulative integrator is responsible not only for the allocation of energy resources but also for the coordination of information processing between the two described channels. Overload of channels' updating information leads to a decreased efficiency of this coordinating block. Our model of attention allows us to use the data from both the single-resource and multiple-resource theories. The notion of resources is tied to the energy aspects of the activity. The notion of regulative integrator is tied to the regulation of information and energy processes.

The model shows the unity of the informational and energetic aspects of the activity. In some situations the same demands are presented for energy resources that have different demands for their allocation and coordination with the information processes. For instance, it is easier to distribute attention between visual- and auditory-processing resources than between two visual resources. This can be explained not just by different modality-defined resources but also, more likely, by the easier coordination of demands of various resources of information processing. Such a coordinating function is related to the regulative integrator functional block. Therefore, the allocation of attention depends not only on the energy resources but also on the complexity of coordination of information processes and their coordination with the energy processes when there is an undifferentiated pool of such resources. The suggested model presents attention as a self-regulative system and includes various functional blocks that are responsible for information processing.

In our model, the first stage of information processing is the sensory-perceptual analysis block. Automatic processes dominate this block. However, this block also includes conscious elements of controlled information processing. Functional blocks regulative integrator and attenuation processes influence the functional block sensory-perceptual analysis through the feedback. This fact is supported by the studies that demonstrate the effect of facilitation of processing of attended information and suppression of unattended information at sensory-perceptual stage (Wickens and Hollands 2000). Therefore, selectivity takes place at the sensory-perceptual level to some extent.

In contrast to Broadbent's (1958) filter model, Treisman (1969) introduced the concept of attenuator as an important mechanism of attention. She proposed that as a filter the attenuator weakens, rather than entirely rejects, the unattended information. There is an inductive relationship between the attenuators (Norman 1976). If one of them becomes more active, the other one becomes less active, and vice versa. In our model, instead of the term "attenuator" we used the term "attenuation processes" block. This means that different psychological and physiological processes are involved in this mechanism. The data obtained in neuroscience also support the existence of attenuation mechanisms. For example, attention operations are associated with the activation of some neural attention networks of the brain and inhibition of the other neural structures related to attention (Sokolov 1969; Fafrowicz and Marek 2008). Hence, the attenuation block is not a separate mechanism of the brain. This block simply demonstrates that there are complex relationships between the activation and inhibition processes in the neural structure of the brain.

The next stage of attention involves the combination of various cognitive processes in the short-term memory. This combination is provided by feedback influences of "regulative integrator" block. Conscious thinking operations are particularly important at this stage. The following functional block is called "information processing in long-term memory." It presents a combination of various cognitive processes in the long-term memory. This block interacts with regulative integrator. The message from the main channel is not attenuated. On the contrary, the information flowing along the additional channel is partly attenuated in the short-term memory and mostly during the transformation to the long-term memory. The information flowing along the additional channel is processed automatically. Hence, a meaningful interpretation of complicated information in the short-term memory is possible only in the main channel. Memorization and efficient manipulation of information in the long-term memory can be performed with attended information. The analysis of the relationship between the functional blocks demonstrates that the regulative integrator activates and regulates the mechanisms involved in information processing from the main channel and ceases the inhibition processes for this channel. At the same time, the regulative integrator turns on the inhibition processes for the additional channel and is not involved in active control of the above functional blocks along this channel. Information processing along the additional channel involves nonvoluntary attention per the activity theory terminology. For this type of attention, the most important features are those of external stimulation, instead of the internal willing processes linked with our conscience and goal-directed activity.

The functional block "regulative integrator" switches information from one channel to the other using attenuators. This functional block continually (1) constructs and revises the expectations and (2) controls and corrects the sensory messages. According to Norman (1976), a conceptually driven analysis should be distinguished from a data-driven analysis. The conceptually driven analysis has a limitation on the number of units of information that can be processed at any given period of time. The regulative integrator is the most important functional block in the conceptually driven analysis. This block creates the conscious goal toward which our attention is directed and compares the system of expectations with input information. Based on this comparison, the functional block generates the feedback influences (depicted by the thin line in Figure 12.2) and is connected with consciousness, language, and speech. Central and peripheral processes are integrated by feedback connections. The presented model demonstrates that the regulative integrator also coordinates the energetic and informational processes. Hence, attention is directed to achieving the established goal in a given period of time through the main channel. At the same time, the additional channel is involved in attaining information based on the existing set of activity. The set can be transformed into a conscious goal and vice versa. There is also a system of expectation and anticipation. The system of expectations and anticipation is connected with the set (Uznadze 1967) and the goal of activity. In the functioning of this system, feed-forward and feedback influences are important. When the set and goal of activity are altered, the system of expectations is altered as well. The specificity of the goal and the set determines the content of the subject's expectations. In perceptual psychology, it is well-known that when people perceive ambiguous pictures, altering the goal of the perceptual process results in modification of their expectations and the result of perceptual process can be different. In other words, once again we see how goals determine the specificity of the selection of information (Bedny and Karwowski 2007). The process of comparison of sensory data with the system of expectations depends on the goal of observation. Feedback that derives from this comparison can influence every functional block involved in the attention process.

As seen in our model, expectations are integrated with the constitution of the goal or set and the feedback processes. In general, the regulative integrator has complicated functions, including formation of the conscious goal or an unconscious set that plays an important role in tuning all other functional blocks involved in the attention process. The information that goes through the additional channel is associated with an unconscious set and can only be partly updated. This information is connected with automatic processing, requiring little conscious attention (i.e., nonvoluntary attention).

The feedback influences for the main channel, depicted by a thin line on the left side of Figure 12.2, should be switched to the additional channel on the right. This happens when the information obtained from the additional channel becomes more significant and the unconscious set is transformed to a conscious goal. At the same time, the regulative integrator activates attenuation processes by feedback influences in the preexisting main channel. As a result, the main and additional channels switch the places. Switching of channels is carried out by the regulative integrator, attenuation processes, and feedback influences. Therefore, our model of attention functions as a self-regulative system.

The regulative integrator has an impact not only on different blocks of a higher level but also on the blocks of a lower level. The blocks at the lower level are responsible for motivation and activation. The regulative integrator governs informational processing and matches it with the energy resources. Due to self-regulation, coordination between the energetic and informational processes is accomplished and is realized most effectively through the main channel. The more coupled and complicated the informational processes are, the more energy resources are required and the less is the ability to allocate the resources necessary for the additional channel. We want to draw attention to the fact that according to modern neurophysiology, excitation and inhibition have the same nature and are represented as active states of the nervous system. Inhibition is seen as an active state of the nervous system involving a waste of energy and requirements for energy supply (Anokhin 1968; Ukhtomsky 1966). Thus, our model of attention emphasizes the interdependence of energetic and informational aspects of the psyche.

The self-regulative model of attention demonstrates that allocation of resources depends not only on constraints imposed by resources' limitation but also on the subject's ability to consciously regulate the strategies of attentions. The notion of "the strategies in allocation of attention" (Navon and Gopher 1979) is important for the support of the self-regulative model of attention. Additional data that prove that attention should be considered a self-regulative system have been derived from the work of Young and Stanton (2002). They introduced the concept of "malleable attention resources pool." The main idea of this concept is that attention capacity can change in response to changes in the task demands to some degree. This can be possible only if the attention system can regulate its functioning and therefore be considered a self-regulative system. According to the self-regulated process, people invest only as much effort as they deem appropriate.

Using the cognitive psychology data provided by other authors as well as the principles of functional analysis of activity developed in SSAT, we created a model of attention that allowed us to describe the performance of cognitive tasks that can interfere with each other. This model allowed us to combine the data of the single-resource and multiple-resource theories. It also showed the relationship between the central and peripheral processes and the relationship between voluntary and involuntary attention. Our model of attention explains why the combination of channels with different modalities (visual and auditory) and spatial and verbal processing makes it easier to perform time-sharing tasks even when the subject has only one undifferentiated pool of attention resources. In such situations, it is easier for the regulative integrator block to coordinate the information processes and to make these processes agree with the energy processes. The existence of the interference of the informational processes with increasing complexity and energy support restrictions makes the work of the coordination block more complicated.

12.7 Conclusion

The goal of this research was to analyze an interaction of complex reactions performed sequentially and to create a model of attention. We analyzed how independent reactions integrate into the whole system of an organized activity. In this study, we used two approaches of SSAT. The cognitive approach was used through parametrical methods of study and the functional approach was used as a systemic method of analysis. The more complicated the previous and/or following reaction are (subtasks), the more they interfere with each other. This interference can be explained using our model. The described experiment is different from prior experiments conducted by other authors. In the experiment, two information channels have different modalities and two messages are sent sequentially. The time interval between the two tasks was the same and was equal zero. The complexity of both tasks was varied. In other words, during a given period of time, a person has to process only one portion of information (perform one task). The more complex the previous and current portions of information (first and second tasks), the more complex is the process of adjustment of the attention mechanisms and the more they influence each other. These experimental data demonstrate that attention can be better explained by single-resource theory.

The obtained result allows us to conclude that the development of a model of attention that can be successfully used in real-work situations should be based on single-resource theory. These data are in agreement with the Kahneman theory. The resource allocation takes place during the performance of interdependent tasks. Coordination of these resources with informational processes can be executed by a specific mechanism we call the regulative integrator. This mechanism is tied to our conscious and verbal-thinking processing to great extent. It also includes unconscious components associated with a set. The existence of this functional block can explain the data revised in the single-resource theory and multiple-resource theory. Intensification of information processing and an increase in the amount of energy used make the functioning of the coordinated block (regulative integrator) more complicated. Difficulty in allocation of attention between two tasks is caused not only by energy restriction but also by the ability to coordinate different informational currents and to match them with energy supply. For example, perceptual modalities can influence the strategy of attention. It is easier to divide attention between visual and auditory channels than between two visual channels. This model shows that coordination may be complicated due to both peripheral and central processes. The task complexity is the main element that determines the possibility for coordinating and matching the energetic and informational processes. Our model of attention emphasizes the interdependence of the energetic and informational aspects of the psyche.

Features of attention put some limitations on the ability to perform time-sharing tasks. In accordance with these limitations, subjects created various strategies of activity to achieve a set of goals. Some subjects paid more attention to the first task, some to the second one, and still others tried to distribute their attention evenly between the two tasks. The subjects varied allocation of their efforts during task performance from trial to trial. All the strategies get feedback based on the instructions given, the subjective understanding of them, individual characteristics, and an estimation of the achieved results. The core is self-regulation properties. Due to self-regulative processes, various reactions are combined into the entire activity, which has a systemic structure. This means that activity cannot be represented simply as a set of independent responses to a set of independent stimuli. Due to self-regulation, the subject develops various strategies, using resources of attention to coordinate cognitive, executive, evaluative, and motivational components of activity. Therefore, the study of the mechanism of attention and their self-regulation process during performance of the time-sharing task is very important. The self-regulative model of attention demonstrates that it is possible to voluntarily regulate the attention strategies. Resources limitation can influence preferable strategies in allocation or switching attention. However, the subjects can be unaware of their resources limitations. The task complexity and the difficulty derived from it are the critical elements of concurrent time-sharing task performance. Increasing the task difficulty increases the unspecified activation of the nervous system.

In cognitive psychology, the ability to perform different elements of cognitive processes simultaneously is considered from the perspective of cognitive processing stages. In SSAT, this problem is considered to be a subject's ability to combine the cognitive and behavior actions and operations. Strategies of activity, which are derived from the mechanisms of self-regulation, determine the specificity of their combination during the task performance. From the activity theory perspective, a subject cannot be considered only a device for updating information with processing resources. Characteristics of attention depend on the goals of the activity, the task significance and the motivational state of the subject, and the strategies of self-regulation; the limitation of processing resources in turn influences these components of activity. Our model of attention considers the functional blocks to be relatively independent stages of information processing, and it also rejects the concept of bottleneck as a possible mechanism of the attention process. We prefer to explain that interference between various elements of activity is a result of activation and inhibition of various structures of the brain.

In conclusion, our model differs from others because it includes three separate subsystems: informational, energetic substructures, and coordination mechanisms. The existence of forward and backward interconnections between the functional blocks allows depicting the formation of a strategy of attention directed toward achieving the conscious goals of activity. The central mechanisms influence the time-sharing process. Attention is considered a complex self-regulative system. The suggested model of attention allows us to explain the data obtained in the single- and multiple-resource theories from the unified perspective and considers a person's ability to voluntarily regulate her or his own activity. This work demonstrates that SSAT and cognitive psychology are considered interconnected approaches. The combination of SSAT and cognitive psychology data help us to develop a more comprehensive model of attention.

References

Aladjanova, N. A., T. V. Slotintseva, and E. D. Khomskaya. 1979. Relationship between voluntary attention and evoked potentials of brain. In *Neuropsychological Mechanisms of Attention*, ed. E. D. Khomskaya, 168–73. Moscow: Science Publishers.

Anokhin, P. K. 1968. *Biology and Neurophysiology of Conditioned Reflex*. Moscow: Medicine Publisher.

Bedny, G., and D. Meister. 1997. *The Russian Theory of Activity: Current Applications to Design and Learning*. Mahwah, NJ: Lawrence Erlbaum Associates.

Bedny, G. Z., and W. Karwowski. 2004a. The functional model of the human orienting activity. In Special Issue, *Theoretical Issues in Ergonomics Science*, ed. G. Bedny 5(4):255–74.

Bedny, G. Z., and W. Karwowski. 2004b. The situation reflection of reality in activity theory and the concept of situation awareness in cognitive psychology. *Theor Issues Ergon Sci* 5(4):275–96.

Bedny, G. Z., and W. Karwowski. 2006. The self–regulation concept of motivation at work. *Theor Issues Ergon Sci* 7(4):413–36.

Bedny, G. Z., and W. Karwowski. 2007. *A Systemic—Structural Theory of Activity. Application to Human Performance and Work Design.* Boca Raton, FL: Taylor & Francis.

Bedny, G. Z., and W. Karwowski. 2008. Time study during vocational training. In *Ergonomics and Psychology. Development in Theory and Practice,* eds. O. Y. Chebykin, G. Z. Bedny, and W. Karwowski, 41–70. London: Taylor & Francis.

Bedny, G. Z., W. Karwowski, and T. Sengupta. 2008. Application of systemic-structural theory of activity in the development of predictive models of user performance. *Int J Hum Comput Interact* 24(3):239–74.

Bedny, G. Z., and M. Seglin. 1999. Individual style of activity and adaptation to standard performance requirements. *Hum Performance* 12(1):59–78.

Broadbent, D. E. 1958. *Perception and Communication.* London: Pergamon Press.

Chebykin, O. Y., G. Z. Bedny, and W. Karwowski., eds. 2008. *Ergonomics and Psychology. Development in Theory and Practice.* Boca Raton, FL: Taylor & Francis.

Dobrinin, N. F. 1958. Voluntary and involuntary attention. In *Scientific Works of the Moscow State Pedagogical University* ed. N. F. Dobrinin, vol. 8, 34–52. Moscow: Pedagogical Publishers.

Fafrowicz, M., and T. Marek. 2008. Attention, selection for action, error processing, and safety. In *Ergonomics and Psychology. Development in Theory and Practice,* eds. O. Y. Chebykin, G. Z. Bedny, and W. Karwowski, 203–20. Boca Raton, FL: Taylor & Francis.

Hick, W. E. 1952. On the rate of gain of information. *Q J Exp Psychol* 4(3):11–26.

Kahneman, D. 1973. *Attention and Effort.* Englewood Cliffs, NJ: Prentice Hall.

Meyer, D. E., and D. E. Kieras. 1997. A computational theory of executive cognitive processes and multiple-task performance: Part 1. Basic mechanisms. *Psychol Rev* 4:3–65.

Navon, D., and D. Gopher. 1979. On the economy of the human processing systems. *Psychological Review* 86:254–255.

Norman, D. A. 1976. *Memory and Attention. An Introduction to Human Information Processing.* 2nd ed. New York: Wiley.

Norman, D., and D. Bobrow. 1975. On data-limited and resource processing. *J Cogn Psychol* 7:44–60.

Pashler, H., and J. C. Johnston. 1998. Attention limitations in dual-task performance. In *Attention,* ed. H. Pashler, 155–90. East Sussex, UK: Psychology Press.

Sokolov, E. N. 1969. The modeling properties of the nervous system. In *A Handbook of Contemporary Soviet Psychology,* eds. M. Cole, and I. Maltzman, 671–704. New York: Basic Books, Inc.

Sternberg, S. 1969. The discovery of processing stages: Extensions of Donders' method. In *Attention and Performance* II ed. W. G. Koster, 276–315. Amsterdam: North Holland.

Treisman, A. 1969. Strategies and models of selective attention. *Psychol Rev* 76:282–99.

Uznadze, D. N. 1967. *The Psychology of Set.* New York: Consultants Bureau.

Wickens, C. D., and J. G. Hollands. 2000. *Engineering Psychology and Human Performance.* 3rd ed. New York: Harper-Collins.

Wickens, C. D., and J. S. McCarley. 2008. *Applied Attention Theory.* Boca Raton, FL: Taylor & Francis.

Young, M. S., and N. A. Stanton. 2002. Malleable attention resources theory: A new explanation for the effect of mental underload on performance. *Hum Factors* 44(3):365–75.

13

Real and Potential Structures of Activity and the Interrelationship with Features of Personality

G. Zarakovsky and W. Karwowski

CONTENTS

13.1 Theoretical Aspects

According to the cognitive approach, any mental activity is a process. However, in systemic-structural activity theory, cognitive activity, as a whole, represents not only a process but also a structure. Activity can be presented as a structure unfolding as a process (Bedny and Karwowski 2007). It is obvious that, to some degree, this structure-organized process, before its realization, has been retained in the memory as some partially organized "template," a basic functional system of activity, which includes the dominant personality features, cumulative knowledge, habits, and skills, in addition to the neuron-level physiological processes. Therefore, activity is not only an active, dynamic entity, but also includes a potential, passive part, which has to be taken into consideration in the further development of activity theory. This idea can be used to further enhance the activity theory into an area that includes the potential forms of existence of functional systems of activity. In this chapter, we attempt to more clearly formulate the theoretical concepts that explain the relationship between the potential and active components of activity, their connection with the individual psychological features of personality, and the possibility of using these data for the analysis of different professions.

Human activity is a process of self-realization of human potentials that have been embedded in a person by nature and have developed during the evolution of mankind. This process includes both physiological functions and activity taking place inside the mind and in the physical and social realms. Therefore, life activity include two components: biological and socially developed. The latter might be spontaneous and seeking, that is, without specific goals, or might be goal oriented. The goal-oriented type is the entity that is generally termed "activity." It is a human process, characterized by a specific, conscious goal, which is oriented toward making changes in some situations of the subject, object, or the person involved.

A person exhibits himself or herself through his or her activity as a rational individual with a conscience, will, and goal-formation capacity; also, a person takes upon himself or herself the responsibilities for the implications of his or her deeds. By human potentials, we mean the intrinsically human propensity for activity. These potentials can be presented as a hierarchical structure as shown in Figure 13.1.

We need to distinguish between nonspecific and activity-related potentials. Nonspecific potentials are formed as a common, nonspecific functional system of an organism, consisting of elements from the psychological, anatomical, and physiological domains. These potentials manifest themselves as a certain level of physical health, as psychological self-effectiveness, and as an adequate or inadequate social behavior in a

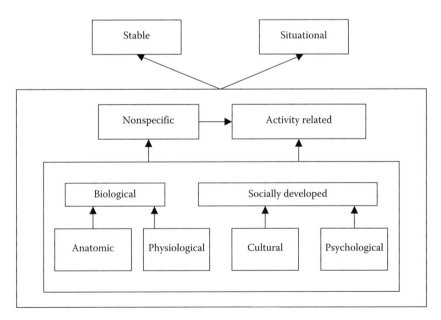

FIGURE 13.1
Structure of a person's human potential.

cultural environment. Activity-related potentials are formed as specific functional systems to perform certain type(s) of activity related to the person's profession, learning, daily life, and other aspects in the public or family domain.

There is a link between these two types of potentials. To function as a support for a specific level of activity-related potentials, the nonspecific (base) potential must have certain properties (i.e., a set of professionally meaningful traits). Some elements of the base potentials might also be a part of the activity-related potential. Both types of potentials (capacities) might have a permanent or situational character. For example, a permanent, stable potential includes well-formed skills to perform some activity, or a good, stable condition of health. The situational potential changes under certain circumstances (i.e., the work capacity is determined by the functional conditions and situational motivation of a person). In general, the structure of the activity potential consists of three connected blocks, with a few subblocks included (Figure 13.2). An activity potential is a "raw" functional system of activity, which can be presented as a psychophysiological model of self-regulation developed by Anokhin (1980) and described in the work of Bedny and Meister (1997). The suggested model describes a dynamic functional system of anatomical, physiological, and psychological elements and processes, selectively merged to achieve some useful result. Psychological functional models of activity as a self-regulative system can be found in the work of Bedny and Karwowski (2007). Together with the functional system of behavior, a human being forms some similar structured types of systems to fulfill partial tasks such as the regulation of homeostasis and adaptation

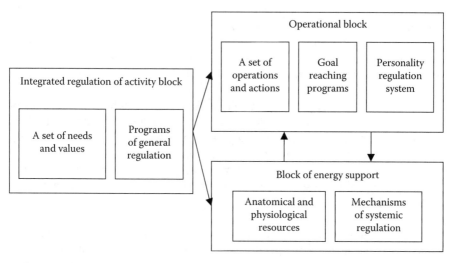

FIGURE 13.2
Structure of activity potential.

of the organism, and tuning some physiological systems to the demands of activity. We have offered a general psychophysiological schema of human activity. This also includes a function that provides a flexible energy support (physiological–biochemical mechanism) for a goal-oriented activity and a function that describes the system's adaptation to external conditions.

The general psychophysiological schema of human activity is presented in Figure 13.3. This schema is much more complex than that of an operational subsystem of activity, because it includes functions that ensure a flow of energy to support the goal-oriented activity and a mechanism of adaptation to the organism's surrounding conditions. Figure 13.3 shows not only the structure of the system but also the mechanism of its transition from a potential or capacity level to an actual, process-realization state. In essence, the general psychophysiological schema of activity represents an ergatic (man–machine) system, as it includes not only an individual, but also the object of his or her activity and the environment (Zarakovsky and Pavlov 1987). The human part of this activity–description system includes five subsystems. The subsystem of spontaneous psychic activity (subsystem 1) includes some processes that can run concurrently with the goal-oriented operational activity (subsystem 2) involved in the work performance. Among these processes, we may name thoughts and images not related to a performed work, some unusual associations caused by such activity, and so on. In addition, we need to take into account subsystem 1 potential impacts on the goal-oriented process involved in the performance of human work.

The operational subsystem of activity (subsystem 2) is primarily responsible for the performance of activity related to work performance; that is, it forms a goal using principal motivations and analyzes the conditions helping to reach it. Based on the results of such analysis, the operational subsystem selects and actualizes an appropriate program (skills, previously used methods, and algorithms) to achieve a goal in similar conditions. If such methods do not exist, the person mentally prepares a new strategy to fulfill a specific task based on past experience. Then, a decision is made to start the activity and subsequently, the process itself is activated, with the incorporation of corrections via feedback channels.

The subsystem dealing with the bioenergetic and biochemical support of activity (subsystem 3) constitutes the physiological subsystem of activity. This subsystem controls various physiological processes, and it works on a subconscious level, when the body itself determines what is important depending on the conditions of activity. The system dynamically adapts and adjusts the physiological processes of an organism to the activity requirements. The system mobilizes the resources of the organism in response to the needs of the aforementioned subsystems.

The subsystem of homeostatic and adaptive control of an organism (subsystem 4) is not as activity-specific as subsystem 3. This subsystem is condition specific. First, subsystem 4 provides for the adaptation of an organism to unfavorable external and internal conditions of a person's work. It is an

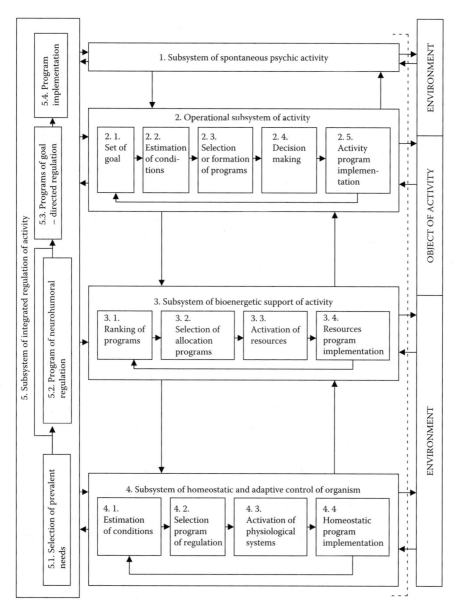

FIGURE 13.3
General psychophysiological schema of a human activity.

adaptive energy-support subsystem. The goal of the subsystem, as a whole, is to ensure an acceptable level of some vital parameters, such as acid-base balance, and tune up other characteristics to compensate for the unfavorable changes in the conditions, for example, those caused by fatigue or high temperature. This system controls the metabolic processes and creates adequate conditions for the biochemical and energy platforms of activity, ensuring the functioning of the anatomic-physiological system of the organism under different work conditions. For example, if the room temperature increases, the system controls the temperature of the body. Thus, physiologically, the system dilates the skin vessels and activates the sweat glands to reduce the body temperature via the skin surface.

The integrated regulatory subsystem 5 synchronizes the work of all the other subsystems at two levels: the visceral (neuronal control of the physiological and psychophysiological functions) and the conscience level (by mental efforts); this subsystem "triggers" the whole activity process based on the prevailing needs and the motivation derived from them. Now let us discuss in detail the functional blocks that control activity.

We start a detailed analysis with the subsystem of integrated regulation (subsystem 5 in Figure 13.3) because it triggers the whole transition from a potential to an actual state of action. The transition starts by the selection of the dominant motivation. A person always possesses a set of motives: aspirations, interests, or needs.

To begin a real activity, a person must activate some dormant factor in particular situational needs, which at the level of subsystem 2 (operational subsystem of activity) are transformed into "dominant motivation," (block 5.1 and its connection with block 5.3). Depending on the external conditions, the central and peripheral nervous systems begin a process of subconscious selection of a program of neurohumoral control of the physiological systems (block 5.2). This tunes up the subsystems of energy support and homeostasis for the expected activity. The program that regulates goal-directed activity (block 5.3) controls the operational system using conscious efforts. In addition, this control can be directed toward other subsystems (such as continuing to work in spite of significant fatigue or overheating of the body). Before a person starts any activity, he or she mentally prepares to overcome some potential impediments. Program implementation (block 5.4) ensures the coordination of all subsystems in case of changing conditions.

Operational activity subsystem 2 includes several blocks. To start the flow of activity, it is not enough to have a strong desire (motivation). It is also necessary to form a goal that can be reached and make a decision to trigger this flow. We need also to estimate the specific conditions of actions (block 2.2). Various goal-achieving programs might potentially already exist in memory as a result of learning, training, or experience in similar activity. In some situations such programs should be constructed based on past experience. Block 2.3 finds the proper goal-reaching algorithms and programs and actualizes them, that is, moves them from the long-term to the short-term memory. The next block (block 2.4) is responsible for selection of the particular program or algorithm and gives a sanction to apply it. Block 2.5 is involved in realization of the activity according to the selected program. Based on the feedback mechanism, the activity-execution process can be corrected.

The subsystem of bioenergetic support of activity (subsystem 3) represents the physiological functions of the whole organism, which includes all organs (breathing, blood circulation, energy generation, and so on). This subsystem activates all the biochemical and energy resources required for activity. As resources are limited, this subsystem includes a part (block 3.1) that ranks the demands of other systems based on their importance and priority. According to these needs, specific programs for resource allocation are selected (block 3.2). For example, if the job includes heavy physical efforts, the flow of blood to the muscles is increased. The part of the physiological processes that is responsible for mobilization of the resources that are usually formed in a potentially ready form is organized as block 3.3. For example, during heavy physical work, glycogen ("carrier of energy") is injected into the blood flow from the liver. Realization of the selected resource-allocation programs (block 3.4) provides all other components with the necessary energy and substances. This is a dynamic process controlled by a feedback mechanism. These blocks function automatically.

Next is the subsystem controlling the homeostatic and adaptive control of an organism (subsystem 4). The first part of this subsystem is block 4.1, a mechanism that evaluates future body demands, depending on the surrounding conditions. Using the evaluated conditions, block 4.2, on a subconscious level, selects a program that is appropriate to these conditions. According to the selected program, block 4.3 initiates the activation of different physiological systems (cardiovascular system, respiratory system, and so on). This is followed by the realization of a selected program (block 4.4), which also has a dynamic character.

The functional psychophysiological system of activity works as the leading component of a bigger, man–machine system. The man–machine system also includes tools and instruments that help to change the object of goal-oriented activity; it also changes, in some manner, the surrounding physical, social, or informational environment.

Figure 13.3 presents the macrostructure of activity. A more detailed description of both the operational subsystem of activity and its relation to physiological and intentional subsystems of activity is presented in Figure 13.4. This figure also includes the personality subsystem as a higher order regulator of activity. This schema shows a more profound organization of the operational subsystem that– an integration of the operations, actions, and algorithms of their realization. At this level, various operational structures, with different psychological and psychophysiological content, are formed, along with the formation of an activity potential or capacity. The activity system includes different subsystems and they are designated as the features or properties of activity. Figure 13.4 demonstrates that each subsystem characterizes certain groups of human properties. There are five basic features or subsystems. The feature of subsystem 1 is a result (outcome) of activity (Zarakovsky 2004).

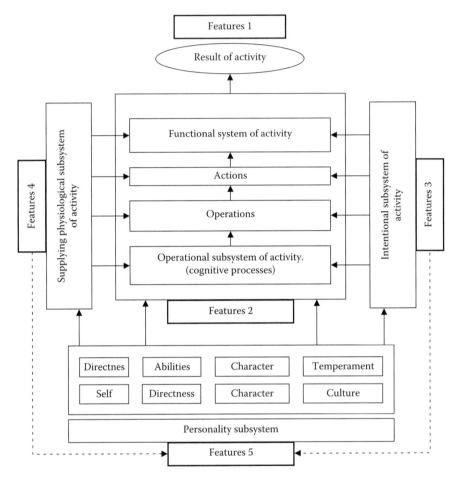

FIGURE 13.4
Activity system and its relation to the personality subsystem.

Knowledge of the result is important for the analysis of activity. In the case of social interaction, the result of activity is evaluated according to solidarity, empathy, subjective judgments and so on. Feature 1, associated with the activity result, focuses on the stages of object transformation according to the goal of activity. Information about the initial state of an object, the intermittent states, and the final state is important for understanding the procedural aspects of activity. The result of an activity is evaluated based on objective and subjective criteria.

The operational subsystem of an activity (feature 2) includes cognitive processes, cognitive and motor actions and operations, and their logical organization. One should distinguish, in this regard, goal-directed and basic functional systems. Subsystem 2 is organized as a goal-directed self-regulatory system. The goal-directed operational subsystem (Feature 2) is directly involved in activity performance. The operational subsystem includes analysis of the situation, the goal-formation process, program formation or selection of a ready program and algorithms, and the execution of actions. This subsystem also includes mechanisms of correction through a proper feedback mechanism. A goal-directed or an operational functional system develops and exists only during the process of planning and realization of the conscious goal. There are two energetic subsystems. One of these subsystems is responsible for the intentional aspects of activity (Feature 3) and the other for the physiological or supplying feature (Feature 4).The latter has a purely physiological role (feature 4). This is the basic functional system that integrates the physiological mechanisms and provides the energetic resources of activity. The personality subsystem (feature 5) is the highest level of activity regulation. This subsystem provides activity regulation at the level of subjective and social significance. All other subsystems are subordinate to this subsystem. This subsystem is the basis for the study of individual styles of performance.

For a more detailed analysis of activity strategies during task performance, a functional analysis of activity is also recommended (Bedny and Karwowski 2007). Functional analysis considers activity as a self-regulative system. The major units of analysis at this stage of activity study are functional blocks. In Section 13.2, we demonstrate the possibility of using these schemas of activity for the classification of various professions and present an example.

13.2 Existing Classifications of Professions

This classification shows that we can build a unified system of identification of various professions by using theoretically supported criteria that are derived from the material in Section 13.1. This classification of professions does not have sufficient theoretical background. Several such systems exist now, but they significantly vary from each other, which causes difficulties in dealing with some issues in personnel departments. During the past few years, these issues became more critical in connection with a mass migration of workers and other employees inside and between countries. In Russia, it became essential also due to a transition from a draft army to contract-based recruitment. A problem has arisen as to how to determine the similarities and differences between military and civilian professions based on various factors, including psychological, psychophysiological, and social-psychological components. To resolve this problem, a single classification of professions and a single nomenclature of professionally important factors (PIFs) became necessary.

In 1965, the constantly updated and enhanced *Dictionary of Occupational Titles* (*DOT*) was introduced in the United States (the fourth edition was published in 1991). In this document, professions are divided into nine categories:

1. Professional, technical, and managerial occupations
2. Clerical and sales occupations
3. Service occupations
4. Agricultural, fishery, forestry, and related occupations
5. Processing occupations
6. Machine trades occupations
7. Benchwork occupations
8. Structural work occupations
9. Miscellaneous occupations

These nine categories are subdivided into 84 groups, and those groups are divided into 603 subgroups. Each profession has been assigned a code number, which defines its position in the classification structure by the following four factors: (1) function of work, (2) interaction with coworkers, (3) performed operations, and (4) occupation area. Some areas of occupation as enumerated by the *DOT*, include art, education, medicine and health, computer-related occupations, engineering, and so on. Each group of professions is accompanied by a qualification profile, which describes the content of activity and psychological, physiological, and anatomical factors (demands), which are important for a profession. Some factors have not only verbal descriptions but also quantifications.

In the former USSR, a three-level classification system was developed:

1. Physical intensity (efforts) of labor
2. Occupational hazards
3. Psychological intensity of activity

The intensity of labor is estimated by an integrated factor, which includes a level of necessary attention and a number of objects (signals, messages) that a worker should consider in the decision-making process. Currently, this classification is used in budget allocations for various groups of employed populations (Bobkov and Polovaya 2002 and *DOT*). This system is also used for developing some official classification documents in the industrial and other spheres.

Klimov (1990) offered a compact hierarchical system using four base tiers.

Tier 1—Type of profession (by the subject of trade)
Man–nature
Man–technical equipment/tools
Man–man system
Man–sign system
Man–artistic images

Tier 2—Classes of professions (by a goal of trade)
Knowledge-oriented professions
Transformation-oriented professions
Exploration or research professions

Tier 3—Divisions of professions (by tools and means of trade)
Manual labor
Machine–manual labor
Work using automated system
Work using mostly functional tools

Tier 4—Groups of professions (by material and social conditions)
Work in an ordinary (room-type) microclimate
Outdoor work
Work in unusual ambient conditions
Work with a significant moral responsibility for health, life, and mental development of other persons, or with responsibilities for the safety of significant material values

The type of working activity is the unit of classification in federal normative documents, and is determined by the qualifications (professional mastership) and professional specialization. To define the similarity of occupations (groups of professions), the system uses the following attributes: the function of work content, particulars of an object and tools, the scale and complexity of managerial organizational work, the conditions of labor, required education, and skills. This classification has a four-level hierarchical structure with nine groups:

1. Heads of state, regional, and local power structures (government bodies), including the heads of companies, factories, and other enterprises
2. High-level specialists
3. Medium-level specialists

4. Clerical and other workers employed in information processing

5. People in sales, services, and maintenance of housing and infra-structure

6. Skilled workers in agriculture, forestry, and fishery

7. Skilled workers in industrial areas

8. Operators, machinists, and drivers

9. Unskilled workers

13.3 Development of a Unified Classification of Professions in the Civilian and Military Areas Derived from an Analysis of the Structure of Activity

As we can see, the existing classifications are based on different methodologies, approaches, and levels of decompositions of characteristics (Zarakovsky et al. 2004). The main drawback of these classifications is that they lack a distinct definition of the psychological and physiological underpinnings of the groups of professions and positions. At the same time, each of them has some positive elements that are used in the proposed unified classification and nomenclature of the PIFs.

Our classification system had a scientific approach and orientation on the practical uses (Zarakovsky et al. 2004, 2005), and we accepted the paradigm of potential and processing structures, which were described in Section 13.1. This is based on the fact that high-level groups of hierarchical PIF nomenclature have been defined using the major components of the functional structure of activity and their connection with different features of personality (Figures 13.3 and 13.4).

The highest levels of the classification group of hierarchical PIF nomenclature are as follows:

1. Value-motivated personality features

2. Personality features that determine the abilities to adapt and self-control

3. General capabilities to perform or operate in productive environment

4. Special abilities to interact with people in productive (business) environment

5. Pliant bioenergetic potential of an individual

6. Abilities of an individual to regulate and control himself or herself

7. Anatomical and aesthetical traits of an individual

Each group includes the nomenclature of PIFs of personality with a code number. For example:

1. Value-motivated features of personality include the following:
    1.1 Values important to a person
        1.1.1 Patriotism
        1.1.2 Pacifism (nonviolence)
        1.1.3 Hedonistic way of life
        1.1.4 Eudaemonism (domination of getting happiness by accomplishing personal goals in life)
        1.1.5 Civilian way of life
        1.1.6 Military way of life
        1.1.7 Altruism
        1.1.8 Selfishness
        1.1.9 Conservatism (following traditions)
        1.1.10 Creativity (innovation)
        1.1.11 Sense of duty and obligation
    1.2 Predispositions of personality toward
        1.2.1 Power
        1.2.2 Leadership
        1.2.3 Conformism
        1.2.4 Cooperation
        1.2.5 Autonomy
        1.2.6 Collectivism
        1.2.7 Risk-taking
        1.2.8 Aggressiveness
        1.2.9 Caution
    1.3 Predisposition in the area of professional self-determination
        1.3.1 Orientation toward manual (physical) work
        1.3.2 Orientation toward contest-type activity that requires physical fitness
        1.3.3 Orientation toward the theoretical and analytical (intellectual) type of activity
        1.3.4 Orientation toward intellectually competitive activity
        1.3.5 Orientation toward creative activity
        1.3.6 Technical-oriented activity

1.3.7   Orientation toward activity related to forestry, plant growing, and so on

1.3.8   Orientation toward work with animals

1.3.9   Activity oriented toward interpersonal interactions

1.3.10 Orientation toward a leadership and organizational type of work

1.4 Motivations of professional activity

    1.4.1 Direct internal motives toward professional activity

       1.4.1.1 Aspiration to master a specific trade

       1.4.1.2 Aspiration for a professional growth (higher qualification)

       1.4.1.3 Career-oriented aspiration (promoting in position, ranking)

       1.4.1.4 Need of self-assertion, self-realization (desire to reach aspired social status)

       1.4.1.5 Social importance of work

       1.4.1.6 Attempts to find positive side in any work

    1.4.2 Direct external motives of professional work

       1.4.2.1 Safety of working conditions

       1.4.2.2 Comfortable conditions of work

       1.4.2.3 Salary

       1.4.2.4 An interest in a package of social benefits

       1.4.2.5 Social protection

       1.4.2.6 Coveting to belong to a certain professional group

    1.4.3 Indirect motives

       1.4.3.1 Aspiration to improve financial state of the family

       1.4.3.2 Aspiration to resolve some family problems (get more time to share with family, upbringing of children, and so on)

       1.4.3.3 Need to get a better house or an apartment

       1.4.3.4 Aspiration to establish useful social connections

2. Regulatory properties of personality include the following:

2.1 Adequate level of self-esteem and self-evaluation

2.2 Independence

2.3 Punctuality, attention to details

2.4 Discipline

2.5 Diligence in work

2.6   Self-discipline, self-organizing

2.7   Industrious, serious attitude to a given job

2.8   Sense of responsibility

2.9   Commitment to work

2.10 Persistence in reaching the goal

2.11 Patience

2.12 Showing initiative

2.13 Self-critique

2.14 Emotional stability

2.15 Optimism, prevailing positive emotions

2.16 Calmness, composure

2.17 Self-control, ability of self-observation

2.18 Alertness, tries to be preemptive

2.19 Tolerance to frustration (in case of failures does not become aggressive or depressed)

2.20 Self-mobilizing reaction to impediments on a path to a goal

2.21 Introvert personality

2.22 Extravert character

2.23 Disposition to reprobate himself or herself

2.24 Disposition to reprobate others

2.25 Ability to plan and schedule his or her work in time

2.26 Ability to organize personal activity in condition of significant flow of information and a variety of tasks

2.27 Ability to make decisions in nonstandard situations

2.28 Ability to behave rationally in extreme situations

2.29 Ability to take responsibility for decision making and actions

2.30 Ability to function effectively in the deficiency of time

2.31 Ability to sustain unpleasant feelings (bad odor, noise, filth, cold water, burns, scratches, electric shock) without losing control

2.32 Ability to argue and defend his or her opinion or position

2.33 Ability to switch from one activity to another

2.34 Ability to overcome fear

2.35 Ability to adjust to new social conditions

3. General properties necessary to perform or operate

   3.1 Properties of senses include the following:

       3.1.1 Strength of vision

3.1.2   Eyesight adaptation to darkness and light

3.1.3   Contrast sensitivity to monochromatic light

3.1.4   Sensitivity to colors

3.1.5   Stability of visual sensitivity in time

3.1.6   Strength of hearing

3.1.7   Contrast sensitivity of hearing

3.1.8   Audio differential sensitivity (to a pitch of sound, interval of sound, its strength, rhythm, background noise)

3.1.9   Endurance to long intervals of irritating sounds

3.1.10 Finger tactility and sensitivity

3.1.11 Sensitivity to vibrations

3.1.12 Sensitivity of muscles and bones to efforts or resistance

3.1.13 Sense of balance

3.1.14 Sense of acceleration

3.1.15 Olfactory sensitivity

3.1.16 Ability to discriminate taste of food

3.2 Cognition properties

3.2.1   Visual estimation of objects' sizes

3.2.2   Discern of distance between objects

3.2.3   Visual distinction of linear, angular, and volume measures

3.2.4   Visual distinction of dynamics (estimate of direction and speed of a moving object

3.2.5   Ability to discern figures (objects, marks, signals, and so on) on a slightly contrasted background

3.2.6   Ability to recognize masked, hidden objects

3.2.7   Ability to comprehend relationship between objects in a three-dimensional (3D) space

3.2.8   Ability to estimate direction of an audio source

3.2.9   Ability to recognize rhythms and their patterns

3.2.10 Aural recognition (cognition of oral speech)

3.2.11 Cognition of time intervals

3.2.12 Comprehension of oral speech

3.2.13 Ability to recognize small deviations of parameters of technological processes from given values by using visual signs

3.2.14 Ability to recognize deviations using audio signals

3.3 Properties of memory (speed to memorize, volume and time of keeping data in memory, the speed of reproducing the stored material) by types of memory:

   3.3.1 Long-term visual memory

      3.3.1.1 For faces

      3.3.1.2 For other images

      3.3.1.3 On symbolic signs (symbols, plans, schemas, graphs, diagrams)

      3.3.1.4 Numbers and dates

      3.3.1.5 Words and phrases

      3.3.1.6 Semantics of the text

   3.3.2 Visual short-term memory

      3.3.2.1 On faces

      3.3.2.2 On natural objects

      3.3.2.3 On signs (symbols, diagrams, graphs)

      3.3.2.4 On numbers and dates

      3.3.2.5 On words and phrases

      3.3.2.6 On the semantics of a text

   3.3.3 Long-term audio memory

      3.3.3.1 On voices

      3.3.3.2 On numbers

      3.3.3.3 On advance-known special signals

      3.3.3.4 On tunes and melodies

      3.3.3.5 On semantics of a message

   3.3.4 Short-term audio memory

      3.3.4.1 On numbers

      3.3.4.2 On semantics of a message

   3.3.5 Kinesthetic (motor) memory

      3.3.5.1 On simple movements

      3.3.5.2 On complex movements

      3.3.5.3 On position and movement of object in 3D space

3.4 Properties of imagination (visualization)

   3.4.1 Ability to visualize

   3.4.2 Ability to imagine in 3D space

   3.4.3 Ability to create images of objects, events, processes

3.4.4 Ability to vividly present something, which a person never encountered before, or something familiar, but in new conditions

3.4.5 Ability to convert an image into a verbal description

3.4.6 Ability to create an image from a verbal description

3.5 Properties of thinking

    3.5.1 Functional properties

        3.5.1.1 Analytical way of thinking (ability to distinguish separate elements of reality and classify them)

        3.5.1.2 Ability to synthesize, summarize, finding links, principles, formulate ideas

        3.5.1.3 Associative thinking, ability to actualize, and use information retrieved from memory

        3.5.1.4 Logical reasoning

        3.5.1.5 Creativity (ability to conceive unusual ideas, get away from traditional way of thinking, resolve problems)

        3.5.1.6 Speed of thinking, intellectual ability

    3.5.2 Using information in the process of thinking

        3.5.2.1 Using concrete objects

        3.5.2.2 Using images (schemas, plans, blueprints, and so on)

        3.5.2.3 Using abstract images and notions

        3.5.2.4 Verbalization (oral and written language)

        3.5.2.5 Using numbers and other quantitative material

3.6 Properties of attention

    3.6.1 The size of attention (number of objects a person can focus on simultaneously)

    3.6.2 Concentration of attention

    3.6.3 Stability of attention in time

    3.6.4 Switching of attention from one object to another

    3.6.5 Ability to share attention among several objects or different types of activity

    3.6.6 Stability of attention to distractions

    3.6.7 Ability to spot changes in the surrounding conditions on a subconscious level

    3.6.8 Ability to spot small changes of an object or in indicators and meters

3.7 Psychomotor properties

    3.7.1 Ability to react on unexpected video signal by performing a certain move

    3.7.2 Ability to react on unexpected audio signal by doing a certain move

    3.7.3 Coordination movements with cognition (complex activity)

    3.7.4 Ability to sensorimotor trekking of a moving object

    3.7.5 Ability to perform small, precision movements

    3.7.6 Ability to perform complex motor acts

    3.7.7 Ability to perform gentle, coordinated motions

    3.7.8 Coordination of the leading hand movements

    3.7.9 Coordination of movements of both hands

    3.7.10 Coordination of hands and feet movements

    3.7.11 Coordination between wrists and fingers

    3.7.12 Stability of hands and wrists (low tremor)

    3.7.13 Ability to write down fast

    3.7.14 Good penmanship

    3.7.15 Significant physical strength

    3.7.16 Ability to quickly acquire sensorimotor skills

    3.7.17 Ability to quickly change sensorimotor skills

    3.7.18 Plasticity and expressiveness of body motions

3.8 Properties of speech

    3.8.1 Good articulation

    3.8.2 Ability to talk for a long time

    3.8.3 Ability to change tone of the voice

    3.8.4 Ability to change the level of the voice

4. Special operational traits in the process of interaction with people include the following:

4.1 Predilection to socialize with people

4.2 Evading social interactions

4.3 Friendliness, compassion

4.4 Modesty

4.5 Tact, diplomacy

4.6 Sense of humor

4.7 Ability to present himself or herself

4.8 Concise, cultural, clear speech

4.9 Emotional stability in a social environment, being composed and undisturbed by aggressiveness or rude behavior of a person you have a dialog with

4.10 Good adaptation to social conditions

4.11 Ability to convince and persuade people

4.12 High personal respect (fulfilling given pledges, duties)

4.13 Demanding to others to fulfill their duties

4.14 Critical view of other people's work

4.15 Ability to obey orders

4.16 Ability to pass to others his or her attitude, mood, emotional charge

4.17 Immunity to inducing influence from other people

4.18 Ability to work in a team and build positive working relationship

4.19 Ability to soberly estimate people around you, to see their strong and weak traits

4.20 Ability to understand and perceive others, feel how sincere they are, what drives them, and so on

4.21 Ability to reach rapport and good contact with people of different psychological profile, cultural, and social level

4.22 Ability to facilitate various types of dialogues (business, personal, as a superior or a subordinate)

4.23 Ability to pass your feelings and views using gestures, mimics, voice manipulation

4.24 Ability to discern potential interpersonal problems in a group, and take proper measures to defuse them

4.25 Ability to resolve conflict situations in a constructive manner

4.26 Ability to control people's behavior properly using different kinds of motivation

4.27 Ability to organize a cooperative activity of a group

4.28 Ability to inspire an aspiration for a goal-oriented activity in other people

4.29 Ability to inspire trust and sympathy

4.30 Ability to delegate authority to subordinates

5. Bioenergetic potential of an individual include the following:

5.1 Level of energy, vitality (activity)

5.2 Endurance to do mental work

5.3 Endurance to a physically intense work

5.4 Endurance to emotional pressure

6. Regulatory properties of an individual include the following:

   6.1   Tolerance to dynamic physical workloads

   6.2   Tolerance to static physical workloads

   6.3   Ability to move quickly from a state of idleness to a state of intensive work

   6.4   Ability to do work despite a lack of sleep

   6.5   Ability to function despite increasing tiredness

   6.6   Ability to be alert in monotonous activity conditions

   6.7   Ability to be alert in conditions of waiting for an event

   6.8   Ability to keep working and be in uncomfortable temperature conditions for a long time

   6.9   Ability to keep working under fluctuating overloads (i.e., air turbulence, high waves)

   6.10 Ability to keep working during vibrations

   6.11 Ability to keep working in case of overloads acting in opposite directions

   6.12 Ability to keep working under increased or decreased air pressure

   6.13 Ability to keep working in condition of decreased partial pressure of oxygen

   6.14 Ability to keep working under decreased partial pressure of carbon dioxide

   6.15 Ability to keep working in limited basic conditions of life (hunger, thirst, and so on)

   6.16 Ability to keep working in various climate conditions

        6.16.1 In optimal conditions

        6.16.2 In the mountains

        6.16.3 In a desert

        6.16.4 In tropical climate

        6.16.5 In high latitudes (Arctic, Antarctic)

        6.16.6 Ability to adapt to new climate conditions

7. Anatomical and outward appearance of a person includes the following:

   7.1   Anthropometric characteristics (in accordance with requirements of the trade)

   7.2   Physique (body structure; in accordance with the requirements of the trade)

   7.3   Presentable outward appearance

In some practical cases, numerical estimates of PIF may be necessary. For these reasons, we recommend the following PIF scale:

+2: The factor is absolutely essential.

+1: The factor is necessary.

  0: The factor does not matter.

−1: The factor has a negative impact.

−2: The factor has especially bad impact.

This PIF nomenclature has been used as a basis of a unified classifier of occupations, which has a multidimensional character. As a primary classification principle, we have used a composite indicator of functions performed by a person. This indicator represents a psychological (psychophysiological, sociopsychological) component of a so-called professional competence. According to Raven (1984), this "competence" is a set of PIFs with knowledge, skills, and experience, which ensures a successful completion of a certain type of activity. Therefore, the first basic principle of classification is "area of professional activity," which is described in functional psychological terms.

Other basic levels of classification are as follows:

- Level and type of organizational control
- Interior conditions of activity (safety, stress, efforts)
- Ambient conditions of activity (temperature, humidity, and so on)
- Level of education (skills)

The developed classification is presented in Table 13.1.

A system of characteristics is used to develop psychophysiological profiles of military personnel positions (all categories) to compare them with the corresponding civilian professions. Some professional profiles of military professions in an infantry division have been tabulated in Table 13.2, which has the following columns:

Column 1: Area of professional activity (consists of three subcolumns, numbered 2, 3, and 4)

Column 2: Level and type of organizational control (numbered 5)

Column 3: Internal conditions of activity (three subcolumns, numbered 6, 7, and 8)

Column 4: Ambient conditions of work (numbered 9)

Column 5: Education (skills, numbered 10)

TABLE 13.1

Occupational Classifications

Classification Structure	Specific Functions	Military Professions	Civilian Professions
1. Areas of professional activity			
1.1 Interaction with people			
1.1.1 Competitive interaction	Interactive activity to reach a person's own goals in conditions of rivalry or animosity from other person (group); game-type activity; activity to detect and predict tactics of a foe	Military commanders, tank crews, snipers, submachine gunners, antitank gunners, and so on	Senior executives, coaches, sport team captains and players, wrestlers, fencers, and so on
1.1.2 Consensus-oriented, organizational type of interactions	Interactions directed to decide organizational issues mutually in a beneficial way, promote constructive dialogues	Deputy commanders in charge of ideology, intelligence officers, heads of military supplies services	Executives and managers in administrations and industries, insurance agents, clerks working with clients, and so on
1.1.3 Informational and emotional influence	Informing about facts and events; aesthetic impact on emotional feelings of a person	Military correspondents and journalists, musicians and actors of military bands	Journalists, reporters, actors, musicians, artists
1.1.4 Psychological impact	Influence a person's motivation, will power, value system to help to adapt to new conditions of activity, to correct behavior, and so on	Deputy commanders in charge of discipline and morale, military psychologists, vocational counselors	Psychologists (personality experts), managers in personnel departments, priests
1.1.5 Educational influence	Impact on people to teach skills, proficiency, knowledge	College professors and school teachers, tutors, instructors	Teachers, educators, instructors, tutors
1.1.6 Medical treatment, rehabilitation	Providing medical help	Doctors in hospitals and in military units, medics, orderlies	Doctors and nurses in hospitals and clinics
1.1.7 Legal and law enforcement activity	Provide normal social behavior people (group)	Military investigators, prosecutors, judges	Judges, prosecutors, detectives, other law enforcement workers

(Continued)

TABLE 13.1 (*Continued*)

Occupational Classifications

Classification Structure	Specific Functions	Military Professions	Civilian Professions
1.2 Interaction with animals			
1.2.1 Work with horses	Training, taking care, using in jobs	Cavalry servicemen	Jockeys, stablemen
1.2.2 Work with dogs	Training, taking care, using in job	Dog-aided guards, landmine searching dog walkers	Dog trainers, dog walkers, dog breeders
1.2.3 Veterinarian area	Preventive practice, treatment and rehabilitation of animals	Military veterinarian personnel	Veterinarian doctors and aides
1.3 Interaction and work with objects of nature			
1.3.1 Ecological control	Control of technical objects and human activity to prevent pollution	Military ecology service specialists	Ecologists
1.3.2 Weather forecasting	Measuring weather conditions, data analysis, weather forecast	Military meteorologists	Meteorologists
1.4 Sensory-cognitive activity			
1.4.1 Recognition and identification of visual signals	Perceptual-analytical activity to identify visual signals from the background images	Radar operators, air reconnaissance data specialists, artillery fire controllers	Radar operators, artists, X-ray images specialists
1.4.2 Recognition and identification of audio signals	Perceptual-analytical activity to identify audio signals from the background images	Sonar specialists, radio operators, electronic counter-measures experts	Radio operators, musicians, sonar operators, singers
1.5 Information-processing work			
1.5.1 Coding and decoding messages	Message translation from one man-made language to another or to a common language	Message-coding clerk, telegraph operator, signalman	Cryptographer, programmer
1.5.2 Calculation involved activity	Performing calculations with or without devices	Command and control centers operators	Accountant, cashier, teller
1.5.3 Language translation	Written and oral translation from one natural language to another	Military translators or interpreters, electronic countermeasures experts	Translators, interpreters

		Programmer	Programmer
1.5.4 Programming	Development and maintenance of computer programs		
1.5.5 Analytical work	Analysis and interpretation of data, research, designing	Staff officers, researchers, officers to test and approve new military equipment	Scientists, economists, sociologists, market researchers, designers
1.6 Interaction with technical objects			
1.6.1 Control of objects moving in space and in the air			
1.6.1.1 Direct control of the moving object	Driving ground, airborne, amphibian vehicles, bulldozers, tractors, and so on	Drivers of autos, tanks, amphibian vehicles, and so on	Drivers of cars, tractors, bulldozers, train engineers, and so on
1.6.1.2 Remote control of dynamic objects	Remote control of dynamic objects	Operators of air-defense systems, operators of antitank guided missiles	Operators of robotic devices
1.6.1.3 Dispatcher-type control	Command and control of a number of objects moving relatively to each other	Air-traffic controllers, fire-control officers in command centers	Air-traffic controllers, train-traffic controllers, ship pilots, road traffic troopers
1.6.2 Use of technical systems, objects, and tools			
1.6.2.1 Using simple tools, no special skills needed	Mostly sensorimotor activity using simple equipment	Auxiliary crew of missile launchers or artillery batteries	Unskilled laborers
1.6.2.2 Using simple tools, some special skills needed	Precisely coordinated sensorimotor activity, needed special skills and training	Weaponry and equipment repairmen, operators of electronic systems	Plant assembly men, watchmakers, electronic lithographic specialists
1.6.2.3 Using devices	Measuring parameters and keeping devices and systems in a work condition	Operators of radio-electronic systems, lab technicians, radioactivity control persons	
1.6.2.4 Using aggregate equipment	Maintenance, starting, work conditions control	Operators of diesel generators, compressors, torpedo launchers	
1.6.2.5 Using systems	Analysis of technical conditions, decision making about normal use or repair	Head of ship's weapon systems, head of a radio or radar station	

(Continued)

TABLE 13.1 (Continued)

Occupational Classifications

Classification Structure	Specific Functions	Military Professions	Civilian Professions
1.6.2.6 Using complex and integrated systems	Analysis of technical conditions, decision making about normal use or repair	Commanders of military engineering and other technical units	
2. Levels of organizational control			
2.1 Individual activity			
2.2. Managing small group (up to 40 people)		Platoon commander, commander of a crew or a special team	
2.3 Control and command of medium-sized unit (up to 150 people)	2.3.1 Command	Commander of a company, battalion, small ship	
	2.3.2 Planning	Head of staff	
2.4 Command and control of a large unit (up to 7000 people)	2.4.1 Command	Brigade commander	
	2.4.2 Planning	Head of staff	
	2.4.3 Control	Commander, head of staff, security service officers	
3. Internal conditions of activity (danger, work pressure, physical efforts)			
3.1 Physical and psychological danger of activity			
3.1.1 During direct contact with a source of threat	3.1.1.1 Does not exist	—	—
	3.1.1.2 Medium		
	3.1.1.3 Significant		
3.1.2 Passively waiting for a threat (constant danger)	3.1.2.1 Does not exist	—	—
	3.1.2.2 Medium		
	3.1.2.3 Significant		
3.2 Pressure or stress during the process of activity			
3.2.1 Stress caused by time deficiency	3.2.1.1 Does not exist	—	—
	3.2.1.2 Medium		
	3.2.1.3 Significant		

(*Continued*)

3.2.2 Stress caused by information deficiency	3.2.2.1 Does not exist	—	—
	3.2.2.2 Medium		
	3.2.2.3 Significant		
3.2.3 Stress caused by the need to be alert in monotonous activity	3.2.3.1 Does not exist	—	—
	3.2.3.2 Medium		
	3.2.3.3 Significant		
3.2.4 Stress caused by sensory deprivation	3.2.4.1 Does not exist	—	—
	3.2.4.2 Medium		
	3.2.4.3 Significant		
3.2.5 Heavy physical load	3.2.5.1 Minimal	—	—
	3.2.5.2 Medium		
	3.2.5.3 Significant		
3.2.6 Deficiency of rest time	3.2.6.1 Minimal	—	—
	3.2.4.2 Medium		
	3.2.4.3 Significant		
4. Environmental conditions of activity			
4.1 Activity in various natural climate conditions			
4.1.1 Comfortable climate conditions		—	—
4.1.2 Beyond the polar circle		—	—
4.1.3 Tropical climate		—	—
4.1.4 Desert climate		—	—
4.1.5 In the mountains		—	—
4.2 Activity on the ground			
4.2.1 Outdoor	4.2.1.1 Comfortable physiological hygiene conditions	—	—
	4.2.1.2 Uncomfortable physiological hygiene conditions		

TABLE 13.1 (*Continued*)

Occupational Classifications

Classification Structure	Specific Functions	Military Professions	Civilian Professions
	4.2.1.3 Extreme physiological hygiene conditions		
4.2.2 Inside moving closed-door objects	4.2.2.1 Comfortable physiological hygiene conditions		
	4.2.2.2 Uncomfortable physiological hygiene conditions	—	—
	4.2.2.3 Extreme physiological hygiene conditions		
4.2.3 Inside stationary closed-door objects	4.2.3.1 Comfortable physiological hygiene conditions		
	4.2.3.2 Uncomfortable physiological hygiene conditions	—	—
	4.2.3.3 Extreme physiological hygiene conditions		
4.3 Activity in water environment			
4.3.1 On the water surface	—	—	—
4.3.2 Under water in a waterproof space	—	—	—
4.3.3 Directly under water	—	—	—
4.4 Airborne activity			
4.4.1 On a plane or other airborne vehicle	—	—	—
4.4.2 Parachuting	—	—	—
5 Educational level (qualification)			
5.1 Basic	—	—	—
5.2 Medium	—	—	—
5.3 Excellent	—	—	—

TABLE 13.2

Profiles of Military Professions in an Infantry Division

Profession		1		2		3		4	5
1	2	3	4	5	6	7	8	9	10
Driver		1.6.1.1	–	2.1	3.1.2.2	3.2.3.2	3.2.6.2	4.2.2.2	5.2
Gun crew commander	1.1.1	1.6.2.2	–	2.2	3.1.1.3	3.2.1.2	3.2.5.2	4.2.1.2	5.1
Tank driver	1.1.1	1.6.1.1	1.6.2.4	2.1	3.1.1.3	3.2.1.2	3.2.5.2	4.2.2.3	5.1
Radio operator	1.4.2	1.5.1	1.6.2.3	2.1	3.1.2.2	3.2.1.2	–	4.2.2.2	5.2
Medic	1.1.6	1.6.2.2	–	2.1	3.1.1.3	3.2.5.3	–	4.2.1.1	5.1
Sniper	1.4.1	1.6.2.2	–	2.1	3.1.2.3	3.2.3.3	–	4.2.1.2	5.2

13.4 Conclusion

In this chapter, we suggested new principles for the classification of professions, derived from analysis of subsystems of activity and its structural organization. The description of the subsystems of activity and their structural relationships became a basis for a multidimensional classification of occupations. As a primary factor, we used an integrated factor of required functions, performed by an individual, in the form of a psychological (psychophysiological, social–psychological) component, which we named "professional competence." In a psychological sense, competence is a set of PIFs that are most significant for a successful performance of functional requirements.

As a scientific base for the classification of professions, we applied the concepts of potential and procedural systems of activity. This approach was expressed by the fact that the top groups of the hierarchical structure have been determined using the basic components of the general psychophysiological schema of an human activity (Figure 13.3); the system itself has been built in accordance with principles or mechanisms presented in activity system and its relation to personality subsystem (Figure 13.4). The high-level classification groups of the PIF nomenclature structure are as follows:

1. Value-motivated properties of personality
2. Regulatory properties of personality
3. General properties to perform or operate
4. Special operational traits of interaction with people
5. Bioenergetic potential of an individual

6. Regulatory qualities of humans

7. Anatomical and external aesthetic qualities of an individual

Groups 1 and 2 represent properties (qualities) associated with the basic processes of psychological regulation and a person's personality (Figure 13.4). They exhibit themselves via the subsystem of an integral activity regulation and operational subsystem (Figure 13.3).

Group 6 describes the properties related to the basic physiological processes of regulation (Figure 13.4). They reveal themselves through the subsystems of bioenergetic support and adaptation—homeostatic regulation. Classification groups 3 and 4 display the features related to basic operational psychological and physiological activity (Figure 13.4). They reveal themselves in the functioning of the operational subsystem of activity (Figure 13.3). In Figure 13.3, this subsystem is depicted as one whole block. When we built the PIF classification, we carved group 4 from this block, which describes special operational features necessary to interact with people. This is explained by the specifics of activity that include interpersonal interactions. Such activities demand the inclusion of not only the basic psychological features and processes but also the properties of personality into the operational subsystem. We included group 7, "anatomical and external aesthetical qualities of an individual," in the classification although this group is not shown in Figures 13.3 and 13.4. This group describes the morphological and culture-related properties of a person, which were not examined when we built the general psychophysiological schema of human activity, but should be considered in the PIF classification.

References

Anokhin, P. K. 1980. *Basic Questions of Functional System Theory.* Moscow: Science Publishers.

Bedny, G., and D. Meister. 1997. *The Russian Theory of Activity. Current Applications to Design and Learning.* Mahwah, NJ: Lawrence Erlbaum Publishers.

Bedny, G., and W. Karwowski. 2007. *A Systemic-Structural Theory of Activity—Application to Human Performance and Work Design.* Boca Raton, FL: Taylor & Francis.

Bobkov, B. H., and E. M. Polovaya. 2002. Determination of minimum norms of earnings by groups of intensity of labor and physical workloads. In *Life Level of Regional Population in Russia,* vol. 8, 24–30.

Klimov, E. A. 2003. *Directory of a Personnel Worker: Managers and Employees, Qualification Characteristics, Tariffs of Earnings.* 2nd ed. Moscow: INFRA-M.

Raven, J. 1984. *Competence in Modern Society: Its Identification, Development and Release.* London: H. K. Lewis & Co.

U. S. Department of Labor, Employment and Training Administration. 1991. *Dictionary of Occupational Titles*, 4th ed. Washington, DC: U. S. Department of Labor. (Orig. pub. 1965.)

Zarakovsky, G. M. 2004. The concept of theoretical evaluation of operators' performance derived from activity theory. *Theor Issues Ergon Sci* 5(4):313–37.

Zarakovsky, G. M., and V. V. Pavlov. 1987. *Law of Functioning of Man–Machine Systems*. Moscow: Soviet Radio.

Zarakovsky, G. M., B. A. Zhiltsov, and I. N. Chunaeva. 2005. Methods of determination similarities of different professions. In *Psychophysiological Research: Theory and Practice*, ed. A. A. Medenkova, 119–41. Moscow: Flight publishers.

Zarakovsky, G. M., I. N. Chunaeva, V. A. Zhiltsov, S. J. Shustova, and A. A. Volokitin. 2004. The proposal of a unified classificator of civilian and military professions, and nomenclature of professionally important characteristics of specialists. In *Human Factor: Problems of Psychology and Ergonomics*, vol. 2, 19–31.

14

Application of Laser-Based Acupuncture to Improve Operators' Psychophysiological States

A. M. Karpoukhina and O. Kokun

CONTENTS

14.1 Introduction

Constant development and an increase in the complexity of equipment due to ongoing technological changes make optimization of the psychophysiological state (PPS) of a person who manages his or her own equipment critically important. The operators' PPS is one of the predominant hindrances to their efficiency, affecting the safety and reliability of their work, and is a prerequisite for their professional upgrade (Karpoukhina 1985, 1990, 2005; Karpoukhina and One-Jang 2003; Kokun 2006). It is well-known that a mismatch between the objectives the operators have to meet and their PPS is often the cause of equipment breakdown, accidents, crashes, operators' occupational diseases, and increasing dissatisfaction with the activity they are charged with. Therefore, optimization of the operators' PPS is directed at the prevention and decrease of the so-called negative praxis conditions induced by factors such as fatigue, strain, and stress and can be considered one of the most important tasks for the modern ergonomics.

We address this task based on functional system (FS) theory, developed by Peter K. Anokhin, and on his comprehensive scheme of structure and functional organization (Anokhin 1978). Of crucial importance is the thesis of leading the "backbone" role that is assigned to the required result or goal. Nothing else but the required concrete result drives the specifics of the structure and functional organization for a concrete integral FS.

From the systematic approach perspective (Karpoukhina 1985, 1990; Kokun 2004, 2006), human PPS is contemplated as a dynamic FS underpinned by a multilayered hierarchical structure, where all the elements interact to achieve the required future result: provision of psychophysiological support for the individual's current motivated activity. The FS acts as an apparatus of self-regulation, which has a cyclic organization with feed-forward and feedback interconnections to present itself as a central-peripheral entity, which includes various components of the neural system. The FS of self-regulation of activity includes important mechanisms such as afferent synthesis, the predictor apparatus (which is called acceptor of "action effect"), program formation mechanism, the decision-making mechanism, and feedback influences. Each of these mechanisms performs particular functions in the regulation of activity (Karpoukhina and One-Jang 2003).

Based on the modern psychological models of self-regulation of activity (Bedny and Meister 1997; Bedny, et al. 2000; Bedny and Karwowski 2007), human PPS can be viewed as a sort of energetic foundation that an individual would rely on to self-regulate his or her activity. On the one hand, PPS is formed in a process under the impact of concrete activity, and on the other, it stipulates its efficiency.

Within the hierarchy of human PPS components (or conventionally speaking, levels), we can distinguish between social, psychological, psychic, physiological, biochemical, biophysical, and bioenergetic constituents (Karpoukhina 1990, 2005; Kokun 2004, 2006). The bioenergetic level is the foundation for the functioning of all superior PSS hierarchical components. Deficient bioenergy affects the implementation of internal information processes and, consequently, interferes with the functioning of superior system hierarchical components. Since a nonspecific activation of PPS bioenergy level promotes specific positive changes on other levels and in PPS as a whole through self-regulation, this level is considered one of the most promising in terms of PPS diagnostics and optimization (Karpoukhina 2005).

The aim of the operators' PPS optimization is to match the PPS parameters with the parameters, tasks, and conditions of their activity (Karpoukhina 1985, 1990. 2005; Kokun 2004). Correspondingly, PPS optimization can be interpreted as the establishment of a complete match between the activity parameters and the operators' PPS parameters (Kokun 2004).

In this chapter, we will study the theoretical, methodological, and practical substantiation to using a bioenergy-level activation method, laser-based acupuncture, to optimize the operators' PPS, and the results of experimental research, which prove the efficiency of this method.

14.2 Theoretical, Methodological, and Practical Substantiation of Applying Laser-Based Acupuncture

Modern science does not dispose of multiple methods and tools to evaluate, control, and optimize the PPS of operators, that is, virtually healthy persons in the process of their activity. Today, mostly pharmacological preparations and methods of psychoregulating training and psychic self-regulation are used to improve the operators' PPS. Sometimes, modifications of such methods use biological feedback to improve the PPS and rheoencephalography (REG) indexes. The common feature of all these methods is that they rely not on increasing the energy level of psychophysiological processes from outside, but rather focus on the redistribution of the internal energy resources of the human body to optimize any individual function.

This leaves open a question related to the possibility of durable repetitive application, harmlessness, and remote consequences of these pharmacological methods (Karpoukhina 1990). The practical application of psychic regulation methods took a considerable lead over their conceptual substantiation. The efficiency of some methods is not permanent and sometimes cannot be prognosticated. The major drawback of the existing psychic regulation methods is the long time and considerable effort they take to produce the effect, as well as the need of high motivation to master them. As a result, only an insignificant percentage of people are capable of utilizing them at the level sufficient to produce a noticeable effect in the optimization of their state and to increase their operational efficiency (Kokun 2004). In addition, these methods are difficult to automate and tend to divert operators from their activity.

Consequently, there is a need to conduct research and develop PPS regulation methods based on the enrichment of energy levels—first of all, the energy of informational (including psychic) human processes, that is, methods that rely on the energy support of the nervous system by virtue of phylogenetic and ontogenetic deterministic algorithms. The latter is all the more essential to provide for a nondoping nature of influence, produced on a human operator, and the elimination of any negative effects.

Such methods include laser-based acupuncture, which has acquired a good reputation in the field of therapeutic treatment of various diseases by using low-intensity laser or "soft laser." By now, we have 30 years of experience applying soft lasers for resolving new tasks such as regulating the PPS of virtually healthy individuals to ensure the efficiency and reliability of their (operator, professional, training, game, daily) activity.

Due to the mechanism of low-energy long-wave laser radiation in red and infrared ranges, this method targets the energy intracellular processes, which are based primarily on the photons functioning at molecular and submolecular levels. Targeting the biologically active points (BAPs), which are characterized by their high concentration of nervous system elements, allows that this effect is being pointed at a particular "address," that is, impacting the energy of physiological and psychic processes. Here, these processes are characterized by energy and informational interrelations, which make it possible through the energy level of nervous and psychic processes to produce an impact on the informational processes occurring in the nervous system and the brain, including perception, coding, transfer, storage, and presentation of information. These processes are mostly responsible for governing the efficiency of the operators' work.

From ancient times, BAPs have been known as acupuncture medicine in the Orient. This method has been widely used for the treatment of various illnesses. BAPs represent small parts of skin (1–3 mm in diameter) and hypodermic tela, which contain a number of mutually interdependent microstructures (vessels, nerves, cells, connective tissue). One individual skin point is distinguished from its neighbor points by an increased oxygen uptake, higher temperature, reduced electrical resistance, asymmetry in current conductivity, considerable electrical capacity, the value of quasistatic and dynamic biopotential, palpatory tenderness, and so on (Karpoukhina 1990; Kokun 2004, 2006). According to oriental medicine, different BAPs are connected to different inner organs and are interconnected into a sort of meridians.

Topographic analysis of BAPs and meridians, which connect them, shows that the path of these channels coincides with the layout of large nerve tracts or blood vessels, densely entwined with nerve fibers. In general, a BAP can be visualized as the localization of energy-labile biomolecules or as an electromagnetic structure (architecture) responsible for powering informational processes inside the human body, which also determines its changes.

From the time low-intensity laser radiation was introduced in medical practice (in the early 1960s) and all the more so in the practice of adjusting PPS in human activity, it was essential to understand its mechanics. Different hypotheses on the nature of its effect on the human body have been developed ever since. Today, we know that the initial mechanism of low-intensity lasers is related to exchange processes and bioenergetics inside the cell, predominantly, the nerve cell (Skulachev 1969, 1972; Rusakov 1988). It is considered to be the most adequate tool affecting the BAP system (Karpoukhina 1985, 1990, 2005). The response of the human body to soft-laser treatment can be explained not only by the specifics of its operation, but also to a considerable extent by the initial functional state of the human body. Since the FS operates at a very low energetic level, the low level of applied energy enhances the power balance within a cell by stirring up self-protective and self-recovery processes as soon as the body's functional state is disturbed. Large quantities of this energy, inhibiting the functional activity and in some cases even capable of destroying tissues, are used in surgery as a laser knife.

Optimization of the cell energetic at the base bioenergy level of the PPS system through energy-information relationships contributes to the optimization of information processes at all hierarchically superordinate PPS system levels. In other words, the reaction to laser-based acupuncture can be represented as a chain of reactions throughout the entire human PPS system. Such a reaction also increases the level of energy channeled to support the body's regulative algorithms, which results in staving off fatigue, enhancing work efficiency, and increasing tolerance to elevated emotional and exercise stresses.

Laser radiation is electromagnetic radiation in the optical band, possessing properties such as monochromaticity, coherence, polarization, and directionality. The radiation source in soft lasers is the electro-optical quantum amplifier or semiconductor oscillator of electromagnetic radiation in the red and infrared bands. The efficiency of such a laser impact depends on the flow rate, duration of the procedure, and length of the respective wave. It has been discovered experimentally that a high biotic level (or affinity with proper biotic process of energy conversion) can be achieved by the red band radiation with a wavelength of 0.6328 µm (Karpoukhina and One-Jang 2003; Karpoukhina 2005). The parameters of the laser radiation used on the BAPs are as follows: a wavelength of 0.6328 µm, a density of energies on the boarder of the skin 0.5–2 µWatt/mm², and a duration between 10–15 seconds and 1–2 minutes. As a rule, laser-based acupuncture is recommended to be applied in treatment courses of up to 12–14 days, but a one-time treatment can also be equally beneficial.

It is necessary to highlight some of the aspects of laser-based acupuncture, such as its normalizing (nonspecific) and nondoping nature. The normalizing effect is produced by soft lasers, which bring the parameters of regulated functions to norm, regardless of initial hypo- and hyperfunctions. This corroborates the fact that a laser operates not through any irritating or stimulating effects, but represents rather an infiltration of energy that underpins the self-regulation processes.

The nondoping nature of lasers was proven through special experimental research, which included the comparison between the efficiency of the signal (selectively) measured by operators of the experimental (exposed beforehand to laser treatment) and control (placebo) groups. They were made to operate in three modes: under no tension, under a certain tension, and with time limitations. In the first mode there were no tangible discrepancies in the groups' performance, but in the second and especially in the third (stress) mode, the performance of the operators who were subjected to laser treatment was considerably better (Karpoukhina 2005). It seems that when working under no stress, the operators had sufficient initial energy and did not need any power "additive" to boost them up; however, while operating in the modes of tension, the power "boost" the operators from the experimental group have enjoyed allowed them to considerably exceed the results achieved by their counterparts from the control group. Significantly, no negative effects were ascertained in any of the multiple experimental series.

The major advantages of laser-based acupuncture used to improve human PPS include the absence of painful sensations and undesirable side effects, the possibility of using cumulative effect, the minimal time required by the treatment, and the preparation procedure, which increases the rate of treatment (up to 60 persons per hour), automation, possibility of combining diagnostics, and optimization.

14.3 Experimental Research on the Optimization of the Operators' PPS Using Laser-Based Acupuncture

The objective of these experiments was to research the possibility of applying laser-based acupuncture for optimizing operators' PPS in the process of their training with a vehicle simulator. The experiment was performed by a group of scientists from the Laboratory of Psychophysiology, Institute of Psychology and Institute of Labour and Professional Diseases (Kiev, Ukraine). The scientific supervisor was A. M. Karpoukhina. Two series of experiments, varying in their structure, were conducted to achieve the established objective: to determine the extent to which laser-based acupuncture influenced the simulator training process of operator. To achieve this objective, the following goals had to be met:

- Study the dynamics of the operator's efficiency, when working with the vehicle simulator under the influence of laser-based acupuncture.
- Study the influence of laser-based acupuncture on the operator's PPS during simulator training, based on the indexes of various FSs of the body, which reflect the hierarchical levels of integral PPS.

The participants were 19- to 21-year–old males, without previous driving skills, who were divided into two groups, each including 14 persons: the experimental group (subjected to laser-based acupuncture) and the control group (with a simulated treatment). The simulated treatment was conducted keeping in place all the exterior attributes of a regular procedure: conducting light to the BAP by light-carrying fibers, fixing the time of influencing each BAP with a stopwatch, and so on. As there are no painful, tactile, or temperature sensations while applying the low-intensity laser beam on the skin surface, the test subjects could not differentiate between the modes, either real or simulated, that they had been working in.

The experiment had five series (including the standard training session), during which the subjects covered 30 kilometers of road shaped in a particular configuration. They were instructed to go at maximum speed, but at the same time try to minimize cases of running into the restrictive lines. Their efficiency was measured by the speed of movement through the route and the number of errors they made (cases of running on the restrictive lines and driving glitches were registered automatically).

The dynamics of achieving the experiment's objective was traced by breaking down the 30-kilometer stretch into six 5-kilometer portions and fixing the performance rates at each of them. This allowed us to represent the obtained experiment data as learning curves with the x-axis representing the duration of learning process, expressed by the ordinal number of task (N) that was assigned, and the y-axis representing the performance efficiency, expressed by the time the subject took to cover the route (t) (Figure 14.1) and the number of errors he or she committed when doing so (n) (Figure 14.2).

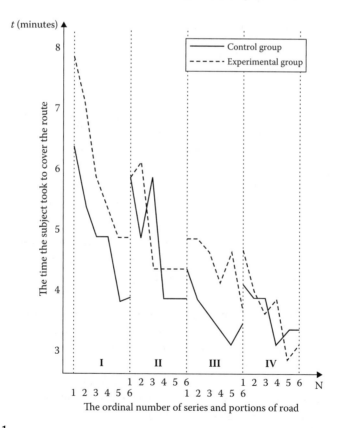

FIGURE 14.1
Influence of laser-based acupuncture optimizing the PPS of the operators on their performance efficiency (by the time the subject took to cover the route, t).

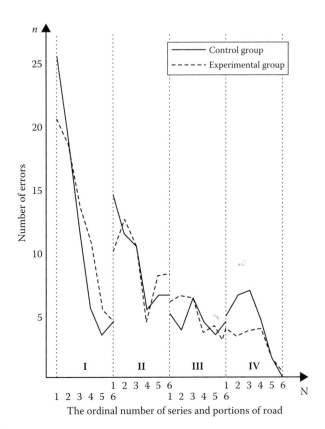

FIGURE 14.2
Influence of laser-based acupuncture optimizing the PPS of the operators on their performance efficiency (by the number of errors, *n*).

While working with the simulator, the optimization of operators' PPS under the influence of laser-based acupuncture was estimated by a number of dynamic parameters, characterizing their body FSs:

- Central nervous system (simple sensor–motor reaction and the mobility of nervous processes by method of A. E. Khilchenko (1960); ability of nervous processes by the method of "critical frequency of light glimpses")
- Cardiovascular system (systolic, diastolic, and pulse blood pressure; heartbeat frequency; rate of cardiovascular system's endurance; cardiovascular integral Shiposh index)
- General health, activity, and mood parameters (by subjective scaled evaluation)
- Reactive anxiety index (Spielberger method)

Quantitative and qualitative comparison of the value and direction, characterizing the psychophysiological shifts in the background and post-stimulation measurements, was made before and after work with the vehicle simulator.

Laser-based acupuncture was performed for 3 days using an LG-12 helium-neon laser with a wavelength of 0.6328 μm and PC86 pliant light-carrying dissector fibers for 10 BAPs placed at the fingertips. Here, the BAP formula was in conformity with the approved international classification (Tabeeva 1994; Luvsan 1986). The exposure time per each BAP was 15 seconds, and the total exposure time was 2.5 minutes. The analysis of the learning curves highlights some specifics of the dynamics of this process.

First, it is evident that the accumulation of information in the course of the learning and training process is accompanied by changes occurring in the (information, algorithmic, graphic) structure of activity. This can be observed from the decrease in the time spent on an individual task, enhanced reliability, or lesser rates of power inputs (a decrease in emotional excitement—the psychophysiological "price associated with the operation") to perform an activity.

Second, the performance improved stepwise. The increase in the efficiency of the operators who were trained at the simulator during one series took place predominantly at the initial stage of this process. The improvement in the operator's efficiency has not occurred over one series but rather from one series to another. Before beginning a new experimental series, the psychological system of the test subject was at a new, higher organizational level as compared to the preceding series. Usually, the achieved level was maintained while performing the new task, and the freshly acquired information was implemented in the next series.

Third, the operator's simulator training was characterized by certain spells of relatively stable levels in terms of efficiency. Each new level was reached by leaps, which makes the graph look more like a staircase. The quantitative characteristic of this process, based on the total data for each temporary period, is represented in Figure 14.3. It is apparent that with the time of the learning process approaching infinity, the efficiency increases step by step (with each step, the efficiency increments become smaller, and the time the efficiency takes to increase gets longer).

When comparing the learning curves built by the fractional exponents of performance efficiency for the control and experimental groups, the similarity in the training processes should be highlighted. The absolute values of the activity's success, both in terms of time and the number of committed errors, do not warrant the conclusion that the subjects from the experimental group have an edge. However, the relative values, that is, the increments of efficiency against the background of the first series' results, speak in favor of the subjects from the experimental group. This can be seen from Figure 14.4, where the x-axis is the learning time and the y-axis is the mean value of the efficiency for each series taken separately (Δt and Δn).

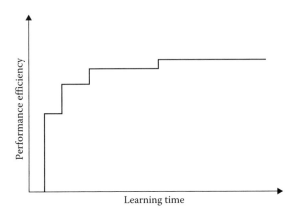

FIGURE 14.3

Features of dynamics performance efficiency during the process of simulator training (in standard units by the average results to cover the portions of road).

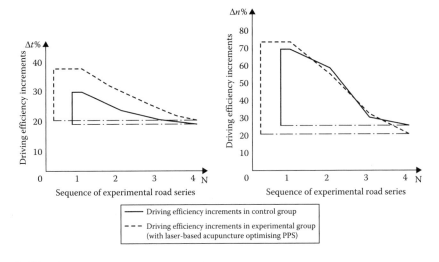

FIGURE 14.4

Comparative estimation of the driving efficiency increments during the process of simulator training in groups with laser-based acupuncture optimizing the PPS and in the control group of the operators.

The analysis of dynamics inherent to the psychophysiological indexes changing under the influence of laser-based acupuncture (in experimental group) and its simulation (for control group) in the process of vehicle simulator training revealed the changes in the principal (registered by us) FSs of the operators' bodies, reflecting various hierarchical levels of integral PPS. The tensity of these systems can be correlated with the efficiency at different stages of the learning process. In the initial stage of training, the subjects showed low efficiency and psychophysiological indexes as registered by objective research methods and demonstrated considerable tension in the functioning of different operators' body systems (central nervous system, cardiovascular systems, etc.).

In the subsequent stage of training, as the subjects accumulated driving skills, the emotional tension (the price of the activity) decreased in both groups, but in the experimental group (against the background of laser-based acupuncture), the changes observed in the psychophysiological parameters were more pronounced and positive. By the end of the 3-day treatment (or its simulation) course, both groups demonstrated increases in the efficiency and decreases in the tension of human body systems, that is, a decrease in the price of the activity. However, in the control group, after work with the simulator as opposed to the background (before work), the psychophysiological indexes either decreased (i.e., got worse) or remained unchanged. In the experimental group, the same indexes either remained at the level of the background quantitative values or improved, even though insignificantly, by 1.5–5%.

Therefore, by experimental research, we revealed a positive effect of laser-based acupuncture treatment on the operators' simulator training process. We made objective records of the changes in the indexes, and their qualitative and quantitative analyses provide the evidence of the operators' integral PPS optimization under the laser-based acupuncture treatment, which eventually resulted in the better performance they demonstrated during the vehicle simulator training. The mean positive changes of the psychophysiological indexes in the experimental group were not so impressive (1.5–5%), probably due to the minimal time (15 seconds) BAP PC86 was exposed to the effecting factor. In further studies, we plan to test for other temporal parameters of laser-based acupuncture treatment to reveal the most suitable modes of impacting BAP PC86, which will ensure attaining a sufficiently higher effect of laser-based acupuncture.

The objective of the second series of experiments was to look into the possibility of improving the operators' driving skills by laser-based acupuncture. REG was used to evaluate the effects, and we also evaluated muscular efficiency. The operators were trained 1 hour per day for 2 days. The high intensity of work was achieved by issuing them 15 orders per minute. The success of the training was controlled by the dynamics of the operators' performance at the simulator. To measure the extent of the laser-based acupuncture treatment, the experimental group (consisting of nine subjects) underwent BAP stimulation before and during their work with the simulator on the second training day. The second group (eight subjects) underwent the simulated treatment.

Laser-based acupuncture was performed by using the following BAP formulas and exposure times:

- Before working with the simulator:
  PC86: 15 seconds per each of 10 BAPs
  TR17 (symmetrical): 10 seconds per BAP

GI4 (symmetrical): 10 seconds per BAP

AP34 (symmetrical): 15 seconds per BAP

AP83 (symmetrical): 15 seconds per BAP

- In the process of working with the simulator (in this case, the flexible light guides were fixed by cuffs to the shin's BAPs):

GI3 (symmetrical)

E36 (symmetrical)

The REG research included the following:

- Study of the influence of the loads on the operators' REG indexes in the process of vehicle simulator training
- Study of the interconnection between the training efficiency indexes and cerebral blood circulation parameters
- Research on the possibility of laser-based acupuncture to produce regulatory impact on the cerebral blood circulation to improve the operators' PPS in the process of their simulator training

The REG method was chosen because of the rather high sensitivity of cerebral blood circulation parameters to the changes in the functional state of humans (Nersesian 1992; Fedorov, et al. 1989). REG is recorded by rheoplethysmograph in the (left) frontomastoidal abduction, which made it possible to study the cerebral blood circulation in the internal carotid artery basin.

The following REG parameters were analyzed:

- Rheographic index (I), which reflects the level of blood in the vessels of the studied brain portion
- Minute rheographic index (I_m), which reflects the blood supply to the studied brain portion
- Dicrotic impulse (DCI), which reflects the tone of small arteries and arterioles and determines the peripheric resistance to the blood flow
- Diastolic index (DSI), which reflects the tone of veins and is the parameter for blood outflow
- Relative time for maximum blood filling of the vessels (α/T), that is, the tone parameter for average and big arteries
- Relative duration of the diastolic period ($T - \alpha/T$)

The results, obtained by research, showed that the work with the simulator on the first day of training provoked an increase in the peripheral resistance. Changes in the vein tone in different groups were insignificant and divergent. A minor (invalid) reduction in the main artery tone was observed. Changes in the pulse blood level and blood supply indexes were divergent: in the control group, I_m increased, and in experimental group, it decreased (Figures 14.4 and 14.5).

On the second day of training with the simulator, the subjects showed an increased peripheric resistance and vein tone. Valid increases of DCI ($p \leq .01$) and DSI ($p \leq .05$) were observed only in the control group. These changes, indicating the augmentation of psychic tension, were more pronounced on the second day of training as opposed to the first day. The lack of valid DCI and DSI changes in the experimental group is evidence of the lesser level of

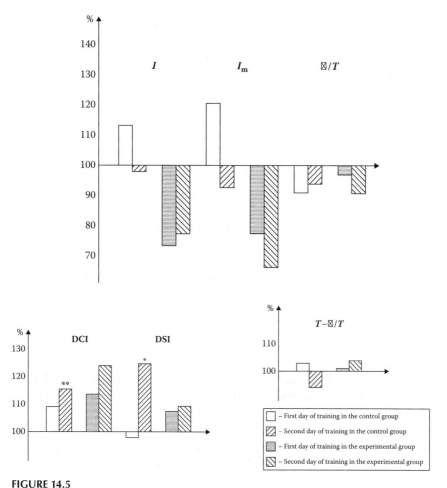

FIGURE 14.5
Influence of simulator training on cerebral blood circulation indexes (according to rheoenceph-alography data). Comparative assessment of indexes in groups with laser-based acupuncture optimizing the PPS and in the control group of the operators. The group's mean values at repose (before training session) are taken to be equal to 100%. * means $p \leq 0.05$; ** means $p \leq 0.01$.

psychic tension in this group as opposed to the control group on the second day of training (this coincided with the laser-based acupuncture treatment).

We have conducted a correlation analysis to determine the interrelation between the indexes of cerebral blood flow and performance parameters while working with the simulator (number of errors, number of kerb touches, duration of touches, number of emergency brakings, speed of emergency braking). The analysis showed no relationship between the REG indexes at rest (before operation) and the efficiency of training at its beginning, that is, on the first day. On the contrary, the after-operation REG indexes correlated with some efficiency indexes. The most constant was the negative correlation between the number of errors and number of touches on the one hand and peripheric resistance and the state of blood outflow (DCI and DSI) on the other. In total, we have observed four significantly valid correlations between the REG indexes and the efficiency of working at the simulator for groups on the first training day (Table 14.1).

On the second training day, two significant correlation links, between the efficiency of working at the simulator and pretest cerebral blood circulation indexes, were observed (in each of the groups), and there were three significant links between the efficiency and after-test REG indexes.

The obtained results lead to the following conclusions:

- The operators' activity, related to the simulator training, are accompanied by distinct changes in cerebral blood circulation.
- There is a link between the cerebral blood circulation indexes and the efficiency of such activity; the highest dependency (in terms of significant correlation rates) can be observed at the initial stage of training.
- The correlation links between the efficiency of working at the simulator and pretest indexes of cerebral blood circulation at the second training stage are an indication of the mobilization processes; that is, the body's setting up for a future activity.
- The decrease in the number of significant correlation links between the after-test cerebral blood circulation indexes and the efficiency of working with the simulator on the second day, as opposed to the first one, reflects the decrease in the dependency of successful activity on physiological supply and indirectly indicates the depreciation of the price of the activity over the training process.
- Indexes of peripheral resistance and outflow can be used as correlators of psychic tension over the period of simulator training.

The analysis of efficiency in the activity related to learning the skills of driving the simulator vehicle shows a slightly better performance in the experimental group: the trainees there had more emergency braking, drove

TABLE 14.1

Correlation Links between Operators' Simulator Training Efficiency Indexes and Rheoencephalography Indexes

Training Efficiency Indexes	Group	I — First Day Before Work	I — First Day After Work	I — Second Day Before Work	I — Second Day After Work	DCI — First Day Before Work	DCI — First Day After Work	DCI — Second Day Before Work	DCI — Second Day After Work	DSI — First Day Before Work	DSI — First Day After Work	DSI — Second Day Before Work	DSI — Second Day After Work
Number of emergency brakings	Exper.	—	—	—	—	—	—	—	—	—	—	—	—
	Contr.	—	—	—	—	—	—	—	—	—	—	—	—
Number of errors	Exper.	—	—	—	—	—	—	—	—	—	—	—	—
	Contr.	—	—	—	—	—	−0.77	—	−0.72	—	−0.79	—	0.69
Speed at emergency brakings	Exper.	—	—	—	—	—	—	—	—	—	—	—	—
	Contr.	—	—	—	—	—	—	—	—	—	—	—	—
Time of scratches	Exper.	—	−0.71	—	—	—	—	—	—	—	−0.73	—	—
	Contr.	—	—	—	—	—	—	—	—	—	—	—	—
Number of scratches	Exper.	—	—	—	—	—	−0.71	—	—	—	−0.76	—	0.69
	Contr.	—	—	—	—	—	—	—	—	—	—	—	—

Training Efficiency Indexes	Group	Im — First Day Before Work	Im — First Day After Work	Im — Second Day Before Work	Im — Second Day After Work	α/T — First Day Before Work	α/T — First Day After Work	α/T — Second Day Before Work	α/T — Second Day After Work	T − α/T — First Day Before Work	T − α/T — First Day After Work	T − α/T — Second Day Before Work	T − α/T — Second Day After Work
Number of emergency brakings	Exper.	—	—	—	—	—	—	—	—	—	—	—	—
	Contr.	—	—	—	—	—	—	—	—	—	—	—	—
Number of errors	Exper.	—	—	—	—	—	—	—	—	—	—	—	—
	Contr.	—	—	—	0.77	—	—	−0.74	−0.72	—	—	−0.74	−0.72
Speed at emergency brakings	Exper.	—	—	—	—	—	—	—	—	—	—	—	—
	Contr.	—	—	—	—	—	—	—	—	—	—	—	—
Time of scratches	Exper.	—	—	—	0.73	—	−0.69	—	—	—	−0.69	—	—
	Contr.	—	—	—	—	—	—	—	—	—	—	—	—
Number of scratches	Exper.	—	—	—	—	—	—	—	—	—	—	—	—
	Contr.	—	—	—	—	—	—	—	—	—	—	—	—

at better speed, committed less errors, and had a significant decrease in curb scratches ($p \leq .01$). The control group had opposite changes: it showed reduction in the rate of training.

This study allows us to draw a conclusion that laser-based acupuncture produces a regulating effect on cerebral blood circulation, which contributes to the optimization of the operators' PPS and, as a consequence, promotes better performance in the simulator training.

The principal indexes of muscular efficiency were the following:

- Experimental study of the operators' muscular efficiency dynamics in the process of 2-day simulator training
- Research on possibility of optimizing the operators' muscular performance in the process of simulator training by applying the laser-based acupuncture
- Comparative evaluation of the operators' muscular efficiency in the process of simulator training between the real and simulated laserpuncture

The research focused on the following indexes of muscular efficiency, which are most critical for the operators' simulator training:

- Coordination of hand movement (by using a Rupp device)
- Statistic muscle endurance (according to V. V. Rozenblat, 1971) by using a manometric dynamometer
- Maximum muscle strength by using a hand-operated spring dynamometer

The static endurance and muscle strength indexes of the test subjects in the experimental (nine persons) and control groups (nine persons) have been registered in the course of 2-day daily training sessions before the session, in the process of training session (30 minutes after driving lessons), and within an hour after the session. The state of hand movement coordination was studied daily before and after working with the simulator.

The analysis focused on the dynamics of results on the first and second days of training. We also performed a comparative analysis with the first day of training, the results of which were taken as the background data, and the second training day after laser-based acupuncture or its simulation. On the first day, the discrepancy between the groups was insignificant. The laser acupuncture or its simulation was carried out only on the second day. Therefore, the intergroup fluctuations of the first training day can be explained by the differences between its participants selected on a random basis.

The analysis of results obtained on the first training day revealed a tendency of all studied indexes in both groups to gradually worsen by the end of the day. For example, the static muscle endurance in the experimental group deteriorated after 30 minutes of training by 28.4% ($p \leq .01$) and by 24.6% ($p \leq .01$) after 60 minutes. In the control group, this index decreased by 34.8% ($p \leq .01$) and 52.1% ($p \leq .01$), respectively. The maximum muscle strength in the experimental group after 30 minutes of driving lessons was 7.9% worse than before, and after 60 minutes, deteriorated by 8.7%. In the control group, the changes also worsened by decreasing correspondingly by 9.5% ($p \leq .05$) in both measurements.

Since the test conditions made it possible to register the hand movement coordination only before the training sessions and after 1-hour work with the simulator, we could establish its downturn at the end of the test in the experimental group by 13.7% ($p \leq .05$) and by 3.4% in the control group. Such unidirectional change of indexes toward their lessening by the end of the test on the first day was caused by the rapidly encroaching weariness induced by activity that was novel for test subjects.

Table 14.2 represents the mean group dynamics of studied indexes (in percentage) over the 2-day training period 30 and 60 minutes after driving the simulator, as opposed to the same indexes registered before the training sessions. Since the operators, while working with the simulators, were subject to static rather than dynamic stress to maintain their posture over a

TABLE 14.2

Dynamics of Changes in Group Mean Values Applicable to Static Muscle Endurance, Hand Movement Coordination, and Muscle Strength in the Process of Operators' 2-Day Simulator Training in Laser-Based Acupuncture and Simulated Laser-Based Acupuncture

Values	First Day			Second Day		
	Before Work	After 30 Min	After 60 Min	Before Work	After 30 Min	After 60 Min
Experimental Group						
Static muscle endurance	100%	71.6% ($p \leq .01$)	75.4% ($p \leq .01$)	100%	108.3%*	130.1% ($p \leq .01$)
Hand movement coordination	100%		86.3% ($p \leq .05$)	100%		95.6%
Muscle strength	100%	92.1%	91.3%	100%	110.9%*	97.7%
Control Group						
Static muscle endurance	100%	65.2% ($p \leq .01$)	47.9% ($p \leq .01$)	100%	67.3%** ($p \leq .01$)	55.6% ($p \leq .001$)
Hand movement coordination	100%		97.56%	100%		98.8%
Muscle strength	100%	90.5% ($p \leq .05$)	90.5% ($p \leq .05$)	100%	94.9%**	88.2% ($p \leq .05$)

Notes: *Laser-based acupuncture influence.
            **Simulated laser-based acupuncture influence.

long time, this was accompanied by increased fatigue. The index of static muscle endurance, to control the operator's physical performance, was the most informative, and consequently, its dynamics over the process of training were the most significant.

While analyzing the results obtained over the second day of simulator training, we discovered that the static muscle endurance in the experimental group 30 minutes after the start of the session improved by 8.3%, and after 60 minutes by 30.1% ($p \leq .01$) as opposed to before-the-session indexes. At the same time, in the control group, as observed on the first day, the static endurance gradually decreased by 32.7% after 30 minutes of operation ($p \leq .01$) and by 44.4% after 60 minutes ($p \leq .001$).

This opposite trend of the changes in the groups was caused by performing laser-based acupuncture in the experimental group and simulated laser-based acupuncture in the control group prior to simulator training on the second day. The higher static endurance in the experimental group after laser-based acupuncture reveals its beneficial impact on muscle performance in the process of simulator training.

These results are corroborated also by the maximum muscle strength of the subjects in the experimental group, which increased on the second training day by 10.9% after 30 minutes of operation, as compared with the indexes taken before the simulator training and receding gradually after 60 minutes of operation down to the initial level. In the control group, muscle strength was smoothly deteriorating: after 30 minutes by 5.1% and after an hour by 11.8% ($p \leq .05$). However, the hand movement coordination of the tests subjects in both groups on the second day of simulator training fluctuated very little as opposed to the initial data. Due to the lack of maximum physical exertions that the subjects would experience while working at the simulator, resulting in a pronounced physical exhaustion, no significant changes in the hand movement coordination indexes have been revealed. Nor, as a consequence, were any significant effects of laser-based acupuncture discovered.

A comparative analysis of the test subjects' results in both groups between the first and the second training days, represented as group means (%) in Table 14.3, showed no considerable difference between the indexes of static endurance and maximum muscle strength registered before the simulator training both in the experimental and control groups on the second day as opposed to the first day because these indexes on both days were registered before laser-based acupuncture or its simulation.

The most pronounced, though, were the changes in hand movement coordination: the experimental group demonstrated a 5.2% progress and the control group showed a 12.6% progress. This can be explained by the learning curve gradually going up as the subjects had to perform the same training routine repetitively. The comparison of this index between the first and second days of training showed that on the second day both groups did better when compared to the first day: in the experimental group, this change was statistically significant ($p \leq .05$) by 16.4%, and in the control group by 14.1%.

TABLE 14.3

Comparative Assessment of Changes Occurring in Group Mean Indexes to Static Muscle Endurance, Hand Movement Coordination, and Muscle Strength of the Operators between the First and the Second Training Days

Values	Experimental Group			Control Group		
	Before Work	After 30 Min*	After 60 Min	Before Work	After 30 Min**	After 60 Min
Static muscle endurance	103%	156.9% ($p \leq .01$)	178.1% ($p \leq .001$)	101%	102.3%	109.9%
Hand movement coordination	105.2%		116.4% ($p \leq .05$)	112.6%		114.1%
Muscle strength	95.6%	115.2%	102.4%	94.5%	89.8%	92.2%

Notes: *Laser-based acupuncture influence.
**Simulated laser-based acupuncture influence.

Further comparison of static endurance and muscle strength indexes between the first and the second training days demonstrated the unilateral direction of changes for the better in the experimental group on the second day as compared to the first day. The statistic endurance registered 30 minutes after working at the simulator on the second day was statistically significant ($p \leq .01$), up by 56.9% as compared to the first day. In 60 minutes, the results increased by 78.1% ($p \leq .001$). The muscle strength on the second training day in experimental group increased correspondingly by 15.2% and 2.4% as opposed to the first day. Consequently, laser-based acupuncture in the experimental group on the second training day produced a positive effect on the studied indexes.

The same data registered in the control group showed that the muscle strength on the second training day both in 30 and 60 minutes of simulator training decreased as opposed to the first day and equaled 11.2% and 7.8%, respectively. However, the static muscle endurance of test subjects from the control group on the second training day, after 30 minutes of simulator training, was the same as on the first day and augmented by 9.9% after 60 minutes of operation. This can be explained by the fact that as the test subjects acquired driving skills in the process of training, static stress was gradually relieved, resulting in a better static endurance index. On the other hand, the operators in the experimental group had significantly higher results after laser-based acupuncture as opposed to the control group operators who underwent only simulated treatment.

Based on the obtained results, we can draw the conclusion that the laser-based acupuncture, by targeting the above BAPs, contributes to and hampers the development of physical exertion. Application of laser-based acupuncture in the process of the operators' simulator training promotes the growth of their static endurance, increases muscle strength, and regulates hand movement coordination. Treating BAPs with laser acupuncture in the process of simulator training contributed to optimization of muscle performance if the subjects showed pronounced physical tiredness at the end of the training session or improvement of static stress if they had to preserve the same posture over long periods of time. If the tiredness was less evident, the impact of laser-based acupuncture on muscle performance was not assimilated and its effect was weak. Therefore, we can conclude that laser-based acupuncture optimizes the operators' PPS in the course of vehicle simulator training by means of improving their muscle performance.

14.4 Conclusion

The studies we have conducted were convincing in showing a rather high efficiency of laser-based acupuncture in optimizing the operators' PPS. We have identified a pronounced positive impact of laser acupuncture on the progress of the operators' simulator training, which was registered by us in an objective manner. Studying diagnostic indexes by quantitative and qualitative analyses shows the optimization of the operators' integral PPS. In particular, we have established that the laser-based acupuncture produced a regulating effect on the state of the operators' cerebral blood flow, improved their static muscle endurance, enhanced their muscle strength, and exercised a regulating influence on hand movement coordination.

Positive shifts in the indexes of psychophysiological provisions of activity systems under the controlling effect of laser-based acupuncture were accompanied by improvements in the operators' training indexes. It enables us to reach a conclusion on the ergonomic significance of laser-based acupuncture in PPS regulation. The interrelation and interaction of two FSs PPS and activity can be construed from the perspective of systematic approach and FS theory as interrelations between hierarchical sub- and supersystems. The results of the system activity are conceived by the PPS system as a component of afferent synthesis block, which, along with other components such as the information on external conditions and initial background PPS, form motivation and participate in making a decision as to what the future result should be; that is, they provide for the "psychophysiological provision of activity" and vice versa. The result of the PPS FS enters as a component into the result-achieving block of activity supersystem.

Further studies will focus on building a single comprehensive method that will make it possible to carry out operators' PPS assessment using BAP diagnostic biophysical indexes in real time and, based on this assessment, perform laser-based acupuncture optimization of their PPS, which will eventually lead to improvement of their performance and efficiency ratios.

References

Anokhin, P. K. 1978. *Contribution to the General Theory of Functional Systems*. Jena, Germany: G. Fischer.

Bedny, G., and D. Mester. 1997. *The Russian Theory of Activity: Current Application to Design and Learning*. Mahwah, NJ: Lawrence Erlbaum Associates.

Bedny, G., and W. Karwowski. 2007. *A Systemic—Structural Theory of Activity: Application to Human Performance and Work Design*. Boca Raton, FL: Taylor & Francis.

Bedny, G., S. Seglin, and D. Mester. 2000. Activity theory: History, research and application. *Theor Issues Ergon Sci* 1(2):168–206.

Fedorov, Б. M., T. I. Sebekina, and E. N. Streltsova. 1989. Cerebral blood circulation during intense mental work. *Hum Physiol* 15(2):48–55 (in Russian).

Karpoukhina, A. M. 1985. *Control and Regulation of Men Psychophysiological State As Method of Increasing Human Productivity*. Kiev: Knowledge Publishers.

Karpoukhina, A. M. 1990. *Psychological Method of Increasing Efficiency of Work Activity*. Kiev: Knowledge Publishers.

Karpoukhina, A. M. 2005. Optimization of relations between man and computer via regulation of the operator's psychophysiologic state. *Relevant Issues Psychol* 5(4):64–7.

Karpoukhina, A. M., and J. One-Jang. 2003. Systems approach to psycho-physiological evaluation and regulation of the human state during Performance. In *Proceedings of the XVth Triennial Congress of the International Ergonomics Association and The 7th Joint Conference of Ergonomics Society of Korea*, 451–54. Seoul: Japan Ergonomics Society.

Khilchenko, A. E. 1960. The mobility of nervous processes: Some research results. *Physiological Journal* 6(1):21–28 (in Russian).

Kokun, O. M. 2004. *Optimization of Man's Adaptative Capacities: Psychophysiologic Aspect of Activity Backup*. Kiev: Millenium.

Kokun, O. M. 2006. *Psychophysiology*. Kiev: Center of Educational Literature.

Luvsan, G. 1986. *Traditional and Modern Aspects of Eastern Reflexotherapy*. Moscow (in Russian): Nauka.

Nersesian, L. S. 1992. *Psychological Aspects of Enhancing Reliability in Controlling Moving Objects*. Moscow: Promedek, Nauka.

Rozenblat, V. V. 1971. *Principles of Physiological Evaluation of Hard Labor and Establishing Time Standards Based on Pulse Measurement Procedures (Methodical Recommendations)*. Sverdlovsk, Russia: Sverdlovski Research Institute of Hygiene.

Rusakov, D. A. 1988. *Comparative Study of Spinal Neurons' Functional and Structural Parameters in Normal State and After Radiation with Low-Intensity Laser*. Unpublished PhD's thesis. Kiev: Institute of Physiology AS of Ukraine.

Skulachev, V. P. 1969. *Accumulation of Energy with in a Cell*. Moscow: Nauka.

Skulachev, V. P. 1972. *Transformation of Energy in Bio-Membranes*. Moscow: Nauka.

Tabeeva, D. M. 1994. *Acupuncture*. Moscow (in Russian): Medicine.

15

Information Processing and Holistic Learning and Training in an Organization: A Systemic-Structural Activity Theoretical Approach

K. Synytsya and H. von Brevern

CONTENTS

15.1 Introduction

Many professions nowadays are related to information processing, which is done with extensive support from computer-based systems. Information workers not only collaborate with each other, but also share tasks with technology systems so that the efficiency of their work depends on the capabilities of each component of the human–computer system, justifiable task distribution, and the quality of the interaction. The changing nature of information workers' professional activity requires them to continuously learn because they must be able to rapidly adapt to the changing problem domains stimulated by their environment. Issues related to human–computer systems are within the focus of ergonomics studies, whereas a combination of activity, for example, training and professional work both performed with computer-based systems, is still hardly explored.

The processes of continuous learning are unique due to their close connection with practical work and the influence of professional activity on the goals and tasks of training. The learners participating in these processes are different from those in a school or university. They have different knowledge, skills, experience, and working habits. They develop their own preferable learning strategies, so they need to be addressed individually. Numerous learning and training (LT) solutions, both traditional and computer-based, ignore the importance of the individual adaptation of the learning content and its tight connection with supporting materials for daily work. Therefore, a general methodology to analyze information processing and learning activity is needed that can support the description of both in a holistic model.

To better understand the needs, goals, and activities of LT and the tasks of information workers, we will present a case study with technical support agents (TSA) in Section 15.2 to illustrate the challenges for research that are not yet answered by technological support. TSAs should have good communication skills; be able to resolve technical problems; know current products and services, customer bases, corporate policies, and procedures; and know how to use support technology. They work under time pressure and bear responsibility not only for the specific solution they offer, but also for the company's image. Therefore, TSAs must know how to understand, anticipate, and meet the customers' needs. All of these requirements make TSAs a perfect object of study and a source for the modeling of technology systems to support both professional information processing and decision making and integrated LT processes.

We first discuss a case study related to the work and training of a TSA. This is followed by an analysis of critical findings that reveals the need for the integration of LT with information support into a professional activity of the TSA. Next, a study of on-the-job training needs is presented with a discussion on the role of instructional theories in shaping LT activity. Due to certain limitations, traditional theories are unable to support the holistic view of the information processing and learning; thus, a new integral approach based on the systemic-structural methodology is introduced in the next section (Section 15.2). Application of this methodology is demonstrated in Section 15.6, which presents a meta-model of a learning technology system based on the triadic schema of activity. In the conclusion, the results are summarized and some directions for future research are outlined.

15.2 Pragmatic Case Study of Technical Support Agents

The following brief case study of TSAs illustrates pragmatic issues that reveal how practical ergonomic and learning rationales fall in place during daily work activity with information systems (ISs). Unfortunately, ergonomic knowledge and methods are still not explored much by corporations, though applying ergonomic principles would greatly maximize the corporate productivity by reducing workers' fatigue and discomfort.

Sarah is an up-front hotline agent who receives and answers incoming calls from external customers of the organization. The job of a hotline agent is stressful, and every agent needs to rest or retire after about 2–3 years of service at the front. Therefore, there is a continuous change of employees, which makes training and learning desirable for these agents.

Every customer has different queries, and hence, each call is different. Therefore, Sarah needs to be acquainted with the latest products and the latest updates. She must be able to categorize product updates from novelties and must know the present and previous versions as well as product features. Moreover, she needs to know the forthcoming, current, and past services and campaigns that the organization has held and is offering and must be able to advise customers' vis-à-vis competition. Since it is impossible to know everything, she must know where to search, how to search, what to search, and what to do in complex situations. Yet, it is not only her knowledge that counts. In view of the diverse issues, she must not only be able to link, connect, and resolve new complexities with the given information meaningfully for each situation, but also have the necessary skills when it comes to practical product functionalities and malfunctions. This adaptive behavior makes her knowledge applicable so that she will be able to understand, classify, and structure inquiries, as well as repair technical devices over the phone successfully.

Sarah works in a modern call center, where she accesses multiple back-end systems and software components like customer relationship management (CRM) systems, enterprise resource planning systems, and the like through one Web front end. In the course of her work, Sarah needs to verify assumptions before she responds to the customer. This verification process must be quick and efficient, so that she needs electronically available information, which she needs to know how to navigate through and operate. Although Sarah works with an IS, defined and interdependent or connected pieces of information are not available for all processes and their tasks in the context of specific problems in the system. Therefore, some pieces of information are predefined and available in the human–computer systems, while others are available only in parts. Some may be partially outdated, while others may not even be available at all. Although some processes and subprocesses and their tasks are defined, numerous processes are neither defined nor have a need to be analyzed.

Apart from face-to-face (F2F) training, Sarah is obliged to read corporate "hot news" every morning. She also does voluntary LT and three compulsory weekly sessions of a minimum of 15 minutes of online LT courses. To understand this myriad of information and gain the urge to continuously learn and to be able to respond correctly to every issue in any situation within milliseconds, some TSAs create their own alphabetically ordered "FAQ card systems," which overcome system constraints.

Sarah's supervisor, Jennifer, believes that she would be able to improve the department's performance if she knew more about each TSA's individual knowledge, skills, abilities, and limitations to offer more subject-relevant and timely training sessions for each one of them. However, the only information available so far is related to the agents' test results from F2F and online training and statistics of their performance. Therefore, Jennifer is unable to reveal the reasons for the difficulties and impasses of TSAs, identify information that should be made available, and formulate individual training plans. She would welcome a multifunctional system that contains and interconnects all information, thus saving time on searching through large documents or whole LT courses, and allows customization and shaping to individual needs. So, Jennifer needs an intelligent system that allows her to access situational fragments of information together with dependent fragments, which customizes information to make it meaningful by adapting to individual LT needs, and allows her to detect where she may need to retrain agents individually.

15.3 Critical Analysis of the Case Study

The duties of TSAs include gathering information about customers and using it to serve the company better. Cognizant answers are not enough any longer because being able to merely solve problems is no longer sufficient. Today, TSAs need to proactively exercise CRM. TSAs are thus a strategic resource to the organization in retaining customers and building loyalty to increase profits. TSAs must be trained to create meaning and resolve inconsistencies by assimilating new knowledge or adapting it to the existing knowledge. In this sense, knowledge and meaning are not fixed, but they are constructed by the individual within the context of meaningful learning. In the work process, meaningful learning is also a result of practical problem solving. Moreover, an active and engaged problem-based learning process continuously forms corporate knowledge.

Corporate LT is subject to cognitive work processes and their activity; this means that LT is competitive on memory, attention, focus, and priorities. Thus, LT is dynamic, yet granular, and therefore, it is situation specific and authentic. Consequently, LT evokes new adaptive behavior and LT content that needs to be adaptively available and accessible during work processes by human beings and ISs. In return, new and changing information and processes stimulate new LT needs. In other words, different sources of information have a competitive effect on the work processes, evolve in real time, and trigger LT on issues such as training new workforces and updating employees on new products, services, promotions, and tools.

Figure 15.1 presents the modeling of the context of technical processes of our analysis. Two systems support a TSA's activity: IS by information support and an LT system (LTS) by arranging the processes of learning and training. Yet, we are challenged to investigate further on the cognitive actions, continuous time events, and the like, because IS and LTS should avoid posing cognitive overload by eliminating irrelevant information that decreases human performance, which in turn affects the work process. Therefore, both systems should facilitate the improvement of human work activity as well as LT performance. Figure 15.1 shows that although LTS and IS share content and knowledge, their behaviors vary with regard to the processes, context, and sequences. Figure 15.2 is a composition of the context shown in Figure 15.1 and provides us a simplified picture from the point of an incoming phone call separated into swim lanes like the automated monitoring system, the human caller, and the subpool of the TSA. Figure 15.2 illustrates connected processes, activity, tasks, events (message, rule-based, timer, and multiple), and decision-making gateways (inclusive, exclusive, complex, and parallel) related to the TSA's performance.

Hence, the changing nature of calls requires TSAs to learn and rapidly adapt to their environment during their work activity. To one individual TSA who is "on call," this may mean that human-computer systems ought to be adaptive to human learning processes and behaviors, while the work activity needs of another individual agent who has been set apart for training may be a human-computer system to detect learning deficiencies for situated LT sessions. Notably, both issues can be put in place provided the organization knows the precise and detailed work activity in their contexts and depths. These aspects, however, are largely unknown, and research in this field is challenging due to the time it takes, the lack of experts who are able to analyze the activity systems, and the fast-moving pace of today's constantly changing economy. Besides, although corporations realize and acknowledge the need for reactive systems to respond adaptively to human activity and better train the staff, they are not willing to invest time and effort to improve machine and human activity systems. Hence, it is now obvious that LT does not take place only in vocational LT sessions, but also during work activity, which is a stimulus for adaptive behavior and learning from decision-making, structured and unstructured processes, activity, and consequent tasks. As we have seen in our practical study, when the organization has not attempted to thoroughly analyze the structured and unstructured activity, the result is no surprise: TSAs suffer from cognitive overload leading to work dissatisfaction. Therefore, a much better understanding of human work activity processes is a critical prerequisite for growth and competitiveness despite today's fast-moving pace.

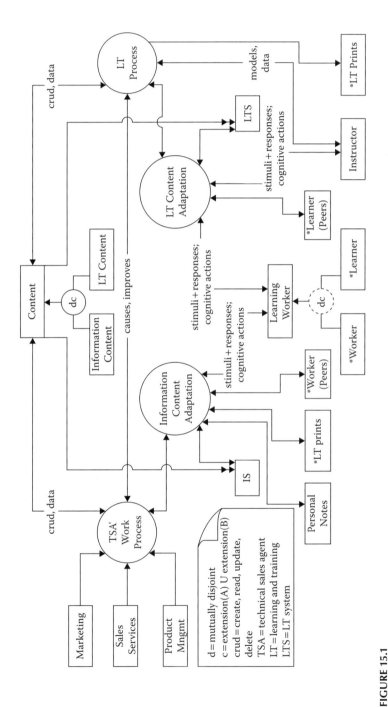

FIGURE 15.1
Context diagram of information processing. (From von Brevern, H., and K. Synytsya. 2006. *Educ Technol Soc* 9(3):100–11. With permission.)

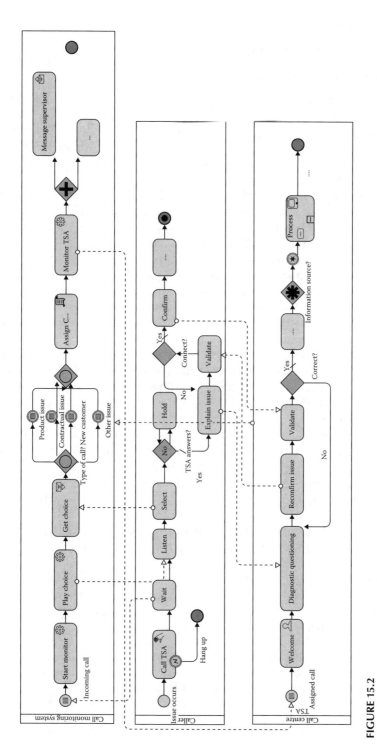

FIGURE 15.2
Processes and tasks.

15.4 Learning and Training

15.4.1 General Methodological Issues

A detailed task analysis of TSAs' activity serves as a cornerstone not only for increasing the efficiency of clients' requests processing, but also for arranging training and learning processes tailored to the individual cognitive needs, that is, specific knowledge and skills to be taught. Although corporate training includes multiple formats and approaches, we will further focus on the specific features of on-the-job training.

On-the-job training is closely related to the professional activity of the learner and addresses issues that were already faced or may be encountered in the near future, thus demonstrating a strong connection among professional activity, information processing, and learning. Both on-the-job training, especially just-in-time training, and performance support focus on filling the gaps in knowledge and skills not covered by traditional education and pretraining. This process is similar to that of repair or re-engineering in contrast to the initial construction. However, most of the instructional theories are dealing with the "construction" process, that is, carefully building knowledge and skills on an empty spot. Thus, all learners are treated the same way, independently of their previous knowledge and experience. Though the time necessary for a particular learner to master new knowledge varies depending on individual abilities, the instruction is usually considered a rather straight path toward learning goals. Classical instructional theories in most cases reflect a "schooling" approach, though recent attention to learner-centered instructional design makes it more appropriate for adults, enabling self-control, setting individual objectives, and supporting reflection. Adult-learning (andragogy) programs accommodate self-directedness and the expectation to take responsibility for decision-making. "In practical terms, andragogy means that instruction for adults needs to focus more on the process and less on the content being taught. Strategies such as case studies, role playing, simulations, and self-evaluation are most useful. Instructors adopt a role of facilitator or resource rather than lecturer or grader" (Kearsley 2007).

As does learning design, andragogy assumes that (1) adults need to know the reason for their learning; (2) they need to learn experientially; (3) they approach learning as problem-solving; and (4) their learning effect is best when their object of learning is of immediate value.

> Andragogy assumes that the point at which an individual achieves a self-concept of essential self-direction is the point at which he or she psychologically becomes an adult. A very critical thing happens when this occurs: the individual develops a deep psychological need to be perceived by others as being self-directing. Thus, when he or she finds himself in a situation in which he is not allowed to be self-directing, he

experiences a tension between that situation and his self-concept. His or her reaction is bound to be tainted with resentment and resistance.

(Knowles 1978, p. 56)

However, on-the-job and just-in-time training require more than that. Relevant training must deal with individuals having partially forgotten, incomplete, and obsolete knowledge. This task is similar to that of reconstruction, restoration, repair, or "building upon"; thus, the instruction plan should address a specific issue in the context of the knowledge structure. So the main questions to ask are "when," "what," and "how" to teach.

15.4.1.1 When to Teach

Training may be triggered by external events, such as appearance of a new product or service in the company, or a call for a scheduled retraining. In this case, the initiation of learning from outside thus forces a TSA (a learner) to switch from solving professional tasks to another activity arranged for him or her. Therefore, a new activity starts with a goal formation (Bedny and Karwowski 2003), which has to be supported by a training system (computer-based or traditional), as training goals are identified by third parties independently of an individual and his or her current activity. To participate in training actively and consciously, learners should be motivated to form their own goals by internalizing the proposed training objectives and transforming them into their own learning objectives. Currently, most corporate training is content-oriented, which means that the role of an individual, and his or her professional tasks related to the individual knowledge taught are ignored. However, to understand the relevance of training and to be able to fully benefit from it, one has to put learning objectives into a context of one's professional duties. The way these issues are addressed may be different depending on the instructional approach selected for the training.

Internally triggered training corresponds to the individual need of a particular person who has to be trained in relation to a specific task. The need may be either identified by the person himself or herself or by an observer (a supervisor or software monitoring personnel capable of identifying shortcomings and impasses in performing tasks). In both cases, training goals are formed dynamically in coordination with a trainee and the first two stages of the activity, goal formation and orientation, are tightly connected.

15.4.1.2 What to Teach

Training content is determined by the training goals, motives, and requirements, which are part of the object of activity, and may be more general in the case of externally scheduled training or more specific, focusing on a product or a task, which a TSA was unable to perform. Whatever the immediate learning objective, it should be considered in the context of overall performance enhancement based on relevant knowledge and skills; thus, filling the gaps should be aligned with support of the general knowledge structure. The actual volume of learning content and depth of training is determined by several factors, including resource limitations, corporate rules, and importance of the topic for professional enhancement.

Here, we would like to stress the difference between the supporting information that is used in TSAs' professional activity, such as a list of current promotions or description of phone models, and learning content. Factual information is useful to avoid unnecessary memorization of details. If properly structured, it facilitates processing requests provided the TSA was taught relevant search and processing procedures. However, the use of supporting information alone does not facilitate building an experience as ad-hoc knowledge or heuristics is even less relevant for systematic understanding of the concepts, relations, and operations of the task. The purpose of learning content is to improve the TSAs' performance through relevant knowledge and skills acquisition or construction. Supporting materials may provide visualization of relationships between fragments of information, but they will not require any circumstantial user activity aimed at knowledge or skills formation. Identification of tasks and information involved in the processing and decision making helps to link supporting information to learning activity.

15.4.1.3 *How to Teach*

The essence and sequence of learning activity are determined by the chosen instructional strategy. The strategy may be considered a specific didactic method for arranging learning events based on some instructional theory. It ensures consistency of training and provides guidance for learning content design. Instructional theories offer sets of recommendations for attaining instructional goals through a sequence of instructional events that are constructed based on some model of cognitive process. Both a teacher and a learner participate in the event, one of them playing an active role. We further consider the application of two theories of instruction, Algo-Heuristic Theory (AHT; Landa 1987) and Components Display Theory (CDT; Merrill 1987), while training a TSA to perform a task according to some procedure.

15.4.2 Application of Algo-Heuristic Theory: The "What" Issue

AHT specifically addresses the cognitive processes of problem solving and decision making and offers instructional methods to teach algorithmic processes at a rather general level facilitating transfer of acquired knowledge and skills to new problems. In other words, the main advantage of the proposed instructional approach is its ability to help a learner develop generalized problem-solving techniques. "The AHT proposes that, in order to be able to effectively teach students how to apply knowledge and solve problems, one should know the composition and structure of cognitive operations involved in the process of knowledge application" (Landa 1987, p. 115). To create relevant training content, an instructional designer, together with a domain expert, analyze the tasks performed by the TSA in advance and identify the typical types of behavior algorithms to be taught. This is exactly the point at which the results of a systemic analysis are needed. By monitoring experts' behavior and extracting explanations about the performance, one can reconstruct the desired algorithm. This is a rather complicated task, as one needs to address a proper level of details and generalization, which may be different for different learners. Therefore, training should address both levels of skills: the motor, sensory, and mental skills important for efficiency of execution of well-defined known tasks (automated skills), and the meta-skills responsible for general problem-solving capabilities. The TSA is given initial information, which may be incomplete, imprecise, poorly formulated, and even contradictory. The TSA's objective is to extract additional information from the client to obtain a description of the case, which is complete and precise enough to deal with. Prescriptive procedures recommended by the company are easy to learn, however, understanding of the process, underlying concepts and algorithm is extremely helpful when new products and services appear, or when a novice TSA faces a new task. Unfortunately

> More complex tasks are nonalgorithmic. The latter class...should be divided into three subgroups: semialgorithmic, semiheuristic, and heuristic. The distinction between the types of tasks is relative, not absolute. We recommend the following major criteria for classification of tasks: (a) indeterminacy of initial data; (b) indeterminacy of goal of task; (c) existence of redundant and unnecessary data for task performance; (d) contradictions in task conditions, and complexity or difficulty of task; (e) time restrictions in task performance; (f) specifics of instructions, and their ability to describe adequate performance and restrictions; (g) adequacy of subject's past experience for task requirements.
>
> **(Bedny and Karwowski 2007, p. 62)**

For instance, if we consider a classification task, the algorithm consists of recursive application of procedure that selects a feature of an object (an attribute) and checks its value. Classes are characterized by a set of values for some features; thus, an object belongs to some class if its attributes have values that are characteristic for that class. Other attributes may be undefined. According to the proposed approach, the learner should: (1) understand the concept of classification and class, that is, be aware of the proposed structure of the domain; (2) understand the role of features in the classification process, that is, understand how knowledge of some values determines a set of possible classes an object can belong to; (3) understand that some features may be related, that is, on knowing the value of one feature one can identify the value of another, or limit the number of variants; (4) know the values specific for each class; (5) be able to identify values of features; (6) know the selection criteria for the features to be checked; and (7) be able to apply the complete algorithm.

AHT focuses on the algorithms, that is, *what* to teach, addressing both specific and generalized procedures. However, the way they are taught depends on the teacher's preference, as the theory accommodates both explanatory and discovery approaches to instruction. Most important for the AHT is knowledge of the "composition and structure of the internal mechanisms of thought" (Landa 1987, p.115), and instructional events should be sequenced to support the development of these mechanisms in learners.

15.4.3 Application of Components Display Theory: The "How" Issue

CDT focuses on the instruction (*how*) and offers guidelines for instructional design depending on the type of performance to be taught. Due to its integrative roots, CDT makes it an excellent choice for instructional strategies. There are three types of instructional events: presentation of learning content, practice (considered to be a monitored and guided performance on learning tasks with instructional feedback), and performance that serves as an assessment of learning results. For each type of learning objective, the theory prescribes a way to design a sequence of typical instructional events, such as "demonstrate a general case of ..." or "ask to provide an example of ..." For instance, in the case of the objective type "teach (a use of) a procedure," the sequence of instructional steps may look as follows: (Step 0) motivation: communicating learning objectives and setting the context of the task; (Step 1) presentation of the general description of the procedure whose purpose is to describe principles and structure of the task performed, and the expected result of the procedure. In case of possible branching, the decision points should be identified, as well as each subtask input–output description provided; (Step 2) for all nontrivial subtasks, continue presentation of their detailed description and explanations; (Step 3) following the explanation of each subtask, its execution is practiced through a sequence of subtasks with similar parameters; (Step 4) individual subtasks are combined into a complete procedure, which is practiced with a declining level of support; and (Step 5) the complete procedure is performed without any external support.

Presentation of material is subject to the basic requirement that it must facilitate each core element of an action that is orientation, execution, and control. With regard to multimedia learning and presentation of information, material presented at computer interfaces is often done in such a way that "the processing demands evoked by the learning task may exceed processing capacity of the cognitive system" (Mayer and Moreno 2003, p. 45), "the learner's visual attention is split between viewing the attention and reading the on-screen text" (Mayer and Moreno 2003, pp. 45–46), and so on. In the first case, the learner is confronted with cognitive overload, while in the second example the learner's attention is split. In both cases, the learner's orientation is disturbed from the beginning already. At its most basic, material and content ought to be presented in such a manner that the most forefront presentation does not disturb any of the core elements of an action. The split-attention principle, spatial contiguity principle, temporal contiguity principle, modality principle, redundancy principle, and coherence principle as addressed by Mayer and Moreno (2003; Moreno and Mayer 2000), and Clark and Mayer (2002) provide a guideline for instructors of how to present multimedia material. In particular, the temporal contiguity and the spatial contiguity principles could be translated into computer-based instructional tasks, for example, "spatial contiguity effect: better transfer when printed words are placed near corresponding parts of graphics… Temporal contiguity effect: better transfer when corresponding animation and narration are presented simultaneously rather than successively" (Mayer and Moreno 2003, p. 46).

With regard to this, the instructional events at each step are selected from the recommended charts that ensure correspondence between the specific learning objectives (such as knowledge acquisition or mastering a skill, dealing with concepts or procedures, and the like), the type of instructional event, and the presentation of an object to be learned.

The approach suggests some guidelines to facilitate memorization and retention of particular elements, such as use of "alternate representations," which may be supported by multimedia. Mastering a procedure starts with a guided execution. The external guidance gradually fades leaving a place to a self-guidance, such as following a plan or a stepwise execution of an algorithm. Finally a learner has to perform the procedure automatically without any support. Instructional control is realized through stepwise guidance and feedback, which can be generated or produced in advance for selected tasks. Immediate feedback may later be substituted by a delayed feedback provided upon completion of some meaningful stage of the procedure execution.

15.4.4 Summary

As one can see, LT theories consider knowledge and skills acquisition in isolation from any other activity, and tend to focus on specific aspects of cognition. Instruction is largely planned in advance and does not take into account specific feedback from the learner—the real state of his or her knowledge is neglected. Finally, the origin of learning content and its structure is not discussed, leaving the design of computer-based instructions to the skills of the author. On the other hand, the need for LT processes tightly coupled with professional activity is obvious. To make LT an integral part of the professional activity from a systemic viewpoint, one needs to apply a sound methodology to analyze cognitive processes taking place while processing and learning information.

15.5 Systemic-Structural Methodology of Analysis

Our analysis of TSAs' activity shows that work processes both originate from, and are equally embedded into, a diversity of human–computer systems, which often lack up-to-date corporate information. Unfortunately, ISs are often just another system in a corporate landscape instead of being the central point of contact, the unique information provider, and also largely fail to respond adaptively to subjective working tasks of TSAs. Moreover, the continuously changing nature of information about new, altered, or obsolete products, services, and promotions within a corporate domain and pressing competition urge TSAs to work proficiently, efficiently, and be more optimized. Hence, such a challenging context requires TSA to continuously pursue LT, but also means that human–computer systems need to change from today's conventional perspective of stimulus and response behaviors to adaptive, cognitive, and reactive behaviors. In other words, we need IS that can grasp human business and learning objectives and tasks and that are capable of responding adaptively within the domain of human–computer systems. The most crucial issues are thus how cognitive human behaviors could be embedded into the boundaries of a structural, systemic, and psychologically valid methodology that allows the use of human–computer systems between humans, one that allows humans to perform activity using human–computer systems, one that achieves results by using human–computer systems, and one that allows human self-regulation within human–computer systems. Notably, this is particularly true for LT. In a previous publication (von Brevern and Synytsya 2005), we argued the need for a holistic viewpoint. In LT, network systems in which both the learner and instructor may reside in different locations no longer suffice because LT cannot be separated any longer from work activity (including its human–computer systems) as illustrated by our case study and therefore need to integrate mediating capabilities (von Brevern 2005). LT impacts an individual's work performance, which in return has an important effect on corporate efficiency and knowledge. Importantly, LT evokes new learning content that needs to be shared by and be made available in IS.

Although comprehensive research is needed to know the requirements for such systems, the systemic-structural activity theory (SSAT; Bedny and Meister 1997) provides a sound methodology and systemic-structural framework that incorporates subjects, mediating tools, the object of activity, goal, result, feed-forward and feedback loops, and methods of experimental and analytical analysis. "In activity theory [AT], the subject is always understood to be socially constituted individual, who is in possession of internal, psychological tools acquired during ontogeny" (Bedny and Karwowski 2007, p. 40). "The two primary types of activity are 'object-oriented' and 'subject-oriented' activity. 'Object-oriented' activity is performed by a subject using tools on a material object, where the subject of activity is the individual or group of individuals engaged in that activity. 'Subject-oriented' activity, also known as social interaction … involves two or more subjects" (Bedny and Harris 2005). The context (e.g., business, corporate, system, local, and international norms and standards) influences the interaction of subject-oriented activity with objects. Object-orientation involves a higher weighting of psychological motivation and behavior, primarily initiated by an individual subject in an internal stimulated act of learning, whereas subject-orientated learning is mediated by group interaction and situated in social learning activity. To determine an object, both goal(s) and task(s) must always be known. Also, the tools that accomplish the integral role mediate between the subject and the object or between subjects. In contrast, a computer can create artificial objects toward which tool-mediated actions may be directed. Under such circumstances, the human–computer mediates human interaction with the external world through the creation of artificial objects and tools. Without knowing a specific task, we cannot precisely determine "the meaningfulness of the tool" (von Brevern and Synytsya 2005, p. 746). Notably, in SSAT, the notion of "objective" relates to the goal rather than to the object of activity so that an object is not synchronous with an objective. A tool is a mediating symbol to the subject and between a subject's work performances. In the course of work activity, the subject remembers how to manipulate the tool. Mediating mechanisms are effectuated through the process of internalization (Bedny and Meister 1997). This process principally transforms external activity on a physical object into an inner plane by mental operations on symbols. Bedny and Karwowski (2007, p. 41) argue that SSAT should respond to the following requirements: "(1) Psychological units of analysis should be expressed in such a way that it will permit identification in real work processes. (2) The qualitative description of an activity should be combined with quantitative measures for the prognosis of the efficiency of the performance. (3) The description of activity should be performed in such a way as to allow us to make an inference or prediction of how we can increase the efficiency of performance." In SSAT, these requirements are satisfied by actions and their microstructure (operation, function block), which are the units of analysis. The composition between activity, task, and action is illustrated in Figure 15.3.

The daily work activity of a TSA that encapsulate LT require us to comprehend and study "the structure of activity, which is a system of actions involved in task performance that are logically organized in time and

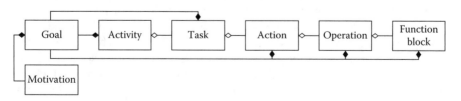

FIGURE 15.3
Units of analysis (activity, task, action, operation, and function block) in the context of goal-oriented activity.

space" (Bedny and Meister 1997, p. 31). Consequently, the critical purpose of "ergonomic design of such tasks is to reduce the degree of objective and subjective uncertainty in problem solving" (Bedny and Karwowski 2007, p. 62). Some examples of tasks during a TSA's work activity revealed in a case study are (1) the task to understand ill-structured problems, (2) the task to search for specific information within time constraints, (3) the task to classify information, (4) the task to master skills consciously, and (5) task switching (von Brevern 2005). This raises the need for task decomposition (cf. example in von Brevern and Synytsya 2006), which is essential to specify and design the subject domain of human–computer LTS. The term "subject domain" is "the part of the world that the messages received and sent by the system are about" (Wieringa 2003, p. 16). SSAT provides us with formal methods of analysis, including morpholo gical, cognitive, parametric, and functional, which leads to exploring concise requirements identification abstracted from formal aspects of task analysis and algorithmic behaviors to thus circumference and decompose the subject domain of human–computer systems.

15.6 Conscious Goal-Oriented Systems for Learning and Training

Albeit praxis proves that LT is an inseparable and holistic part of daily work activity requiring us to rethink human–computer system design for which no single conjecture has been accepted yet as a paradigm of LT for psychology for the use in the human–computer systems. Moreover, contradictory LT methods and theories challenge us as to how to design the holistic human–computer LTS on one hand and how to integrate heuristic and algorithmic behaviors of a subject into the systemic and structural activity system described by SSAT on the other hand. As demonstrated and discussed in Section 15.4.2, AHT provides us with a noncontradictory paradigm that can be encapsulated into SSAT. Yet, neither does a holistic LT human–computer system exist that can cope with the needs as derived from our case study, nor do we know of any human–computer LT framework that could incorporate holistic and goal-oriented activity as described by SSAT or AT.

Since the subject in AT is the principal object of study, cognition and along with it also situation awareness become crucial elements. Thus, cognition is not merely a process of actions as shown in Figure 15.2, but also "a system of actions that combine in a particular way and emerge as building blocks of holistic, cognitive activity. An individual does not passively receive information, but also actively selects information from the environment, using different actions or operations" (Bedny et al. 2000, p. 181). This dynamic process and system require us to include and interpret the context.

In interpreting a situation, we extract from it a certain content which reflects aspects of that situation, and then fixate this content in the mind in sign form. The subject extracts from the objectively set situation those aspects and properties which are connected to a particular system of meanings and senses actualized at that moment in time. As a result, a presented objectively situation can be associated with several methods of interpretation. The character of interpretation and comprehension depends not only on cognitive aspects of activity, but also on the significance of the situation and motivation which affects the specifics of activity goal formation.

(Bedny and Karwowski 2004, p. 14)

Cognitive and situational tasks of a subject within an activity system reflect the subjects' heuristic behaviors, which are thus a subject's algorithmic processes. "Algorithmic tasks completely define the rules and logic of actions to be performed and guarantee successful performance if the subject follows the prescribed instructions" (Bedny and Karwowski 2007, pp. 61–62). In LT, this requires us to effectively develop *"specific* cognitive abilities and problem-solving skills in students, through teaching algorithms and algorithmic processes, is only one *of* the tasks of the AHT. Another task—and not less important—is to develop *most general, content-independent,* cognitive abilities. The AHT provides tools and techniques to accomplish this" (Landa 1987, p. 123). In addition to such cognizance, the significant tasks of a subject within an activity system include a complex system of self-regulative actions, shown in simplified form in Figure 15.4.

However, from the perspective of SSAT, the very underlying root of tasks or actions is the goal of activity. "In [AT] a goal is a conscious mental representation of humans' own activity in conjunction with a motive. Goals are considered cognitive, informational components of activity. In contrast motives or motivation in general, are treated as energetic components of activity. The more intense the motive is, the greater the effort to reach the conscious goal" (Bedny and Karwowski 2007, p. 23). Hence, in SSAT, the goal is purposeful, significant, and conscious to the subject and is inseparable from the object of activity as shown in the triadic schema of the activity system of SSAT by Bedny and Karwowski (2007, p. 40).

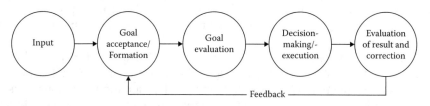

FIGURE 15.4
Simplified model of action as a one-loop system. (From Bedny, G., and W. Karwowski. 2007. *A Systemic-Structural Theory of Activity: Applications to Human Performance and Work Design.* Boca Raton, FL: Taylor & Francis. p. 41. With permission.)

Leont'ev (1978) argued that another

> important aspect of the process of goal formation consists in the concret-
> ization of the goal, in isolating the conditions of its achievement ... Every
> purpose, even one like the 'reaching of point N,' is objectively accom-
> plished in a certain objective situation. Of course, for the consciousness
> of the subject, the goal may appear in the abstraction of this situation,
> but his or her action cannot be abstracted from it. Therefore, despite its
> intentional aspect (what must be achieved), the action also has its opera-
> tional aspect (how, by what means this can be achieved), which is deter-
> mined not by the goal in itself, but by the objective object conditions of
> its achievement.

In view of a holistic LT activity system, we have illustrated and discussed a possible triadic schema in our previous publication (2005, pp. 747–8; 2006, p. 103).

However, when it comes to the concrete design of a human–computer LTS, the triadic schema raises research issues of how to formally translate infor-mal events (feedback, feed-forward loops, instructional events, and learn-ing events), entities (objects, tools, goal, object of activity, and result), and dynamic processes on a mental plane into identifiable human–computer LTS that make up part of the subject domain.

Figure 15.3 is the result of a synthesis of the triadic schema of a holistic LTS based on SSAT, learning and instructional events based on the AHT, and some tasks from the TSAs' work activity translated into a human–computer Unified Modeling Language (UML) meta framework (Object Management Group 2009). Figure 15.3 reflects the adaptive negotiation, learning, and instructional "strategy" behaviors of the instructor's role and the respective learner's role, as we introduced (von Brevern 2004) based on the strategy and composite patterns by Gamma et al. (1994). These patterns allow behaviors derived from AHT. Giest and Lompscher (2003) talk about the zone of actual performance and argue that instruction has to be organized and structured in such a way that learners become the subjects of their own activity, that is, they must be conscious of the goals, tasks, and actions.

Figure 15.5 is a further granularity of the task of Figure 15.6, while it illus-trates the taxonomy of the task of SSAT that is synchronous with the tax-onomy of the task according to AHT. This schema therefore allows studying the action that is critical to a goal-oriented activity as well as in view of the task analysis.

Therefore, the method of LT does *not* exclude the instructor from LT, but challenges the instructor on goal-oriented algorithmic guidance to operate on the environment and on objects to make a learner and peers consciously discover hidden associations and relationships (Bedny and Meister 1997). Goal-oriented learner's guidance by the instructor is critical and is contrary to "trial-and-error" LT, non-goal-oriented self-study, or the like because of the critical aspect of individualization. A further requirement that needs to

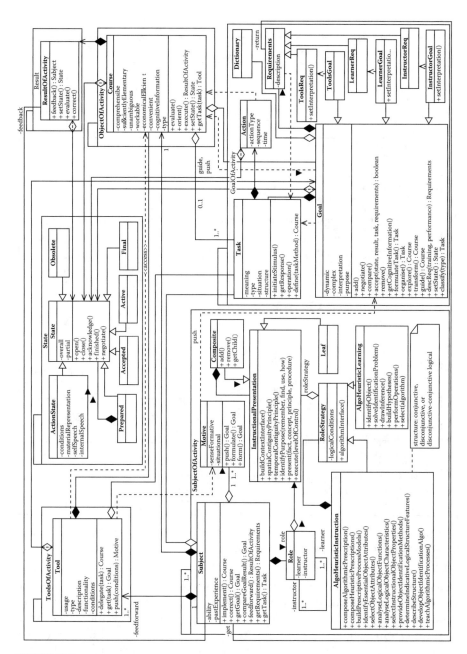

FIGURE 15.5
Task taxonomy of the systemic-structural activity theory and task implementation.

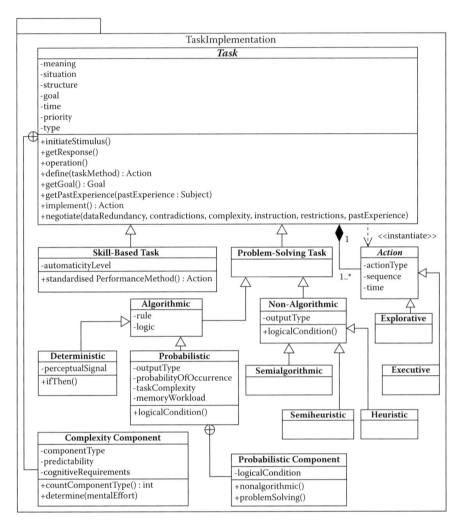

FIGURE 15.6
UML model of the learning and training system.

be included alongside the "strategy" behaviors of the roles is the concept of reference points. Gal'perin's work (1969) on the stages in the development of mental acts demonstrates that learning includes various psychic stages in the process from "understanding" to "mastery." According to Gal'perin (1969), understanding is on a mental plane where a learner is able to mentally trace the relationships between concepts or the conditions of tasks, while mastery requires understanding at a higher level. A challenging issue is the subjective perception of the objective object of activity. This not only implies the need to explore how to detect the discrepancy between subjective perceptions versus the objective object of activity and how LTS can help an instructor to detect and (self-)correct the subjective perceptions, but also how instruction should actively shape learning. An interesting approach also suitable for adult learning has been posited by Gal'perin's (1969) reference points.

The principle UML packages—subject, object (of activity), goal, result, state, and tools—encapsulate the major components of SSAT. Although Figure 15.6 does not explicitly visualize, but rather encapsulates, the concept of self-regulation of SSAT, it demonstrates that motives push a subject to reach the goal, which in return are the cause of tasks; sources of requirements are tools and subjects, and that the abstract class goal and its derivates, task, and requirements are part of one package, that is, the goal. Bedny and Karwowski (2007, p. 23) argue that "in order to understand what is task or action, it is essential to understand the goal of activity. In cognitive psychology goal is considered a combination of cognitive and motivational components." However, in any human–computer system design, entities must be identifiable by the system to make part of the subject domain. Thus, the metamodel shown in Figure 15.6 explicitly implies that the goal, the motive, and the task are identifiable parts of the subject domain. These identifiable and conscious classes by the subject domain can be implemented in human–computer LTS design based on the concept of verbalization. Verbalization is crucial in the process of internationalization and in the course of analyzing activity systems. "Orienting the basis of action and control components of action can be performed in perceptual form based on audible speech or mental form. Mental actions also have their orienting basis of action, control, and executive parts. Orienting and control components of action can exist in perceptual, external verbal, and mental form" (Bedny and Meister 1997, p. 325). Furthermore, verbalization is critical in the course of performing actions, internalization, analyzing actions, and the like.

Consequently, if a human–computer LT is able to grasp dynamic goals and implement identifiable dynamic negotiation and task processes, LT will result in a product. "Product is a result of the transformation of an object of activity. Product may be material, spiritual, aesthetic, and so on. Indeed, the subjects themselves may be the objects of change as a result of activity. This is why in AT; instead of the term 'product' one may find the notion of result. The result does not always match the goal of activity" (Bedny and Karwowski 2007, p. 24).

The case study of TSA activity demonstrated the need for various forms of LT to support the efficient performance of agents. Initial training prepares them for solving typical problems, whereas the constantly changing professional environment requires new skills, procedures, and knowledge to be acquired in context of existing professional duties. To create meaningful on-the-job LT processes harmonized with professional duties, a holistic view on these activity is essential. Information processing taking place while working with IS and LTS has certain similarities in algorithms and data, and thus a combination of them may increase the efficiency of agent behavior. Hence, we applied SSAT and looked at the TSAs work through the prism of solid methodology, which suggests a stepwise increase in the level of details when describing mental processes related to problem solving and knowledge or skills acquisition.

Following SSAT guidelines, we focused on the task level as the one that should be addressed by the LT processes and further emphasized the goals as a core mechanism for leading the agent's self-regulated activity. To further elaborate the triadic schema, we had to focus on certain instruction theories that determine the content and context of interactions between agents as learners and those who arrange training, whether they are subjects (tutors, supervisors or peers) or LTSs containing learning content and instructional algorithms. AHT was chosen for its relevance to the learning of procedural knowledge and skills that are essential for TSAs. The suggested approach facilitates a focus on the essence of learning and training and on determining specific tasks and methods, whereas the assignment of instructional tasks to a human tutor or LTS may be done at later stages.

The taxonomy of the TSAs' tasks and variety of specific tasks within each node of the taxonomy determine the overall training program for the agents. Knowing critical and typical tasks, one can create training in a way the learned skills can be developed through the professional performance on the tasks of the same type. The agents may benefit from scheduling the training in such a way that most frequent tasks are addressed first, enabling construction of automatic procedures in the course of work. Automatic procedures support an agent's confidence in his or her skills and enable the agent's cognition for acquiring new skills and mastering more complex tasks. A systemic approach in this case keeps the training scheduled in small portion from disintegration into unconnected chunks by supporting a framework for mastering knowledge by the agents. The suggested analysis of the TSA's tasks demonstrated a value of the SSAT approach for the design of the training system in support of the daily work. However, the results of this paper could be considered only a sketch for the LTS implementation.

15.7 Conclusion and Future Work

This study elaborates on the works of von Brevern and Synytsya (2005, 2006), where applications of the SSAT to the holistic modeling of the LTS were first introduced. This demonstrates how abstract notions of subjects, tools, tasks, and goals may be filled with specific meaning to implement certain instructional theories like the AHT. AHT allows the development of algorithms at the level of elementary operations and the use of prescriptions and provides structured and algorithmic means for teaching. The results presented in this chapter are just the first steps toward the understanding of how LT processes may be integrated into the professional activity and what kind of mediating tools might be necessary for this purpose.

A more in-depth study of tasks, events, and goals is necessary to elaborate the metamodel proposed in this chapter and to identify the requirements to specific mediating tools as system components of the human–computer LTS. The metamodel presented in this chapter is a formal UML interpretation of the informal triadic schemata of SSAT. The metamodel serves as a basis for the development of the human–computer LTS and is reactive to identifiable human goal-oriented activity. On the basis of the concepts of verbalization and reference points, we posit that the goal, motives, and tasks that need to be part of and be integrated into a human–computer LTS. On the positive side, connections between subjects, goal, tasks, actions, mediating tools, the object of activity, and the result provide systematic means for task analysis and allow for studying actions, which underlie activity systems. On the negative side, more research and further decomposition are required in view of individual classes and their dependencies within the metamodel.

Specification of the subject domain as an abstract model that describes key concepts related to human–computer LTS in their interconnection is critical not only for fostering theoretical results, but also for practical implementation of LTS components. The holistic view emphasized in this chapter supports the creation and enhancement of adaptive, reactive, and cognitive systems that are developed in harmony with ergonomics principles. At the theoretical plane, SSAT provides the researcher with a valid framework, a methodology that is psychologically sound, systemic, and structural, and one that includes formal methods of analysis that are so much needed in system research. In the system design dimension, SSAT offers scalable solutions enabling a developer to focus on particular system features at a time.

Increasing the efficiency of organizational performance requires a systemic approach, taking into account both human factors and machine components and thus ensuring a holistic view on the processes within the human–computer systems. Although no single method would ensure a complete solution consistent with human activity and engineering, SSAT builds a backbone for the integration of findings in psychology, didactical strategies and experience, and engineering practice and trends, including object-based content construction and its interoperability among subsystems.

References

Bedny, G. Z., and S. R. Harris. 2005. The systemic-structural theory of activity: Applications to the study of human work. *Mind Cult Activ* 12(2):1–19.

Bedny, G. Z., and W. Karwowski. 2003. Functional analysis of orienting activity and study of human performance. In *Proceedings of the Ergonomics in the Digital Age*, 443–6. Seoul: The Ergonomics Society of Korea.

Bedny, G. Z., M. H. Seglin, and D. Meister. 2000. Activity theory: History, research and application. *Theor Issues Ergon Sci* 1(2):168–206.

Bedny, G., and D. Meister. 1997. *The Russian Theory of Activity: Current Applications to Design and Learning.* Mahwah, NJ: Lawrence Erlbaum.

Bedny, G., and W. Karwowski. 2004. Meaning and sense in activity theory and their role in the study of human performance. *Ergon Int J Ergon Hum Factors* 26(6):121–40.

Bedny, G., and W. Karwowski. 2007. *A Systemic-Structural Theory of Activity: Applications to Human Performance and Work Design.* Boca Raton, FL: Taylor & Francis.

Clark, R. C., and R. E. Mayer. 2002. *e-Learning and the Science of Instruction.* Hoboken, NJ: John Wiley.

Gal'perin, P. Y. 1969. Stages in the development of mental acts. In *A Handbook of Contemporary Soviet Psychology (1st ed.).* eds. M. Cole, and I. Maltzman, 249–73. New York: Basic Books, Inc.

Gamma, E., R. Helm, J. Vlissides, and R. Johnson. 1994. *Design Patterns: Elements of Reusable Object Oriented Software.* Reading, MA: Addison Wesley Longman.

Giest, H., and J. Lompscher. 2003. Formation of learning activity and theoretical thinking in science teaching. In *Vyogotsky's Educational Theory in Cultural Context,* eds. A. Kozulin, B. Gindis, V. Ageyev, and S. M. Miller, 267–88. Cambridge: Cambridge University Press.

Kearsley, G. 2007. *Andragogy (M. Knowles),* from http://tip.psychology.org/knowles. html (accessed April 10, 2010).

Knowles, M. S. 1978. *The Adult Learner: A Neglected Species.* Houston, TX: Gulf.

Landa, L. N. 1987. A fragment of a lesson based on the algo-heuristic theory of instruction. In *Instructional Theories in Action: Lessons Illustrating Selected Theories and Models,* ed. C. M. Reigeluth, 113–60. Hillsdale, NJ: Lawrence Erlbaum Associates.

Leont'ev, A. N. 1978. *Activity, Consciousness, and Personality.* Englewood Cliffs, NJ: Prentice Hall.

Mayer, R. E., and R. Moreno. 2003. Nine ways to reduce cognitive load in multimedia learning. *Educ Psychol* 38(1):43–52.

Merrill, M. D. 1987. Lessons illustrating selected theories and models. In *Instructional Theories in Action: Lessons Illustrating Selected Theories and Models,* ed. C. M. Reigeluth, 201–44. Hillsdale, NJ: Lawrence Erlbaum Associates.

Moreno, R., and R. E. Mayer. 2000. A learner-centered approach to multimedia explanations: Deriving instructional design principles from cognitive theory. *Interact Multimed Electron J Comput Enhanced Learn* 2(2) (online journal available at http://imej.wfu.edu/articles/2000/2/05/index.asp; accessed April 10, 2010).

Object Management Group. 2009. UML® Resource Page. http://www.uml.org (accessed 13th April, 2010).

von Brevern, H. 2004. Context aware e-learning objects and their types from the perspective of the object-oriented paradigm. In *Proceedings of the Fourth IEEE International Conference on Advanced Learning Technologies (ICALT 2004),* eds. C-K. Looi. Kinshuk, E. Sutinen, D. Sampson, I. Aedo, L. Uden, and E. Kähkönen, 681–3. Finland: IEEE Computer Society Press.

von Brevern, H. 2005. Support of cognitive processes for corporate learning and training. In *Proceedings of the UNESCO Workshop on Knowledge Society Building for Youth through 21st Century Technologies,* ed. K. Synytsya, 32–7. Kiev: International Research and Training Center.

von Brevern, H., and K. Synytsya. 2005. Systemic-structural theory of activity: A model for holistic learning technology systems. In *Proceedings of the Fifth IEEE International Conference on Advanced Learning Technologies (ICALT 2005)*, eds. P. Goodyear, D. G. Sampson, D. J.-T. Yang, Kinshuk, T. Okamoto, R. Hartley, and N.-S. Chen, 745–9. Kaohsiun, Taiwan: IEEE Computer Society Press.

von Brevern, H., and K. Synytsya. 2006. A systemic activity based approach for holistic learning & training systems. Special issue on next generation e-learning systems: Intelligent applications and smart design. *Educ Technol Soc* 9(3):100–11.

Wieringa, R. J. 2003. *Design Methods for Reactive Systems: Yourdon, Statemate, and the UML*. San Francisco, CA: Morgan Kaufmann.

16

Effort, Fatigue, Sleepiness, and Attention Networks Activity: A Functional Magnetic Resonance Imaging Study

T. Marek, M. Fafrowicz, K. Golonka, J. Mojsa-Kaja, H. Oginska,
K. Tucholska, E. Beldzik, A. Domagalik, W. Karwowski, and A. Urbanik

CONTENTS

16.1 Visual Perception and Attention

The majority of information that reaches an operator in the work process is of a visual character. The sense of sight is the most developed of all the senses. Consequently, there is a considerable interest in visual perception and the relationship between this perception and attention processes. A specific synergy between the attentional systems and the system controlling the movement of eyes is observed.

The visual reaction of attention shifting is usually connected with fixation shifting from one area or object to a new one. The shifting operation of the fixation point assures that the new object or area is perceived by the part of the retina called the "central retinal fovea," encompassing an area up to two angular degrees, which offers the best vision. This amounts to one ten-thousandth of the whole field of vision. In comparison with an image that reaches this part of the retina, images reaching the remaining areas have significantly lower visual resolution and acuity. The peripheral areas of the retina are adapted to detect movement or, more broadly speaking, sudden, rapid changes in the field of perception. By contrast, the analysis of complex information from the visual perception field is executed by the central retinal fovea. Because of a relatively small visual field, the analysis of an area greater than one ten-thousandth of the visual field requires moving the areas in the field vision which are being analyzed into the region of acute vision. The fixation shifting is executed eye movements. In classic literature devoted to the subject, the so-called oculomotor cycle, which consists of saccadic eye movement and fixation pause, is discussed (Rayner 1978). During eye movements (shifting of the fixation point), perception is switched off; otherwise, a blurred image would occur.

In the 1970s, fixation shifting was still identified with attention shifting. Eye fixation shifting was also identified with visual orienting reactions. Posner was the first scientist who postulated the separation of the mechanisms underlying fixation shifting and attention shifting (Posner 1978, 1988). He introduced the concept of so-called discrete covert attention shifting. Posner's discovery was confirmed in many further studies conducted in various research centers (Remington 1980; Posner and Cohen 1984; Fischer and Breitmeyer 1987). The research showed beyond any doubt that covert (discrete) attention shifting is independent of eye movements. It was also confirmed that a reverse interconnection exists—the attention shifting constitutes an indispensable foundation for programming and executing eye movements. The direction and the amplitude of the movement are estimated on the basis of the difference between the point of present fixation and the new point of attention engagement. The effectiveness of visual perception depends on the systems (in the sense of underlying mechanisms) functioning independently of one another: the attentional system (mainly the system responsible for the attention shifting) and the system responsible for eye movements (fixation shifting). Efficient cooperation of the two systems determines the field of perception and the effectiveness of visual perception (Fafrowicz 2006; Fafrowicz et al. 2008).

16.2 Diurnal Variability of Human Cognitive Processes

By the turn of the twentieth century, researchers had already started studies concerning diurnal variability in the performance of various cognitive tasks (Folkard 1996; Fafrowicz 2006). The first attempts in the 1890s, which were strictly connected with the names of Dressler and Bergstrom, confirmed what can be observed in everyday life: that the performance of various cognitive tasks is characterized by variability depending on the time of day.

In 1934, Freeman and Hovland (Folkard 1996) analyzed the findings of earlier studies of diurnal task performance. The results of this analysis showed that simple sensory and motor tasks performance reaches its peak in the afternoon. However, the scientists encountered a problem while trying to determine the highest performance level for complex cognitive tasks. In this case, the results were inconclusive and sometimes even contradictory. Freeman and Hovland stated that based on the analyzed tasks, it is difficult to decide which time of day would offer the greatest performance level for complex mental tasks.

The next researchers who concentrated on diurnal variability of the task performance were Nathaniel Kleitman in the United States and Peter Colquhoun in the United Kingdom. Kleitman discovered that the daily rhythm of speed for performing simple psychomotor tasks is parallel to the rhythm of body temperature. Thus, he concluded that a higher temperature (activation) causes a better performance of cognitive tasks. Kleitman's generalized conclusions were not confirmed in further studies, because he focused on simple cognitive processes and the performance was studied only from the point of view of the speed involved in the task performance. He did not take into account other indicators, such as precision. Later, Folkard and Monk (1980, 1983) revealed that the precision of performing sensory and motor tasks decreases during the day. Thus, a higher temperature is accompanied by a decrease in reaction time, but, simultaneously, the precision of simple cognitive tasks drops. The value of Kleitman's studies was, above all, to confirm the existence of variability in the performance of human cognitive functions depending on the time of a day. The finding that late afternoon and evening hours are characterized by the highest level of performance of simple cognitive tasks proved that a simple model, which refers to the tiredness increasing during the day as the main factor that lowers the task performance, cannot be accepted.

Colquhoun's (1971) findings concerning simple cognitive task performance were similar to Kleitman's results; the differences concerned early afternoon hours, in which Colquhoun observed a significant decrease in the performance of tasks ("post-lunch dip"). Another exception was the result of the study on memorizing a sequence of digits, in which case the highest performance level was observed in the morning hours (at around 10:00 hours), and from then on it decreased to reach its lowest level in the evening. A concept resorting to the so-called basal arousal, defined in opposition to sleepiness experienced, emerged as a consequence of the research by Colquhoun (1971). According to Colquhoun, it was the 24-hour rhythm of basal arousal, which was parallel to the internal body temperature, that caused the diurnal variability of cognitive tasks performance. In other words, an increase in basal arousal during the day caused an increase in simple task performance. To explain the results of the research on memorizing digit sequences, Colquhoun introduced a model of optimum level of arousal, which had an inverted U-shaped course. He assumed that different activation levels condition the optimum level of various task performances. The absence of an independent method of measuring the basal arousal was the most significant disadvantage of Colquhoun's theory.

Scientists exploring the diurnal or circadian variability of cognitive processes underline the complexity of the processes studied (Fafrowicz 2006). According to Folkard and Monk (1983) and Owens et al. (1998), the diurnal course of cognitive task performance is connected with the type of the task. Other research emphasized the role of the information-processing strategy (Folkard 1979, 1983).

The complexity of studies on diurnal variability of various cognitive task performances entails a difficulty in interpreting the results. Folkard and Monk (1983) stressed the necessity to investigate the elementary cognitive processes. According to them, research tasks should constitute a strict, clear measure of the studied processes, the ways of motor reaction should be as simple as possible, and the task should be made up of precisely measurable reactions.

16.3 Neuronal Attentional Networks and Saccadic Eye Movements

Functional neuroimaging data have supported the presence of three independent brain networks related to different aspects of attention (Fan et al. 2005; Posner and Rothbart 2007). These networks carry out the functions of alerting, orienting, and executive controls.

The *alerting* network is responsible for maintaining a state of high sensitivity to incoming stimuli and is associated with thalamic, frontal, and parietal regions of the cortex. The *orienting* system is related with selection of information from sensory input and is located in the regions of the posterior brain: the superior parietal (SP) lobe and the temporal parietal lobe. Research findings suggest that the right hemisphere plays a crucial role in the orienting process of attention shifting (e.g., Heilman et al. 1993) and is associated mainly with location-based attention. This was confirmed by research carried out with the use of positron emission tomography and functional magnetic resonance imaging (fMRI). The *executive* attention, which involves mechanisms for conflict monitoring and resolving discrepancies among thoughts, feelings, and responses, is linked with activity of the anterior cingulate cortex (ACC) and the lateral prefrontal cortex.

One of the widely studied forms of attention is visual orienting, which is carried out by two neuronal frontoparietal systems: the dorsal and ventral attentional systems (Corbetta and Shulman 2002). The dorsolateral is bilateral and consists of the intraparietal sulcus and the frontal eye field (FEF). This structure is considered to be involved in a top-down (voluntary and endogenous) orienting system that exhibits a higher activity in response to the presented cues referring to where (space), when (time), and to what objects the subjects should direct their attention (Fox et al. 2006).

The ventral system is lateralized to the right hemisphere and is composed of the temporal-parietal junction (TPJ) and the ventral frontal cortex. It is linked with bottom-up (involuntary and exogenous) mechanisms and shows an increase of activity in response to unexpected and salient stimuli (Shulman et al. 2004; Fox et al. 2006).

Recent research suggests that the model of functional networks responsible for executive attention is even more complex (Sridharan et al. 2008). The first network, called the "central executive network" is comprised of the dorsolateral prefrontal cortex and posterior parietal cortex and is considered to be crucial for maintenance and manipulation of information in working memory, as well as for decision making in goal-oriented behavior. The second, called the "salience network," includes the ACC and frontoinsular cortex, which combines of the ventrolateral prefrontal cortex and the anterior insula. The salience network's activity reflects the degree of subjective salience.

These networks remain in the dynamic interaction, which is modified by the task specificity and the natural rhythm of activity. Many research works state that disorders of the effective functioning of attentional processes are connected with the states of fatigue and sleepiness (Folkard 1983; Moray 1984; Aston-Jones et al. 1999; Marek et al. 2004).

The oculomotor cycle is one of the fundamental processes involved in exploring the surrounding environment. A previous study indicated that saccadic eye movements are controlled by a frontoparietal network including FEF, the presupplementary motor area (pre-SMA), SP, and the inferior parietal (IP) lobes. This network is considered to play a crucial, different role in the control of attention and is implicated in planning saccadic eye movements (Matsuda et al. 2004). The SP appear to be involved in representing the locations of interest, whereas IP areas appear to play the role of reorienting attention whenever a novel stimulus enters awareness. The FEF is thought to be involved in preparatory set in terms of coding both readiness and intention to perform a specific movement. It is involved in preparation and triggering of intentional saccades. The FEF contributes to the allocation of covert attention and overt orienting through gaze shift.

Additionally, Grosbras et al. (2001) suggest that FEF is involved in the selection of saccadic parameters in a behavioral and visual context, whereas pre-SMA is associated with elaboration of motor plans in the context of spatiotemporal set of movements. Therefore, pre-SMA is more engaged in the preparatory period, whereas FEF is more involved in the stimulus response stage of the saccadic movement. Many authors suggest the role of pre-SMA in updating motor plans for subsequent temporally ordered movements.

16.4 Workload, Mental Fatigue, and Effort

Mental fatigue is a decrease in performance level or a condition evoking feelings of fatigue. Grandjean and Kogi pointed out that "sensations of fatigue have a protective function similar to those of hunger and thirst. They force us to avoid further stress, and allow recovery to take place" (Grandjean and Kogi 1971). It is the signal for recovery. Mental fatigue develops due to a long-lasting workload. The phenomenon is highly correlated with attention impairment. The impairments of functioning of the three attentional networks are crucial symptoms of mental fatigue.

Depending on work demands, the model of an action is built for all the three attentional networks based on adaptation effort, which is related to task demands and level of difficulty as well as novelty and processing complexity. Long-lasting workloads (phenomenon of fatigue) cause changes in some parameters of this model. Due to these changes, the compensatory effort, which is related to the control of state, appears. In the state of developing fatigue, the impairment of the initial model accomplishment has to be compensated. Because of the decrement in the functioning of neuronal networks, a new model fitted to new internal conditions (developing fatigue state) is built.

From many research works, it is known that the magnitude of glucose utilization by the brain neuronal networks is a function of task complexity (e.g., Marek 2006). On the one hand, a high level of effort is correlated with increased glucose utilization; on the other hand, there is no evidence that any energetic limitations could occur in the brain even due to the most intense load. It seems that mental fatigue is not a consequence of a depletion of the brain metabolic energy. It is a process that is switched on to prevent the depletion of metabolic energy at the active processing sites in the brain (Marek 2006). Mental fatigue "may be best characterized as the inability to continue to exert executive resources control" (Mulder 1986).

16.5 Fatigue and Sleepiness

A *fatigue state* encompasses many physical and psychological symptoms, including lack of concentration, periods of inattention, and reduced alertness. Its effects on performance are reduced efficiency, slower response time, impaired decision making, and increased error rates. Operator fatigue is, therefore, one of the most important causes of accidents (Hockey 1983; Marek et al. 2004).

Fatigue results from the amount of time spent on a task and the amount of effort expended as well as other factors such as sleep deprivation and circadian disruptions. It is also related to the nature of the task, such as novelty, complexity, and variability, and to situational context, such as time pressure. Objective causes of the fatigue state, which can explain the anticipated level of tiredness, are modified by individual factors such as the level of motivation, experience of a particular task, and efficiency in cognitive and motor skills. Individual factors influence the degree of mental overload, with increasing discrepancy or divergence between task demands and the level of the worker's efficiency and skills as the workload increases (e.g., Sanders 1979; Moray 1984). Apart from the objective indicators of fatigue, which are mainly connected with lower levels of performance, the important indicator of fatigue is a subjective feeling of being tired, exhausted, overloaded, and the feeling that the task exceeds an individual's capabilities (Nijrolder et al. 2009; Leone et al. 2007).

A large body of research has shown that fatigue is rapidly emerging as one of the greatest safety issues, for example, in transportation and industry. The effects of fatigue and sleepiness on human performance are profound. Some researchers have shown that fatigue-related impairment may be compared to the effects of moderate alcohol intoxication (e.g., Dawson and Reid 1997; Lamond and Dawson 1999). Fatigue slows down the reaction times, affects logical reasoning and decision making, and impairs visual-motor coordination—all critical safety issues in the workplace.

Fatigue and sleepiness are potential sources of human errors. They may be considered a time-related phenomenon to a large extent. Operator functional state, that is, one's capacity for sustaining effective task performance under constraints imposed by environmental factors, fluctuates with the time of day and the time spent on a task. The demands of managing complex control systems force the operator to sustain a high-level performance independently of temporal circumstances and actual workload.

Fatigue and sleepiness are difficult to differentiate in terms of the symptoms and consequences. In normal circumstances, they interact resembling the two-factorial model of sleepiness by Alexander Borbely (1982), in which the homeostatic mechanism (total time awake modified by the time on task) coincides with the time of the day, that is, the circadian component.

According to a review by Mathis and Hess (2009), prevalence rates of excessive daytime sleepiness up to 15% were reported in young adults and elderly people. The major causes included sleep insufficiency syndrome, irregular sleep-wake rhythm (shift work, jet lag), sedative drugs, sleep apnea syndrome, narcolepsy, and idiopathic or nonorganic hypersomnia. Our own research indicated an even larger prevalence of daytime drowsiness—over 30% of investigated medical doctors ($n = 162$) revealed an increased level of sleepiness as measured by Epworth Sleepiness Scale (Johns 1991). The percentage of medical students ($n = 649$) who complained of excessive somnolence "often" and "very often" exceeded 60% (Oginska 2003).

16.6 Effort, Fatigue, and Sleepiness Study

The aim of the study was to observe the diurnal changes in the functioning of neural attentional subsystems considering the time of the day and the time spent on the task, analyzing the subjective (energy levels, tiredness, anxiety, calmness) and objective (fMRI) measurements. Subjective outcomes were correlated with the activity of neuronal structures engaged in the attentional subsystems. One of the main research questions was whether subjective assessments of sleepiness and activation do reflect the objective measures of neuronal activity registered in brain structures by means of fMRI.

16.6.1 Participants

The participants of the experiment were 23 healthy, male, paid volunteers with a mean age of 27.9 years (SD = 4.9). All of them were right-handed, had driver's licenses, and met the criteria for magnetic resonance (MR) scanning (normal color vision, no neurological disorders, no history of head injury, sleep-related disorder, nonsmokers, and drug-free). The subjects were informed about the procedure and goals of the study and gave their written consent. The study was approved by the Bioethic Commission at the Jagiellonian University.

Participants' chronotype (subjective phase and amplitude, according to a chronotype questionnaire; Oginska 1996) and daytime sleepiness level (according to the Epworth Sleepiness Scale; Johns 1991) were controlled before participating in the experiment to exclude extreme circadian types and subjects with excessive or pathological sleepiness. Before the experiment, participants had been trained in performing tasks to exclude the learning and the novelty factors.

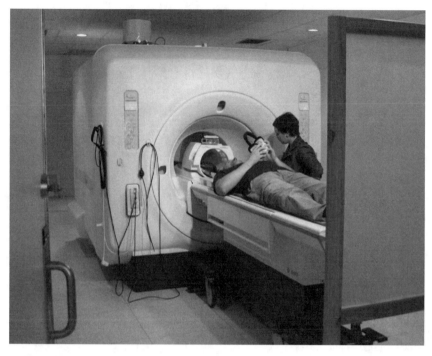

FIGURE 16.1
Magnetic resonance scanner used in the study.

16.6.2 Experimental Procedure

The experiment lasted 18 hours and consisted of five successive sessions, starting at 06:00, 10:00, 14:00, 18:00, and 22:00 hours. The times of measurements were chosen according to specific points of the hypothetical diurnal curve of speed and accuracy in performing cognitive tasks (Folkard 1979). Each session included the subjective assessment of sleepiness and fatigue and fMRI examination (Figure 16.1) during the performance of the saccadic task.

Between the MR registrations, there were four sessions of driving tasks, each lasting about two-and-half hours. Driving sessions were conducted in a simulator equipped with a car chair, 900°-rotation wheel with a dual-motor force feedback transmission, six-speed stick shifts, gas, brake, and clutch pedals (Figure 16.2). In the simulator, the subjects executed chosen assignments from three car games: the 18 Wheels of Steel Haulin' computer game (2005 SCS Software), Test Drive Unlimited (2006 Atari Inc.), and Driver (2000 Infogrames). Some sample shots of the games are shown in Figures 16.3 and 16.4. To augment the realism of the situation, visual effects were projected on a screen (234 × 176 cm), and sounds were transmitted by the stereo speaker

FIGURE 16.2
Workstand in the driving simulator.

FIGURE 16.3
Sample shot of 18 Wheels of Steel Haulin' game.

set with five speakers and a subwoofer (Logitech S500 5.1 speakers). The subjects completed, in total, about 10 hours of driving with three approximately one-hour breaks. To sum up, the participants spent approximately 18 hours in the lab in controlled conditions (type of cognitive activity, diet, stable temperature, isolation from daylight, and external noises).

FIGURE 16.4
Sample shot of Test Drive Unlimited (2006 Atari Inc).

16.6.3 Subjective Assessments

The methods used in the study included:

- The Karolinska Sleepiness Scale (KSS; Åkerstedt and Gillberg 1990) is a popular, nine-point, self-reporting measure of alertness, with 1 referring to "extremely alert" and 9 to "very sleepy, a great effort to keep awake, fighting sleep."
- The Activation-Deactivation Adjective Checklist (AD-ACL) by Thayer (1989) is a multidimensional test of various transitory arousal states, including energetic and tense arousal. Its short form lists 20 items referring to four dimensions: energy, tiredness, tension, and calmness (five items in each subscale). Scoring is based on four possible points for each adjective. The energy and tension scales seem to be good indicators of energetic and tense arousal, respectively, according to many studies analyzed by Thayer (1989).

16.6.4 Saccadic Task Description

The covert attention paradigm to measure object-based attention shifting was used. The mechanisms of the attention shifting involving the right and left hemispheres were studied. A fixation point was displayed in the centre of the screen for 30 seconds; then, a target stimulus was randomly presented at one of the five right- or five left-side situated squares, while the fixation point remained continuously visible (overlap condition). The subjects were instructed to direct their attention and gaze straight ahead toward the fixation point and, when the target appeared to the left or right side, to execute a saccadic eye movement and then to move back to the fixation point. The trial sequence and timing are shown in Figure 16.5.

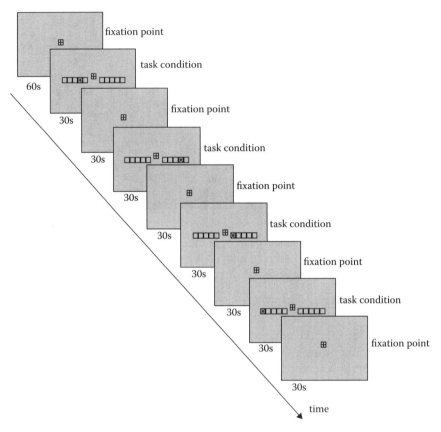

FIGURE 16.5

Schematic description of the task. Target stimulus was randomly presented at one of the five right- or five left-side situated squares, while the fixation point remained continuously visible (overlap condition). One session was combined with nine blocks: five blocks of fixation point presentations and four blocks of targets presentations. Blocks of targets consisting of 18 stimuli (each presented for 1500 milliseconds with a gap of 500 milliseconds between the targets) were tested, each block taking 30 seconds (total time of the task: 5 minutes).

16.6.5 Data Acquisition

Imaging was performed with the 1.5 T General Electric Signa scanner (GE Medical Systems). High-resolution anatomical images of the whole brain were acquired using 3D T1-weighted spoiled GRASS (SPGR) sequence. A total of 60 axial slices were acquired (voxel dimension = $0.4 \times 0.4 \times 3$ mm^3; matrix size = 512 $\times$ 512, TR = 25.0 seconds, TE = 6.0 milliseconds, FOV = 22×22 cm^2, flip angle = 45°) for coregistration with the fMRI data. Functional T2*-weighted images were acquired using a whole-brain echo-planar imaging (EPI) pulse sequence with a TE of 60 milliseconds, matrix size of 128 × 128, FOV of 22 × 22 cm^2, spatial resolution of $1.7 \times 1.7 \times 5$ mm^3, and flip angle of 90°. Each functional session was composed of 50 images acquired for each of 20 axial slices taken in an interleaved fashion with a TR of 3 seconds. To ensure the stability of the magnetic signal, the first three images of each session were not included in the functional analysis.

16.6.6 Data Analysis

The images were analyzed with Analysis of Functional NeuroImage (AFNI) software (Cox 1996). Each 3D image was first time-shifted so that the slices were aligned temporally. After head-motion correction, the functional EPI data sets were zero padded to match the spatial extent of the anatomic scans and then coregistered with them. The anatomical and functional images were transformed into a coordinate system of Talairach space (Talairach and Tournoux 1988). The functional data were then smoothed using a full-width at half-maximum isotropic Gaussian kernel of 8 mm. During the scaling procedure, low signal intensity voxels corresponding to voxels located outside the brain were excluded from the functional images by a clipping function.

Standard procedure was applied for each general linear model (GLM) analysis; that is, fitting the baseline to a second-order polynomial and using movement parameters as regressors. For each voxel, the beta coefficient, which represents the general linear model's estimate of the fMRI activity as a function of time, was calculated. Two procedures have been applied during GLM analysis. First, all five scans of each time of day were joined, and their general maps of activation due to the task were created. These maps were averaged across the subjects with corresponding t test. Second, the same procedure was applied to each time of the day separately; however, with the false discovery rate (FDR) algorithm instead of the t test, two-way mixed effects analysis of variance (ANOVA) was performed on single-factor level maps to achieve F statistical map ($p_{uncor} < .05$, cluster size > 15 voxels) for further conjunction analysis. The maps obtained after conjunction of F-map to t-map had been corrected for multiple comparisons with the FDR algorithm ($p_{cor} < .01$, cluster size > 15 voxels). Mean beta parameters were extracted from each cluster that survived threshold, and they were compared across each time of the day without any bias. Their significance was found by ANOVA.

16.7 Results

16.7.1 fMRI Measurements

Performing the saccadic task activated the neuronal structures belonging to the orienting system: right TPJ, bilaterally SP, and bilaterally FEFs. The activations of brain structures belonging to the alerting system—IP, BA 40, bilaterally and executive system, pre-SMA—were also observed. An overall analysis of the neuronal activity level evoked while performing the saccadic task is illustrated in Figure 16.6. Statistically significant differences between activity levels in different times of the day were analyzed with the ANOVA model (Table 16.1).

The neuronal structures located in the right hemisphere and forming the orienting network—TPJ, SP, and FEF—show a similar diurnal profile of neuronal activity (measured five times during the day) with a significant

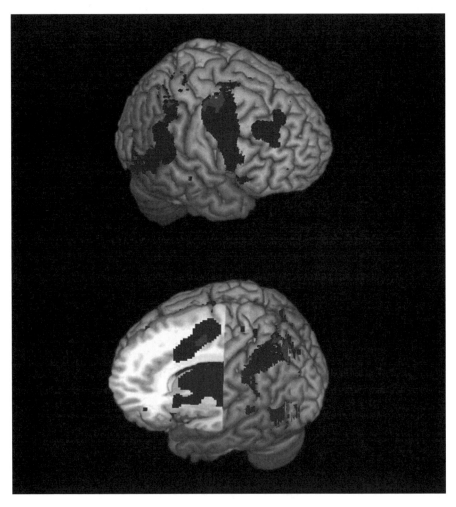

FIGURE 16.6
Brain regions activated while performing saccadic task (fMRI, block design)—a group map for all subjects at five times of the day.

TABLE 16.1

Diurnal Differences in Activity Levels of Considered Neuronal Structures Based on ANOVA Model

Attentional System	Structure	Hemisphere	F	p	Post Hoc Test (hours)
Orienting	TPJ	R	5.04	.001	6:00 > 10:00, 14:00, 18:00, 22:00
	FEF	R	3.34	.014	6:00 > 10:00, 14:00, 18:00, 22:00
	FEF	L	3.23	.016	6:00 > 14:00, 22:00
	SP	R	3.38	.013	6:00 > 10:00, 14:00, 18:00, 22:00
	SP	L	2.83	.030	6:00 > 10:00, 14:00, 18:00
Alerting	IP	L	3.10	.019	6:00 > 10:00, 14:00
	IP	R	2.83	.030	6:00 > 10:00, 14:00, 18:00, 22:00
	IP	R	3.23	.016	6:00 > 10:00, 14:00, 18:00, 22:00
Executive	pre-SMA	–	2.50	.050	6:00 > 10:00, 22:00

TPJ = temporal-parietal junction; R = right; FEF = frontal eye field; L = left; SP = superior parietal; IP = inferior parietal; pre-SMA = presupplementary motor area.

dominance of the activity level measured at 06:00 hours in comparison with the activity levels observed at other times of the day (TPJ: $F = 5.04$, $p = .001$; SP: $F = 3.38$, $p = .013$; FEF: $F = 3.34$, $p = .013$; Figure 16.7).

In the case of the left SP lobe, a significantly increased neuronal activity level was observed at 06:00 hours compared with activity levels measured at 10:00, 14:00, and 18:00 hours (ANOVA, $F = 2.83$. $p = .029$). Therefore, the neuronal activity level measured at 06:00 hours did not differ significantly only from the activation level measured at 22:00 hours. Both these times of the day are associated with increased sleepiness.

The left FEF showed an increased activity at 06:00 hours compared with the activity level measured at 14:00 and 22:00 hours (ANOVA, $F = 3.23$, $p = .016$). This indicates that the neuronal activity level at 06:00 hours, associated with "pure" sleepiness, differs from the activity level at the times of day that are associated with afternoon (postlunch dip) and evening sleepiness (fatigue and sleepiness).

The presented task also activated neuronal structures belonging to the alerting system (Figure 16.8). In the case of the right IP, two clusters had significantly higher activity at 06:00 hours compared with other times of the day (ANOVA, $F = 3.23$, $p = .016$; $F = 2.83$, $p = .030$). Additionally, the neuronal activity level observed in the left IP at 06:00 hours was higher than the activity level measured at 10:00 and 14:00 hours (ANOVA, $F = 3.10$, $p = .019$). Performing a saccadic task activated the pre-SMA, which belongs to the executive system with higher activity observed at 06:00 hours compared with the activity levels measured at 10:00 and 22:00 hours (ANOVA, $F = 2.50$, $p = .048$; Figure 16.9).

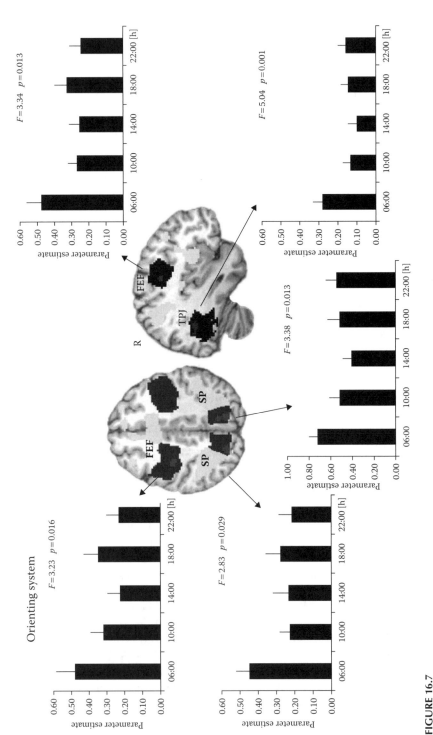

FIGURE 16.7
Brain regions showing different temporal activations in saccadic task (fMRI, block design)—the orienting system. The bar graphs indicate the mean parameter estimates of clusters, which survived conjunction analysis.

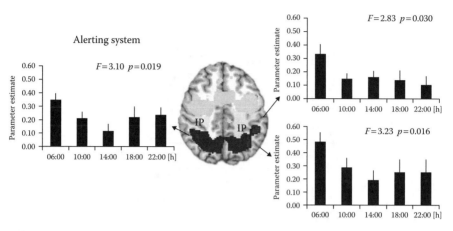

FIGURE 16.8

Brain regions showing different temporal activations in saccadic task (fMRI, block design)—the alerting system. The bar graphs indicate mean parameter estimates of clusters, which survived conjunction analysis.

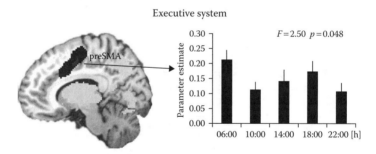

FIGURE 16.9

Brain regions showing different temporal activations in saccadic task (fMRI, block design)—the executive system. The bar graphs indicate mean parameter estimates of clusters, which survived conjunction analysis.

16.7.2 Subjective Reports on Fatigue and Sleepiness

Thayer's adjective list allows for subjective estimation of activation level in four dimensions, intercorrelated in pairs: "energy–tiredness" and "tension–calmness." Only the first two dimensions show the diurnal variability (ANOVA: $F = 2.972$, $p = .023$ and $F = 5.387$, $p = .001$, respectively). The level of energy decreases, while the level of tiredness increases noticeably in the evening hours (Figure 16.10).

The level of tension–calmness was constant during the experiment day (Figure 16.11), and this may be due to the primary qualification procedure when the subjects were assessed and were selected according to their reactivity and tension levels, which had to be below the average level (because of the uncomfortable position of the subjects and long-lasting nature of MR examination).

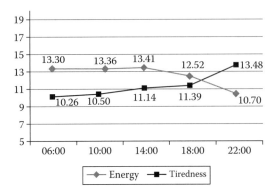

FIGURE 16.10
Results of Thayer's AD-ACL—components of "energy" and "tiredness."

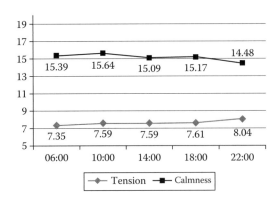

FIGURE 16.11
Results of Thayer's AD-ACL—components of "calmness" and "tension."

The subjective sleepiness assessed with KSS showed a continuous increase during the day, which was reported both before (Figure 16.12) and after the MR scanning. The differences were statistically significant, as shown with ANOVA method ($F = 2.481$, $p = .048$; $F = 4.024$, $p = .004$, respectively).

16.7.3 Subjective Reports on Fatigue and Sleepiness and fMRI Measurement—Correlation Analysis

The subjective and objective measurements of the states of sleepiness and tiredness were compiled in the correlation analysis (Table 16.2).

The results of the analysis indicate some statistically significant correlations between the subjective and objective measurements. The objective measurement refers to the activation level of the neuronal structure in the two attentional subsystems, which were significantly activated in saccadic tasks—executive and orienting control.

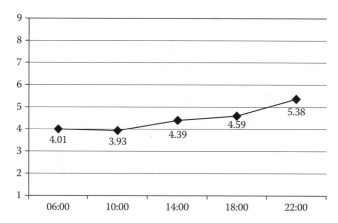

FIGURE 16.12
Self-assessments of vigilance-sleepiness (KSS).

TABLE 16.2

Pearson Correlation Coefficients Matrix—Correlation between Objective
Measurements (the Levels of Activity in Neuronal Structures, fMRI Data)
Registered during the Saccadic Task Performance and Subjective
Measurements: KSS and AD-ACL Scales (Energy, Tiredness, Tension, Calmness)

Neuronal Structure	KSS	AD-ACL Energy	AD-ACL Tiredness	AD-ACL Tension	AD-ACL Calmness
TPJ–R	−0.00	0.14	−0.16	0.07	−0.06
SP–R	0.07	0.13	−0.14	0.14	−0.01
FEF–R	0.02	0.15	−0.13	−0.04	0.13
IP–L	0.00	0.16	−0.11	0.04	0.09
IP–R	0.00	0.11	−0.09	0.12	0.01
SP–L	−0.03	0.20*	−0.18	0.12	−0.08
IP–R	−0.03	0.14	−0.04	−0.05	0.06
FEF–L (1)	−0.14	0.33*	−0.29*	0.01	0.03
pre-SMA	0.13	−0.00	−0.02	0.05	0.01

KSS = Karolinska sleepiness scale; AD-ACL = activation-deactivation adjective checklist; TPJ = temporal-parietal junction; SP = superior parietal; FEF = frontal eye field; IP = inferior parietal; pre-SMA = presupplementary motor area. * means $p < .05$.

Interesting relationships were found between the subjective outcomes and the FEF activity in the left hemisphere (FEF–L) and SP (SP–L), also in the left hemisphere. The FEF is linked to the attentional orienting subsystem, and it shows a statistically significant correlation with two subscales of AD-ACL: a positive correlation with subjective feeling of energy ($r = 0.33$; $p < .05$; Figure 16.13) and a negative correlation with a feeling of tiredness ($r = -0.29$; $p < .05$; Figure 16.14). In the SP areas of the left hemisphere, which are also related to

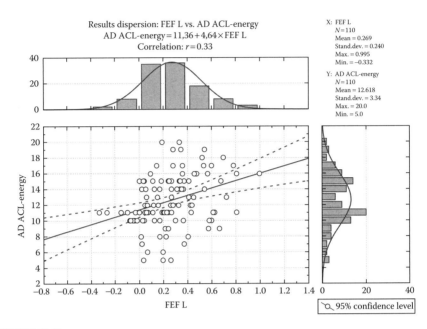

FIGURE 16.13
Diagram of results dispersion: activity of the orienting system's structure—frontal eye field in the left hemisphere (FEF–L) and the subjective level of energy.

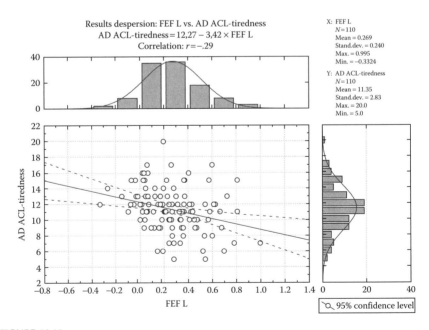

FIGURE 16.14
Diagram of results dispersion: activity of orienting system's structure—frontal eye field in the left hemisphere (FEF–L) and subjective level of tiredness.

neuronal system of the orienting control, a positive correlation with energy was found ($r = 0.20$; $p < .05$).

Other correlations between the subjective and objective measurements were not found. Data analysis shows that only the orienting control demonstrates relations with the subjective measurements, specifically with the activation level. The positive correlation with the level of energy indicates a higher activation of neuronal structures of the orienting subsystem when a subject feels energetic. Additionally, the negative correlation with the tiredness subscale confirms that the orienting system is less activated when a person feels tired and exhausted, which may have a considerable impact on the speed and precision of operator reactions.

16.8 Summary

The efficiency of attentional networks plays a crucial role in operator reliability. Different configurations of attentional networks are actuated under different task demands. Functional MR registrations of brain activity during saccadic tasks obtained in the study showed a specific pattern of attentional subsystems performance. Significant activations in the regions linked with orienting attentional system—bilaterally FEFs, SP, and right TPJ—were observed. The regions of alerting attentional system showed neuronal activity of IP lobe (bilaterally IP). Finally, significant activations of pre-SMA were observed in the executive network.

An overall analysis of the data revealed a significantly higher level of activation of all the analyzed structures belonging to attentional subsystems in the early morning hours (06:00–07:00 hours). This can be explained by an adaptive effort in a state of circadian sleepiness caused by waking up too early as well as by task demands. In the case of the right hemisphere, the lack of differences in the remaining times of the day indicates a relatively high stability of all attentional subsystems, which seems to enable an effective performance at the level of elementary cognitive processes in the context of diurnal rhythm. The structures of the orienting and alerting systems located in the right hemisphere have a similar profile of diurnal variability, with the highest activity observed at 06:00 hours when compared to the other times of the day, while the structures in the left hemisphere exhibit more variability in relation to the time of day (prolonged workload).

A diurnal profile of pre-SMA activity shows no difference between the results obtained at 06:00 hours and at 14:00 and 18:00 hours. This may suggest an occurrence of the similar activation effect, but one evoked by different neuronal mechanisms of morning adaptation, that is, the adaptive effort (06:00 hours) and afternoon/evening compensation, that is, the compensation effort (14:00 and 18:00 hours). The extraordinary function of pre-SMA in the compensatory effort is consistent with the role of pre-SMA in updating motor plans for subsequent temporally ordered movements.

Correlations between the results of the subjective and objective measures revealed a statistically significant relationship between the subjectively perceived state of energy and tiredness and the neuronal activity level of the orienting network in the left hemisphere (i.e., FEF). It could be hypothesized that the subjective assessment of fatigue adequately refers to the orienting processes. It is visible only in the left hemisphere. In the case of the right hemisphere, it is probably hidden by the ongoing process of attention shifting related to localization. According to current knowledge, in such a case, the right hemisphere is the leading one.

Analyzing the results, we must keep in mind that in addition to the time-of-day and time-on-task effects, many other factors may contribute to operator fatigue. Individual factors include preceding sleep length and quality, health status, family or social pressures as well as individual personality-temperamental traits modifying coping with stress. The non-work–related factors are highly changeable between individuals and between days, therefore, they are difficult to control.

The efficiency of the attentional system plays a crucial role in every moment of an operator's work. New technologies put special demands on the operator's neuronal systems. The awareness of diurnal profiles of activity of attentional system subdivisions under the constant demanding workload may be of importance from the perspective of the risk-prevention programs (Fafrowicz and Marek 2008).

Acknowledgments

This research was supported by the grant from the Polish Ministry of Science and Higher Education (N106 034 31/3110) (2006–2009). We would like to thank Justyna Kozub, Barbara Sobiecka, Adam Swierczyna, and Izabela Gatkowska for their assistance in the data acquisition.

References

Åkerstedt, T., and M. Gillberg. 1990. Subjective and objective sleepiness in the active individual. *J Neurosci* 52:29–37.
Aston-Jones, G. S., R. Desimone, J. Driver, S. J. Luck, and M. I. Posner. 1999. Attention. In *Fundamental Neuroscience*, ed. M. J. Zigmond, F. E. Bloom, S. C. Landis, J. L. Roberts, and L. R. Squire, 1385–1409. San Diego, CA: Academic Press.

Borbely, A. A. 1982. A two process model of sleep regulation. *Hum Neurobiol* 1:195–204.

Colquhoun, W. P. 1971. Circadian variations in mental efficiency. In *Biological Rhythms and Human Performance*, ed. W. P. Colquhoun, 39–107. London: Academic Press.

Corbetta, M., and G. L. Shulman. 2002. Control of goal-directed and stimulus-driven attention in the brain. *Neuroscience* 3:201–15.

Cox, R. 1996. AFNI: Software for analysis and visualization of functional magnetic resonance neuroimages. *Comput Biomed Res* 29:162–73.

Dawson, D., and K. Reid. 1997. Fatigue, alcohol and performance impairment. *Nature* 388:235.

Fafrowicz, M. 2006. Operation of attention disengagement and its diurnal variability. *Ergonomia Int J Ergon Hum Factors* 28:13–31.

Fafrowicz, M., K. Golonka, T. Marek, J. Mojsa-Kaja, K. Tucholska, H. Ogińska, A. Urbanik, and T. Orzechowski. 2008. Diurnal variability of attention disengagement process—EOG and fMRI studies. In *Conference Proceedings. 2008 Applied Human Factors and Ergonomics Conference (AHFE)*, ed. W. Karwowski and G. Salvendy.

Fafrowicz, M., and T. Marek. 2008. Attention, selection for action, error processing, and safety. In *Ergonomics and Psychology: Developments in Theory and Practice*, ed. O. Y. Chebykin, G. Bedny, and W. Karwowski, 201–16. Washington, DC: Taylor & Francis.

Fan, J., B. D. McCandliss, J. Fossella, J. I. Flombaum, and M. I. Posner. 2005. The activation of attentional networks. *NeuroImage* 26:471–9.

Fischer, B., and B. Breitmeyer. 1987. Mechanisms of visual attention revealed by saccadic eye movements. *Neuropsychology* 25(IA):73–84.

Folkard, S. 1979. Time of day and level of processing. *Mem Cognit* 7:547–92.

Folkard, S. 1983. Diurnal variation. In *Stress and Fatigue in Human Performance*, ed. G. R. J. Hockey, 245–72. New York: John Wiley & Sons.

Folkard, S. 1996. Effects on performance efficiency. In *Shiftwork. Problems and Solutions*, ed. W. P. Colquhoun, G. Costa, S. Folkard, and P. Knauth, 65–87. Frankfurt am Main, Germany: Peter Lang.

Folkard, S., and T. H. Monk. 1980. Circadian rhythms in human memory. *Br J Psychol* 71:295–307.

Folkard, S., and T. H. Monk. 1983. Chronopsychology: Circadian rhythms and human performance. In *Psychological Correlates of Human Behavior*, ed. A. Gale and J. Edwards, 57–78. London: Academic Press.

Fox, M. D., M. Corbetta, A. Z. Snyder, J. L. Vincent, and M. E. Raichle. 2006. Spontaneous neuronal activity distinguishes human dorsal and ventral attention systems. *Proc Natl Acad Sci U S A* 103:10046–51.

Grandjean, E., and K. Kogi. 1971. Introductory remarks. In *Methodology in Human Fatigue Assessment*, ed. I. Hashimoto, K. Kogi, and E. Grandjean, 17–30. London: Taylor & Francis.

Grosbras, M., U. Leonards, E. Lobel, J. Poline, D. LeBihan, and A. Berthoz. 2001. Human cortical networks for new and familiar sequences of saccades. *Cereb Cortex* 11:936–45.

Heilman, K., R. Watson, and E. Valenstein. 1993. Neglect and related disorders. In *Clinical Neuropsychology*, ed. K. Heilman and E. Valenstein, 279–336. New York: Oxford University Press.

Hockey, R. 1983. *Stress and Fatigue in Human Performance*. New York: John Wiley & Sons.

Johns, M. W. 1991. A new method for measuring daytime sleepiness: The Epworth Sleepiness Scale. *Sleep* 14:540–5.

Lamond, N., and D. Dawson. 1999. Equating the effects of fatigue and alcohol intoxication on performance. In *XIV International Symposium on Night- and Shiftwork*, Wiesensteig, Germany, Abstracts, p. 23.

Leone, S. S., M. J. H. Huibers, J. A. Knottnerusm, and I. J. Kant. 2007. Similarities, overlap and differences between burnout and prolonged fatigue in the working population. *Int J Med* 100:617–27.

Marek, T. 2006. Mental fatigue and related phenomena. In *The International Encyclopedia for Ergonomics and Human Factors*, ed. W. Karwowski, 798–99. London: Taylor & Francis.

Marek, T., M. Fafrowicz, and J. Pokorski. 2004. Mechanisms of visual attention and driver error. *Ergonomia Int J Ergon Hum Factors* 26:201–8.

Mathis, J., and C. W. Hess. 2009. Sleepiness and vigilance tests. *Swiss Med Wkly* 139:214–9.

Matsuda, T., M. Matsuura, T. Ohkubo, H. Ohkubo, E. Matsushima, K. Inoue, M. Taira, and T. Kojima. 2004. Functional MRI mapping of brain activation during visually guided saccades and antisaccades cortical and subcortical networks. *Neuroimaging* 131:147–55.

Moray, N. 1984. Mental workload. In *Proceedings of the 1984 International Conference on Occupational Ergonomics*, Toronto, Canada.

Mulder, G. 1986. The concept and measurement of mental effort. In *Adaptation to Stress and Task Demands: Energetical Aspects of Human Information Processing*, ed. G. M. Hockey, A. W. K. Coles, and M. G. H. Gaillard, 175–98. Dordrecht, The Netherlands: Martinus Nijhoff Publisher.

Nijrolder, I., D. van der Windt, H. de Vries, and H. van der Horst. 2009. Diagnoses during follow-up of patients presenting with fatigue in primary care. *Can Med Assoc J* 181:683–7.

Oginska, H. 1996. Typ rytmu okołodobowego aktywacji a podatność na zmęczenie chroniczne (Type of diurnal rhythmicity of activation and susceptibility to chronic fatigue). PhD dissertation. Krakow, Poland: Jagiellonian University.

Oginska, H. 2003. Hypersomnia and drowsiness at work: Individual and situational circumstances. Research report 501/PKL/146/L. Krakow, Poland: Collegium Medicum UJ.

Owens, D. S., I. Macdonald, P. Tucker, N. Sytnik, D. Minors, J. Waterhouse, P. Totterdell, and S. Folkard. 1998. Diurnal trends and performance do not parallel alertness. *Scand J Work Environ Health* 24:109–14.

Posner, M. I. 1978. *Chronometric Exploration of Mind*. New York: Erlbaum.

Posner, M. I. 1988. Structures and functions of selective attention. In *Master Lectures in Clinical and Brain Function: Research, Measurement and Practice*, ed. T. Boll and B. Bryant, 171–202. Washington, DC: American Psychological Association.

Posner, M. I., and Y. Cohen. 1984. Components of performance X. In *Attention and Performance*, ed. H. Bouma and D. Bowhuis, 531–56. New York: Erlbaum.

Posner, M. I., and M. K. Rothbart. 2007. Research on attention networks as a model for the integration of psychological science. *Annu Rev Psychol* 28:1–23.

Rayner, K. 1978. Eye movements in reading and information processing. *Psychol Bull* 85:618–60.

Remington, R. 1980. Attention and saccadic eye movements. *J Exp Psychol Hum Percept Perform* 6:726–44.

Sanders, A. F. 1979. Some remarks on mental workload. In *Mental Workload*, ed. N. Moray, 41–78. New York: Plenum Press.

Shulman, G. L., S. V. Astafiev, and M. Corbetta. 2004. Two cortical system for the selection of visual stimuli. In *Cognitive Neuroscience of Attention*, ed. M. I. Posner, 114–23. New York: The Guilford Press.

Sridharan, D., D. J. Levitin, and V. Menon. 2008. A critical role for the right fronto-insular cortex in switching between central-executive and default-mode networks. *Proc Natl Acad Sci U S A* 105:12569–74.

Talairach, J., and P. Tournoux. 1988. *Co-Planar Stereotaxic Atlas of the Human Brain*. New York: Thieme Medical.

Thayer, R. E. 1989. *The Biopsychology of Mood and Arousal*. New York: Oxford University Press.

Index